DYNAMIC
ASTRONOMY

DYNAMIC
ASTRONOMY

THIRD EDITION

ROBERT T. DIXON

Associate Professor of Astronomy
Riverside City College

Prentice-Hall, Inc., Englewood Cliffs, New Jersey 07632

Library of Congress Cataloging in Publication Data

DIXON, ROBERT T
  Dynamic astronomy.

  Includes bibliographies and index.
  1. Astronomy. I. Title.
QB43.2.D58 1980   520   79-24257
ISBN 0-13-221267-6 pbk.

Printed in the United States of America

10 9 8 7 6 5 4 3 2 1

Editorial/production supervision by Zita de Schauensee
Interior and cover design by Mark A. Binn
Line illustrations by Herschel Wartik, Inc.
Manufacturing buyer: John Hall
Cover photographs (JPL–NASA)
Front cover: Great Red Spot of Jupiter as seen by Voyager.
Back cover: Io, moon of Jupiter.

Prentice-Hall International, Inc., *London*
Prentice-Hall of Australia Pty. Limited, *Sydney*
Prentice-Hall of Canada, Ltd., *Toronto*
Prentice-Hall of India Private Limited, *New Delhi*
Prentice-Hall of Japan, Inc., *Tokyo*
Prentice-Hall of Southeast Asia Pte. Ltd., *Singapore*
Whitehall Books Limited, *Wellington, New Zealand*

CONTENTS

TWO
METHODS OF ASTRONOMY 45

THREE
THE EARTH 103

FOUR
THE MOON 141

SIXTEEN
EXTRATERRESTRIAL LIFE 483

GLOSSARY 491

APPENDIXES 501

INDEX 517

PREFACE

Like the first and second editions of this text, which has been used in over two hundred colleges and universities in the United States and elsewhere, the third edition is designed to help the liberal arts student and general reader to gain an understanding of how the astronomer studies his subject and to appreciate the grandeur of the universe in which we live. No prior background in science is presumed. Concepts are developed in the light of common experiences, and what little mathematics is needed is developed as the need arises. Relationships that are usually presented only as mathematical expressions are also stated verbally in order to dispel any anxiety on the part of the reader as to the mysteries of mathematics. The real beauty of astronomy lies not only in the facts that surround the subject but also in the development of an understanding of relationships within the universe. The background of thought that has brought us to our present view of the cosmos is presented here in conjunction with the methods whereby we make pertinent observations.

This revision represents not only the updating of information relative to a fast-moving science but also a move in the direction of a more comprehensive treatment of astronomy. Significant portions of the text have been completely rewritten to reflect the current state of the art. Two new chapters have been added, covering the topics of relativity and extraterrestrial life. Within the coming years, we in North America will be moving gradually toward use of the International System of Units (*Système International*, SI), and this is the system adopted for the present edition. In the first two chapters the SI units are followed by English units in parentheses, in order to familiarize the reader with the system; in subsequent chapters the SI units are used exclusively. An explanation of the system, together with the conversion factors, may be found in Appendix 2.

While the text is designed primarily for a one-semester course, it lends itself as well to a natural division for the two-quarter or two-semester program. Chapters 1 through 6 emphasize the methods of astronomy and deal with the solar system in particular, while the remaining chapters treat of the stars and galaxies.

To a large extent, the astronomy student's understanding is dependent upon his ability to visualize the motion of whatever is being discussed in terms of both space and time. To assist the student in visualizing certain rather intricate motions of objects in the universe, a unique kind of illustration is presented in the margins of this text. By flipping the successive pages, the illustrations appear to move.

This book is dedicated to my wife, Marian. Without her constant encouragement and the patience of my children, this book would not have been written.

Riverside, California   ROBERT T. DIXON

Key to Flip Pages

The eight flip-page sequences that occupy the upper margins of the pages of this text are a unique instructional tool dynamically illustrating the basic motions of astronomy. To use the flip pages on the right-hand margins, grasp the desired section with the right hand, bending these pages as illustrated here and allowing the pages to flip, one by one, from beneath your thumb. Left-hand flip pages may be handled in a similar fashion, using the left hand. Remember, flip pages starting on an odd-numbered page (right-hand) are flipped toward higher page numbers; those starting on an even-numbered page (left-hand) are flipped toward lower page numbers. (See the illustrations here in the left margin, which make this clear.)

A brief introduction to each flip sequence follows, together with an indication of its location in the text.

The Ptolemaic system (Right-hand flip pages beginning on page 25) Ptolemy envisioned the earth to be the center of the universe. In order to explain the apparent retrograde (westward) motion of the planets among the stars, he utilized the concept of the epicycle. Each planet was

Odd-numbered pages

Right-hand flip pages

Even-numbered pages

Left-hand flip pages

thought to revolve on its epicycle as the epicycle revolved on its primary orbit, called the deferent. By assigning the proper speed to each of these motions, he was able to create a retrograde motion. This motion also explains the fact that at certain times a given planet is closer to the earth. Since it had been observed that Mercury and Venus always remained close to the sun, the centers of their epicycles must always align with the earth and the sun. (This sequence of flip pages will be useful in your study of Chapter 1.)

Retrograde motion of Jupiter (Left-hand flip pages beginning on page 114) This sequence shows Jupiter moving first in a direct (eastward) motion among the stars, then apparently stopping and moving in a retrograde (westward) motion among the stars. Later the planet appears to stop again and resume its direct motion. By careful observation over an extended period of time, this apparent retrograde motion of Jupiter or of any other planet may be observed in the real sky. (This sequence will be helpful in your study of Chapters 1 and 5.)

The Copernican system (Right-hand flip pages beginning on page 115) In this sequence, the planets will be seen to move at their proper speeds in relation to the earth's motion. A period of approximately one year is depicted. The period of revolution of Mercury is 88 days, hence it can make four complete revolutions in one year. Within the period of 116 days, it returns to inferior conjunction with the earth. This is called its synodic period. Venus, on the other hand, makes a complete revolution in 225 days but does not return to inferior conjunction in the period shown. This planet requires 584 days for one synodic revolution. Jupiter needs approximately 12 years for one revolution, hence it is seen to move only about 30° during the period shown. Other configurations such as maximum elongation, superior conjunction, conjunction, quadrature, and opposition may be seen on certain pages individually. See if you can find all possible configurations. (This sequence will be helpful in your study of Chapters 1 and 5.)

A binary system (Left-hand flip pages beginning on page 182) This sequence demonstrates the fact that the earth and the moon form a binary (two-body) system and that it is the barycenter of the system which follows a smooth elliptical orbit. The barycenter of the system is located approximately 3000 miles from the center of the earth. As the moon revolves about this barycenter, the earth also deviates up to 3000 miles on either side of the system's orbit. In a very similar way, two binary stars, stars which lie in each other's gravitational field, orbit around a barycenter. The position of the barycenter is determined by the way in which the material (mass) is distributed in the two stars. (This sequence of flip pages should be helpful in your study of Chapters 4 and 9.)

A comet in motion (Right-hand flip pages beginning on page 235) This sequence depicts Halley's Comet, moving in its highly elongated elliptical orbit about the sun. This comet returns to the region of the sun every 76 years, its next expected return being in 1986. Halley's Comet travels in a retrograde direction that carries it beyond the orbit of Neptune. When a comet is at such a great distance from the sun, it possesses no coma (head) nor tail but exists only as a swarm of frozen gas bodies. As the comet approaches the sun, however, the warmth of the sun vaporizes a portion of the gas, thus producing the coma and tail. The outflow of particles from the sun (the solar wind) continually pushes the gases of the tail in a direction away from the sun. The speed with which the comet travels increases as it approaches the sun, hence only a relatively short time is spent in the vicinity of the sun. (This section of flip pages will be helpful in your study of Chapter 6.)

The proper motion of stars (Left-hand flip pages beginning on page 284) Over a 100,000 year-period, the familiar constellation of the Big Dipper will change in appearance until it no longer resembles a dipper. This change results from the fact that stars are in constant motion. The stars that make up this constellation are moving in different directions. The apparent change in the position of a star in 1 year is very small, perhaps in the order of 1 second of arc per year. This is called the proper motion of the star. (This sequence of flip pages will be helpful in your study of Chapter 8.)

Motion of globular clusters (Right-hand flip pages beginning on page 345) The upper view in this series shows the motion of globular clusters that form the halo of the Milky Way galaxy. Each globular moves along an elliptical path which causes it to periodically "dip" into the nucleus of the Galaxy; however, it spends a relatively short time there. (In this sequence, it is best to choose a particular globular and follow its motion, say, the one marked by the double circle.) *Rotation of the Milky Way galaxy* The lower half of the pages shows the revolution of the Galaxy. The sun participates in this revolution and makes a circuit around the center of the Galaxy in 200 million years. From the sun's position, we see one spiral arm beyond and two arms toward the center of the Galaxy. (This sequence of flip pages will be helpful in your study of Chapter 13.)

An eclipsing binary system (Left-hand flip pages beginning on page 410) The top view in this section shows the motion of a binary system, seen from above. The brighter star (light in color) is about five times as massive as its cooler (darker) component.

The middle view shows the same system, seen from our position on earth. One star is seen periodically to eclipse the other, for the earth lies very nearly in plane of their orbit.

The lower view shows the light curve which is generated as these stars move in their orbit. When the cooler star almost completely eclipses the hot star, the lowest light output is apparent. When the hotter (brighter) star is in front, only a slight dip occurs. While you are seeing the motion of these stars and the light curve generated simultaneously, the astronomer usually observes only the light curve and he must infer the actual motion from that curve. (This sequence will be helpful in your study of Chapter 9.)

Chapter-opening Illustrations

Chapter 1 Stonehenge, in England. (British Tourist Authority)

Chapter 2 The 64-m Goldstone radio antenna, near Barstow, California. (NASA-JPL)

Chapter 3 The earth photographed from Apollo 10 in 1969. (NASA)

Chapter 4 Astronaut James Irwin of Apollo 15 walking on the moon, August 1971. (NASA)

Chapter 5 Surface of the planet Mercury, photographed by Mariner 10 from a range of 86,800 km, March 29, 1974. (NASA-JPL)

Chapter 6 Comet Ikeya-Seki (1965f). (Lick Observatory)

Chapter 7 Solar prominence 64,500 km high, photographed in red light of H_α at Big Bear Solar Observatory, March 31, 1971. (Hale Observatories)

Chapter 8 Region of the Orion Nebula. (Lick Observatory)

Chapter 9 An open cluster in Cancer (M67). (Hale Observatories)

Chapter 10 The Small Magellanic Cloud. (Mount Stromlo and Siding Spring Observatories, The Australian National University)

Chapter 11 The Horsehead Nebula in Orion. (Hale Observatories)

Chapter 12 The Crab Nebula, the remains of a supernova first seen in 1054 A.D. (Hale Observatories)

Chapter 13 Portion of the Milky Way. (Hale Observatories)

Chapter 14 Spiral galaxy (NGC 628; M74).(Hale Observatories)

Chapter 15 Relativity

Chapter 16 The Cyclops proposal for a radio telescope array. (NASA-Ames)

DYNAMIC
ASTRONOMY

ONE

HISTORY OF ASTRONOMY

The history of astronomy is characterized by humans' ever-expanding concept of the cosmos. Your study of astronomy will no doubt expand your own awareness. Through their observations of celestial objects, humans have been able to relate the rhythms of their lives to those of the visible universe. In order that you may understand the development of this harmonious relationship, place yourself backward in time about 6000 years and try to imagine what was taking place then. The very early humans had been nomadic—constantly searching for a new supply of food and a shelter from the weather and unfriendly animals. Even as nomads, humans surely recognized several obvious cycles of nature: day and night, the monthly phasing of the moon (see Figure 1.1), and the seasonal changes which made hunting more difficult at certain times. These cycles became far more important, however, when humans turned from nomadic patterns to become farmers—domesticating animals and planting crops.

The most natural locations for agriculture lie within fertile river valleys, which provide a ready supply of irrigation water. Out of such humble beginnings arose at least four early civilizations: that of China, along the Hwang Ho river; that of India, along the Indus river; that of Babylonia and Sumeria, along the Tigris and Euphrates rivers; and that of Egypt, along the Nile. Dating from these early cultural beginnings, we have clear evidence that early humans watched the skies and charted the motion of the sun, moon, and planets among the stars. In some cases these objects were thought to possess godlike qualities and they became objects of worship, which motivated the watchers to record their rhythmic motions.

Figure 1.1 *The phases of the moon: (a) "new moon"; (b) waxing crescent, 4 days old; (c) first quarter, 7 days old; (d) waxing gibbous, 10 days old; (e) full moon; (f) waning gibbous; (g) last quarter; (h) waning crescent; (i) "new moon" again. (Lick Observatory)*

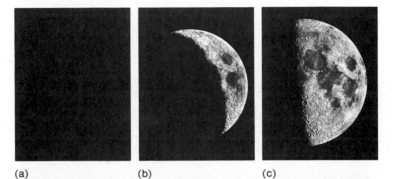

(a)   (b)   (c)

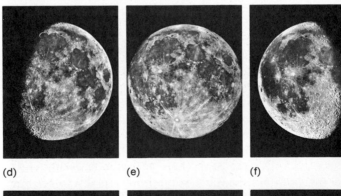

(d)   (e)   (f)

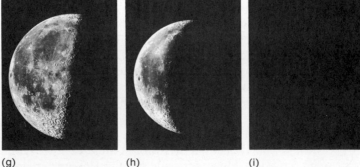

(g)   (h)   (i)

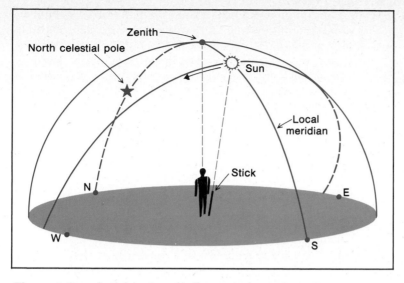

Figure 1.2 *A stick is placed in the ground so as to cast no shadow when the sun appears highest in the sky.*

The people of China produced one of the more accurate determinations of the length of a year. Their findings were recorded by means of a number of circular objects divided into 365.25 parts. What kind of experiment could they have performed in order to determine the number of days in a year so accurately? Visualize yourself standing under the imaginary dome of the sky, facing south. Let an imaginary curved line run on that dome directly north and south over your head. This line is called your *local meridian*. The point directly over your head is called your *zenith*.

Choose a time of year when the sun passes closest to your zenith, and place a stick in the ground so that at noon it points directly toward the sun as it crosses the meridian and therefore has no shadow. (Figure 1.2). Note that if the sun were located east or west of the meridian, the stick would cast a shadow, and the time would not be noon. You can then easily tell when a full day has passed, because at that moment the stick will cast no shadow again. As the days go by, however, it will become evident that the sun is crossing your meridian lower in the sky and as a consequence the stick casts a shadow even at noon. But count the number of days until the sun again passes close to your zenith and the stick casts no shadow at noon and you will have determined that a year is approximately 365 days. After several years of counting, you may make a more accurate determination by averaging the counts for those years, revealing that a year contains 365.25 days. (Even this is not exact and must be corrected periodically. We will consider this further adjustment later.)

Legend tells us that the Chinese were also very adept at predicting eclipses, a feat that becomes possible only after very accurate observations have been made of the motions of the sun and moon in relation to the background of stars. The story relates that two observers named Hi and Ho became lax in their priestly duties and failed to predict a given eclipse, which was unexpectedly viewed by the citizens of their province. Because an unheralded eclipse was thought to be a bad omen, the two priests were beheaded.

The great temple to the sun at Peking is oriented so as to point toward the rising point of the midwinter sun (the winter solstice). As you will see, sun alignments are quite common among the temples and monuments of many cultures. Even today, St. Peter's in Rome is oriented toward the rising sun at the time of the equinox (approximately March 21 and September 21). At this time sunlight floods the entire length of the basilica and illuminates the high altar. At the setting of the sun, a similar phenomenon illuminates the window of the Holy Ghost, producing a spectacular, almost mystical appearance.

Early Indian culture is recorded in a work called the Vedas. Included are hymns to the sun which are at least 3500 years old—hymns which recognize human dependence upon the sun for life and light.

Babylonian "Astronomy"

The Sumerians were very early inhabitants of the region we refer to as the "fertile crescent," a region dominated by the Tigris and Euphrates rivers. These people produced a written record using cuneiform script. In these records one finds a very early recognition that the cycle of seasons, as it relates to the sun's apparent position in the sky, is also directly related to the sowing and reaping of crops. This relationship is expressed in a calendar which has for its fourth month a name that is a contraction of the words for hand and seed—a time of sowing. The name of its eleventh month indicates a time of reaping, and that of its twelfth month indicates a gathering of the corn into storage.

In constructing their calendar the Babylonians faced the same problem we still face today: that of having months that closely reflect the cycle of the moon's phases and a year that accurately reflects the seasonal changes. To grasp this problem, suppose one were to count the number of days required for the moon to go through six complete cycles of phasing, starting with its appearance as a new crescent. This would require 177 days. The average lunar cycle then is $29\frac{1}{2}$ days ($177 \div 6$). Next suppose one charted the seasonal cycle, as illustrated in Figure 1.2, for four complete cycles and found that this required 1461 days. The average cycle of seasons is then $365\frac{1}{4}$ days ($1461 \div 4 = 365\frac{1}{4}$). One might construct an ideal calendar based on moon phases, something like the one on the next page.

This method of placing first 30, then 29, days in a month would make a given phase of the moon fall on the same day each month, because the average of $29\frac{1}{2}$ days per month would be preserved. But note that the

MONTH	NUMBER OF DAYS
1st	30
2nd	29
3rd	30
4th	29
5th	30
6th	29
7th	30
8th	29
9th	30
10th	29
11th	30
12th	29
	354

12-month total of 354 days is $11\frac{1}{4}$ days short of the $365\frac{1}{4}$-day cycle of seasons, so that the planting season would not fall on the fourth month in successive years. Even if an approximate lunar month of 30 days were used, 12 such months would only total 360 days, and the calendar year would be $5\frac{1}{4}$ days short of a seasonal year.

The Babylonians solved this problem by simply inserting an extra month in certain years whenever it was necessary to bring the calendar back in tune with the seasons. This they did, not according to a regular plan but sporadically as they realized that the eleventh month did not coincide with the ripening of the corn.

About 800 B.C. the Assyrians became dominant in the land of Babylon, and they employed a more accurate method for keeping their calendar in tune with the seasons. They noted that as the year passed, certain stars were visible in a winter evening, others in the spring evening, and still others in the summer and fall (see Figure 1.3). By noting the return of the stars which were characteristic of a given season, they were able to keep their calendar more closely synchronized with the year of the seasons.

A new motivation for celestial observations grew out of the Assyrian culture, based on a belief that the events of nations and individuals could be predicted by events in the sky. We see the rise of a religion called *astrology*, with its priests of Zoroaster. Because the location of the sun, moon, and planets against the background of stars became such an important factor in this new religion, observations became much more precise. The 12 constellations through which the sun appears to move in a year are called the signs of the zodiac; besides these, many more constellations were named by the observers of this time and place. The religion of astrology and the science of astronomy had a long marriage, parting ways only in very recent times. Certainly without the strong motivation of astrology the modern astronomer would not have the wealth of ancient observations he has today.

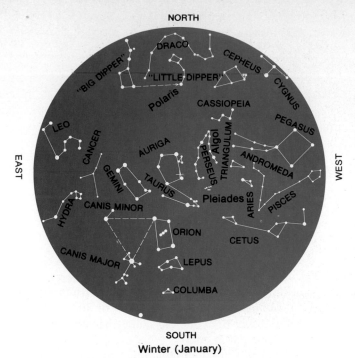

NORTH

DRACO
"BIG DIPPER"   "LITTLE DIPPER"   CEPHEUS
                                    CYGNUS
        Polaris   CASSIOPEIA
                                    PEGASUS
LEO   Algol
    CANCER   AURIGA   PERSEUS TRIANGULUM ANDROMEDA
EAST   GEMINI   TAURUS   WEST
                                            ARIES PISCES
        CANIS MINOR   Pleiades
HYDRA   ORION   CETUS
                                        ARIES
    CANIS MAJOR   LEPUS

                    COLUMBA

SOUTH
Winter (January)

(a)

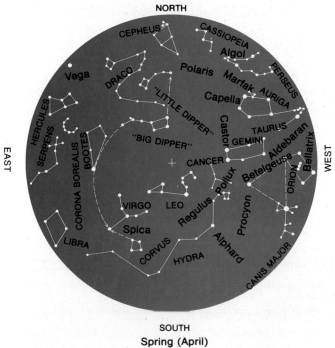

NORTH

CEPHEUS   CASSIOPEIA
                    Algol
        DRACO   Polaris Marfak PERSEUS
Vega   AURIGA
                "LITTLE DIPPER" Capella
HERCULES   Castor TAURUS
SERPENS   BOOTES "BIG DIPPER" GEMINI Aldebaran Bellatrix
EAST   CORONA BOREALIS   Betelgeuse ORION WEST
                    CANCER Pollux
                VIRGO   LEO Regulus Procyon
            Spica   Alphard CANIS MAJOR
LIBRA   CORVUS   HYDRA

SOUTH
Spring (April)

(b)

Figure 1.3 *Evening constellations of winter, spring, summer, and fall. (Griffith Observatory)*

8 Chapter One

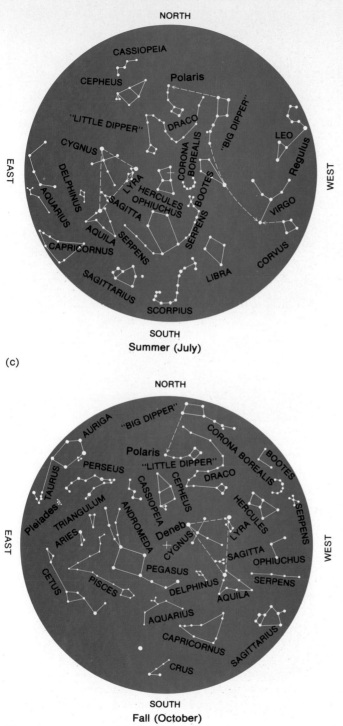

NORTH

CASSIOPEIA

CEPHEUS   Polaris

"LITTLE DIPPER"   DRACO

CYGNUS   CORONA BOREALIS   "BIG DIPPER"   LEO

EAST   DELPHINUS   LYRA HERCULES   BOOTES   Regulus   WEST

AQUARIUS   SAGITTA   OPHIUCHUS

AQUILA   SERPENS   SERPENS   VIRGO

CAPRICORNUS

                LIBRA   CORVUS

SAGITTARIUS

SCORPIUS

SOUTH
Summer (July)

(c)

NORTH

AURIGA   "BIG DIPPER"   CORONA BOREALIS   BOOTES

       Polaris

PERSEUS   "LITTLE DIPPER"   DRACO

TAURUS   CASSIOPEIA   CEPHEUS   HERCULES   SERPENS

Pleiades   TRIANGULUM   ANDROMEDA   Deneb   LYRA

EAST   ARIES   CYGNUS   SAGITTA   OPHIUCHUS   WEST

CETUS   PISCES   PEGASUS   SERPENS

                DELPHINUS   AQUILA

AQUARIUS

CAPRICORNUS   SAGITTARIUS

CRUS

SOUTH
Fall (October)

(d)

Figure 1.4   *The zodiacal constellations, as depicted on the ceiling of the Temple of Dendera, now in the Louvre. (This drawing, first published in 1822, is by an unknown author.)*

Egypt, Land of the Nile

The Nile river played a dominant role in the life of the Egyptians, largely because it rose each year to overflow a large delta region. This action of the river deposited fertile silt in which crops could be grown, and the flood waters were retained behind a series of check dams, to be released as needed. Because the food supply of the entire nation depended upon the proper utilization of this natural occurrence, it was vital that the people be forewarned as to when the flooding would begin.

The Egyptians had already made observations which led to a very accurate calendar. They were well aware of the signs of the zodiac through which the sun appears to move in a year's time. See how many of the zodiacal constellations you can find in Figure 1.4, a reproduction of the ceiling painting from the temple of Dendera. How can one observe the position of the sun, in relation to the background of stars, when one

Figure 1.5 *The Great Pyramids of Giza are so accurately aligned with the cardinal points (north, south, east, west) that we can be certain that their builders must have observed the stars. (Robin Rector Krupp)*

cannot see the stars in the daytime? The Egyptians observed the star or constellation which appeared about one hour ahead of the rising sun, when dawn was just "breaking." Each morning they noticed a slight change in star positions, relative to the sun. Only on one day per year would a given star rise exactly one hour before the sun. For example, the Egyptians noticed that when the bright star Sirius rose exactly one hour before sunrise, the Nile would soon begin to rise, hence this celestial event became a useful predictor. Certainly, after the Egyptians had observed this cycle for a number of years, it would have been clear that the seasons reoccur in a cycle of $365\frac{1}{4}$ days, yet their calendar contained only 365 days. Perhaps we can understand this inaccuracy when we realize that it was the priests who observed such cycles. By purposely leaving the calendar in error, they retained for themselves the secret of the proper correction to make—a secret that enhanced their power.

A Megalithic Legacy

No other culture has left so many huge and enduring reminders of its astronomical observations as did the Egyptians. Perhaps the most obvious are the pyramids. As shown in Figure 1.5, the pyramids of Giza are so accurately oriented with the cardinal points (north, south, east, and west) as to force the assumption that the alignment could only have been made by observing the stars. The greatest deviation is less than $\frac{1}{12}°$.

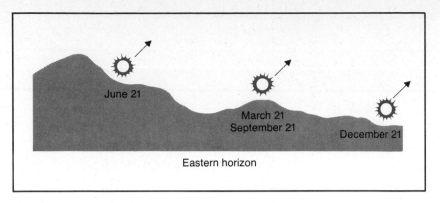

Figure 1.6 *During the course of one year the sun appears to rise over different points along the eastern horizon.*

The temples of the Egyptians reveal the importance of the sun in their religious beliefs, as in those of so many other cultures also. The temple builders took advantage of one fact that the sun follows a predictable pattern, throughout the year, in terms of its rising points and setting points. As illustrated in Figure 1.6, the sun rises directly out of the east on the dates we call the equinoxes (usually March 21 and September 21); it rises considerably to the north on June 21, known as the summer solstice, and to the south on December 21, known as the winter solstice. These dates may vary from year to year by up to a day, but within this degree of accuracy they do return year after year. The reason is that we keep our modern calendar in tune with the seasons by adding a day every four years, making our average year $365\frac{1}{4}$ days. The Egyptian calendar of 365 days allowed the date of the summer solstice to move through the calendar, occurring 25 days earlier every century. Only the priests knew the correction to make. If the temples of Egypt had merely been oriented along the cardinal direction, as were the great pyramids of Giza, this would not necessarily indicate a fascination for the sun; however, we find many temples oriented toward one of the solstice points. Note the plan at Karnak, illustrated in Figure 1.7. A long series of doorways was definitely oriented toward the setting sun at the time of the summer solstice. On the one day of the year when the sun appeared to travel across the sky in an arc farthest to the north, the sunset event produced a most spectacular flash of light upon the holy of holies. You can imagine the mystical quality of such an event, especially if the king stood in that darkened location just before the flash.

Further evidence of Egyptian sun worship is the written hieroglyphic record left as inscriptions on temples and other monuments. The name of their sun-god was Ra, who brought life to all things. Each day Ra would appear and travel in his barge up over the back of the sky goddess, Nut (pronounced "Noot"). We might call this concept, as illustrated in Figure 1.8, an Egyptian cosmology—their concept of the universe.

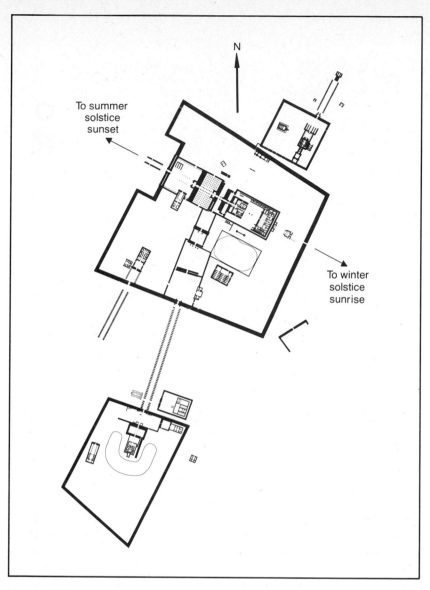

Figure 1.7 *The Great Temple of Amen-Ra at Karnak, Egypt, shows an alignment of many corridors and doorways with the setting sun at the time of the summer solstice. On the longest day of the year, the sun's last rays would seem to flash upon the holy of holies, creating a rather mystical phenomena. (Griffith Observatory, Joseph Bieniasz)*

Stonehenge

Far away from all the "cradles of civilization" we have mentioned, yet dating to a similar period of time (some 4000 to 6000 years ago), is Stonehenge, a megalithic monument to human intelligence, set out on

Figure 1.8 *An Egyptian concept of the sun god traveling in his barge over the back of a starry goddess and down into the underworld. (Yerkes Observatory)*

the Salisbury Plain of southern England (Figure 1.9). Here massive stones were aligned so as to point to the extreme rising or setting positions of the sun at the summer and winter *solstices;* that is, they were the means whereby the shortest and longest days of the year could be noted and the length of the year accurately determined. If the megaliths of Stonehenge marked only the extremes of the sun's apparent motions, then we might surmise that it served primarily a religious function; however, this structure also marks the extremes in the rising and setting points of the moon—hence it was a lunar observatory. Whenever we think of the moon and sun simultaneously, we think of eclipses. Eclipses occur on any occasion in which the sun, moon, and earth are aligned. The subject of eclipses is treated more fully in Chapter 4; suffice it to note here that eclipses occur in cycles. The period of one such cycle is 18.61 years. In that length of time the circumstances which produced a particular eclipse are repeated. Three such cycles (3 × 18.61) yield a number very close to 56 (55.83). In that length of time a given eclipse would occur again at the same season of the year. Some observers feel that this is the significance of the 56 Aubrey holes which mark the perimeter of Stonehenge (see Figure 1.9c). By moving a marker one hole per year, such a seasonal cycle of eclipses may be charted. At a number of other sites in England and in Brittany (in northwestern France) similar megalithic structures were erected, although of somewhat less massive stones. It is interesting to contemplate the degree of intelligence and sophistication which the builders of such monuments possessed.

(a)

(b)

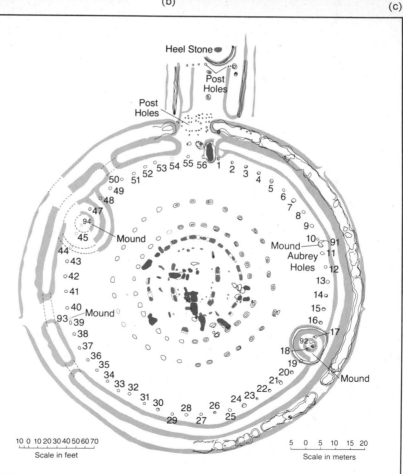

(c)

Figure 1.9 (a) and (b) Megaliths at Stonehenge, in England, possibly used by early man for celestial observations. (c) Plan view of Stonehenge. The alignment of certain positions (for example, Heel Stone, center of circle, mounds 91, 92, 93, 94) pointed to key rising and setting points of the sun and the moon. The 56 Aubrey Holes may have been a counting device for predicting eclipses. (a,b, British Tourist Authority; c, Controller of Her Britannic Majesty's Stationery Office. British Crown Copyright)

Early Concepts of the Universe

Man's concept of the earth on which he abides appears to influence greatly his concept of the universe, and logically so. Can you imagine your own interpretation of the earth if you had never traveled farther than you were able to walk in a day? Undoubtedly you would be convinced that the earth was flat and that the stars simply formed a tent or canopy overhead. It might appear as though the sky were a huge brass shield stretching over your head with a fire on the outside and hundreds of small holes punched for the light to shine through—an idea of the ancient Greeks.

To explain the apparent motion of the sun, the Greeks held that Helios, the sun god, drove his chariot across the sky each day from east to west and sailed round the northerly stream of Ocean each night to arise in the east again at dawn. We have already seen that the Egyptians favored a scheme in which the sun god (Ra) sailed in his barge up over the back of his mother, Nut (the sky), who arched her body over the earth (Figure 1.8). As you can see, the flat-earth concept caused men to create explanations for the motion of the sun which might appear absurd to us today but which reflect their personal feeling for objects in the sky.

Figure 1.10 depicts the early Greek concept of the universe in general terms. Various ideas were set forth by major schools of Greek philosophers. Thales of Miletus (636–546 B.C.) said that water was the first principle of all things, and he visualized the earth as a flat disk, floating on water; Anaximander (611–547 B.C.) thought of the world as infinite and viewed the earth as a cylinder floating free in space; and Anaximenes (585–526 B.C.) pictured the earth as a disk supported by air.

It is interesting to note that from the same school of training which produced these imaginative schemes came a man whose ideas were so advanced for his day that they were branded as sacrilegious and he was exiled for his teachings. This was Anaxagoras (499–428 B.C.), who more accurately pictured the moon and planets as earthlike in nature—having a solid, crusty surface and shining by reflected sunlight. Furthermore, he rightly explained lunar eclipses as the result of the moon's moving into the earth's shadow.

Another school was established about the same time in southern Italy by the philosopher-mathematician Pythagoras of Samos and his followers (about 582–507 B.C.). It is thought that Pythagoras was the first to recognize the shape of the earth as spherical (ball-like). Perhaps his observation of ships led him to this conclusion. A ship appears to "sink" into the water as it sails away from an observer and appears to "rise" from the water as it sails toward an observer, indicating the curvature of the earth's surface. Pythagoras may also have noticed that during a lunar eclipse, as the moon passes into the earth's shadow, the curvature of the edge of that shadow is suggestive of the curved nature of the earth itself (Figure 1.11).

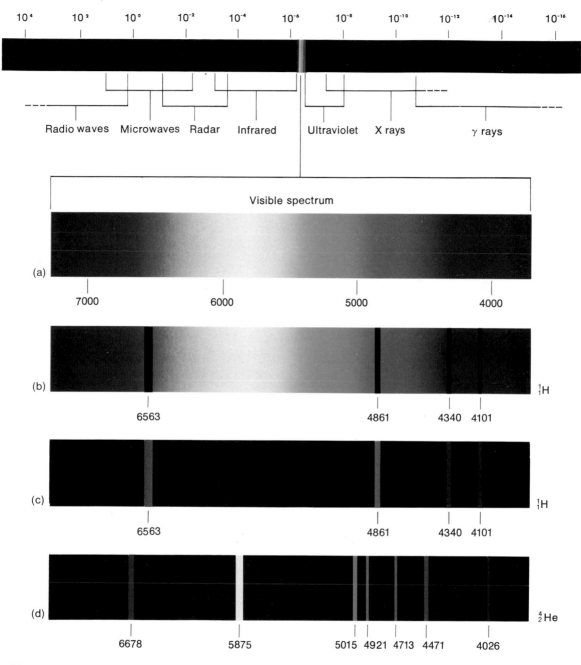

Wavelength (meters)

| 10^4 | 10^2 | 10^0 | 10^{-2} | 10^{-4} | 10^{-6} | 10^{-8} | 10^{-10} | 10^{-12} | 10^{-14} | 10^{-16} |

Radio waves   Microwaves   Radar   Infrared   Ultraviolet   X rays   γ rays

Visible spectrum

(a)

7000   6000   5000   4000

(b) $^1_1$H

6563   4861   4340 4101

(c) $^1_1$H

6563   4861   4340 4101

(d) $^4_2$He

6678   5875   5015 4921 4713 4471   4026

Plate 1   *Spectrograms of (a) white light—a continuous spectrum;*
(b) hydrogen—an absorption spectrum; (c) hydrogen—an emission
spectrum; (d) helium—an emission spectrum. Numbers on the
scale under the spectrograms indicate wavelengths corresponding to
each color and are given in angstroms (Å); 1 Å equals 10^{-10} m. See
pages 79 to 83 of the text.

Plate 2 *The planet earth as seen from Apollo 11 at a distance of about 98,000 nautical miles, or 178,000 km. (NASA)*

Plate 3 *Apollo astronaut Aldrin places the seismic station on the moon. Visible to the left of the lunar lander is the laser reflector and the American flag. (NASA)*

Plate 4 The first color photograph taken by Viking 2 on the Martian surface, showing the rocky reddish surface. Note the pink sky which is characteristic of Mars. The horizon is actually level although it appears tilted due to the 8° tilt of the lander. (JPL–NASA)

Plate 5 Jupiter, as seen by Voyager 1 while 33 million kilometers from the planet. Vigorous swirling activity surrounds the Great Red Spot. Objects as small as 600 km across can be identified—the best resolution of Jupiter ever achieved. (JPL–NASA)

Plate 6 (a) Io, as seen by Voyager 1 at a distance of 900,000 km. Surface activities identified on this moon of Jupiter include cratering and erupting volcanoes. This moon is bathed in the intense radiation of its mother planet. (JPL–NASA)

Plate 6 (b) Ganymede, Jupiter's largest satellite, as seen by Voyager 1 from a distance of 2.6 million kilometers. Ganymede is probably composed of ice and rock. Some of its surface markings remind one of the maria and the rayed impact craters of the moon. The small colored dots (blue, green, and orange) are not present on the satellite but are simply camera markings (JPL–NASA)

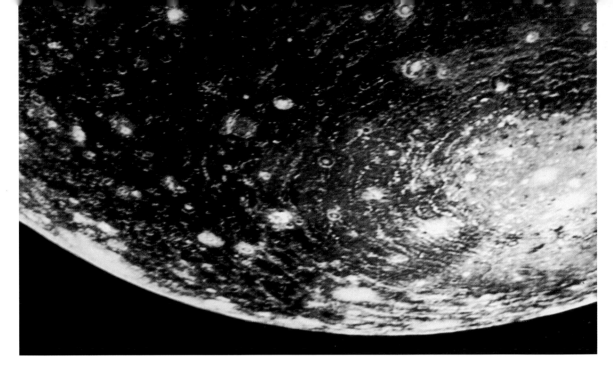

Plate 6 (c) Callisto, a little smaller than Ganymede, is thought to have a similar composition; however its darker appearance may be due to "dirty ice" on the surface. Visible near the upper left is a large basinlike structure, the center of which is much brighter than its surroundings. The bright spots may represent clean ice. (JPL-NASA)

Plate 6 (d) Europa, smallest of Jupiter's four Galilean moons, appears bright due to a mantle of ice perhaps 100 km thick. Cracks that developed appear to be filled with dark material from below. (JPL-NASA)

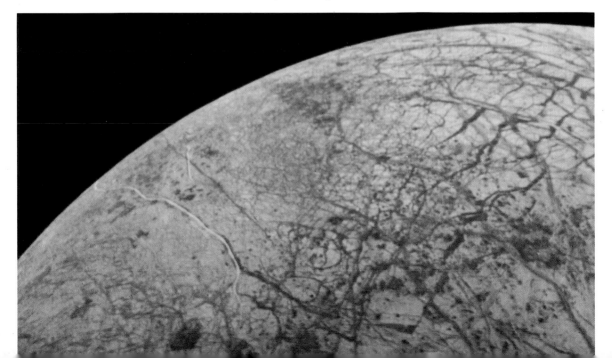

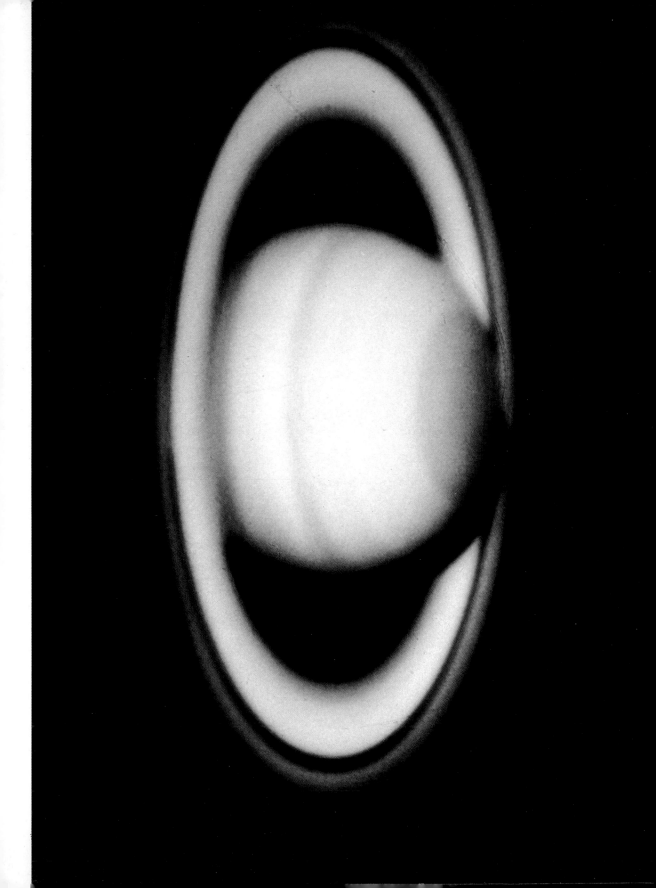

Plate 7 *(a) A ground-based view of Saturn, showing its banded structure and rings. This image was produced by combining sixteen original color images. (Ames–NASA) (b) The rings of Saturn, as seen by Pioneer on September 1, 1979, representing the first passage of a probe near this planet. (Ames–NASA)*

(b)

(a)

Plate 8 *Comet West (1975n) showing a great amount of detail in its dust and gas tails. (Clifford Holmes, Riverside Astronomical Society)*

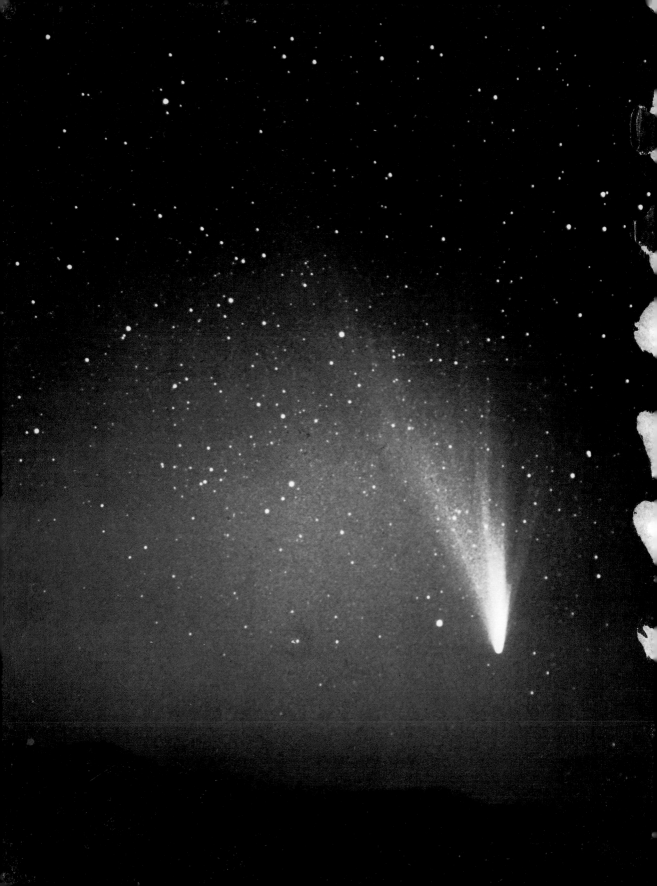

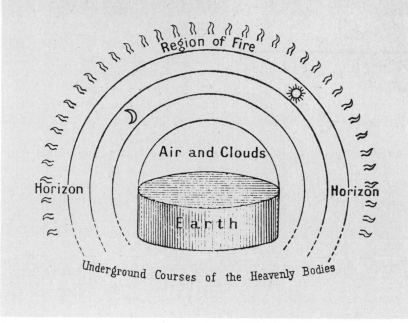

Figure 1.10   *An early Greek concept of the universe. (Yerkes Observatory)*

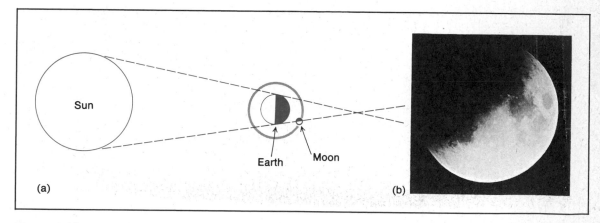

(a)

(b)

Figure 1.11   *(a) The moon passing into the shadow of the earth. (b) A partially eclipsed moon.*

The concept of a spherical earth was a great step forward, for it changed humans' entire concept of the universe. In place of a canopy of stars, they could now view the stars as surrounding the earth in all directions. Some thought of them as fixed to a vast crystal sphere. Now, too, the sun could be visualized as moving in a smooth orbital motion around the earth.

This is an example of how a single discovery may topple an entire conceptual structure that is based upon a false premise. Such discoveries serve to produce a scientific revolution. As you progress through this book, see if you can identify other such discoveries that produced major revolutions of thought.

In the sixth century B.C. there were still no indications that the earth is in motion (*rotating* on its axis or *revolving* around the sun). If the earth does not move in space, then all apparent movement of objects in the sky must be explained in terms of their own motion. It was not until about 450 B.C. that another Greek scholar, Philolaus, made an attempt to loosen the earth from its fixed position in the minds of his contemporaries by suggesting that there existed a "central fire" (not the sun) around which revolved the earth, the moon, the sun, and the planets. The earth was supposed to make one trip around this central fire each day, producing the apparent motion of all objects around the earth in that period of time. When asked if he had ever seen the central fire, he replied in the negative, indicating that Greece was always turned away from it. While the system of Philolaus did not find many adherents, it served the significant purpose of suggesting that the earth is in motion. Today we realize that the apparent daily motion of all objects in the sky is due to the rotation of the earth on its axis.

Also of the Pythagorean school, Democritus (about 450 B.C.) recognized that the fuzzy band of light we call the Milky Way is in reality numerous distant stars which appear close together. It is interesting to compare the wildly imaginative scheme of Philolaus, on the one hand, and the very keen insight of Democritus, on the other. Perhaps it is only our advantage of hindsight that allows us to judge between these ideas.

Plato's Academy, the world's first university, was founded in Athens in the fourth century B.C., and one of its most famous pupils was Aristotle (384–322 B.C.). Aristotle, who became the predominant philosopher of his age and tutor to the young Alexander of Macedon (Alexander the Great), was to influence astronomical thinking for approximately 2000 years, having firmly established the idea that the spherical earth was the center of the universe and stood stationary in that position, with the sun, moon, and planets moving around it in circular orbits. Aristotle's contemporary, Heraclides (388–315 B.C.), suggested that the apparent daily motion of these objects was due to the rotation of the earth, a very logical idea, but because it went counter to the thinking of the dominant academic group, it was cast aside.

In 332 B.C. Alexander the Great founded the city of Alexandria on the Mediterranean coast of Egypt. This city was established as a cultural center, a place where theoreticians and practical observers alike could find a level of support unheard of in other lands. Here Aristarchus (about 270 B.C.) challenged the teachings of Aristotle by asserting that the sun is the center of the solar system and that the earth and other planets revolve around the sun. He explained the apparent daily motion of all objects to be the result of the earth's rotation, and he believed the stars to be very

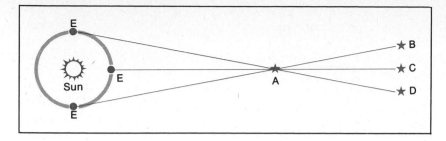

Figure 1.12 *The apparent shift of a nearby star (A) against the background of more distant stars, due to the revolution of the earth, as suggested by Aristarchus.*

distant. These ideas were far too radical for his contemporaries. Because the erroneous theory of Aristotle had become so deeply ingrained by that time, the challenge failed. It might have been argued by the Aristotelians that if the earth traveled around the sun, then nearby stars should appear to shift in their alignment with more distant stars—a phenomenon called *stellar parallax*. No such shift could be seen with the naked eye. The distances are too great. But had observers of that day been able to measure (as we can with our modern instruments) the very slight shift in the apparent position of nearby stars against the background of distant stars, as the earth revolves around the sun, the validity of Aristarchus's *heliocentric* (sun-centered) model would have been apparent. One can visualize this shift in the alignment of a star (A) with respect to more distant stars (B, C, or D) as the earth moves to various positions as shown in Figure 1.12. Even in the case of the nearest star, α (Alpha) Centauri, the apparent shift is only $\frac{1}{4800}$ of one degree (1°) to either side of its apparent central position.

Hipparchus, also of the Alexandrian school, made several very significant contributions to astronomy toward the end of the second century B.C. His methods of observation were very advanced for his time. He is often called the father of positional astronomy, for he constructed one of the first systematic catalogues of stars. He accurately stated the position of more than 1000 stars and ranked them according to their apparent brightness. He placed them into six categories: the brightest were called *first magnitude stars;* the dimmest (to the naked eye) were called *sixth magnitude stars.* This cataloguing project appears most remarkable when we realize that Hipparchus had no telescope or other optical aids with which to work. Hipparchus had observed that the sun appeared to move faster through the stars during one part of the year and slower during the other part. He reasoned that when it moved more slowly the sun must be farther from the earth and when it moved more quickly the sun must be nearer the earth. He concluded that its orbit could not be circular and devised the model depicted in Figure 1.13. The sun moves on a second, smaller orbit called an *epicycle.* His model, you

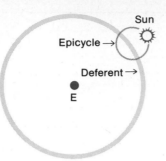

Figure 1.13
The epicycle of Hipparchus.

might say, is a wheel turning on a wheel. By choosing the proper speed of turning for the epicycle and for the larger orbit, the *deferent*, upon which it moves, the apparent variations in the sun's rate of motion and distance from earth could be explained.

Planets in Motion

The model conceived by Hipparchus for motion of the sun could also be applied to the moon with only minor adjustments, but the apparent motions of the sun and moon were really quite simple compared to those of the planets. Not only did planets vary in their rate of travel against the background of stars, but they even appeared to stop and reverse their direction among the stars at certain intervals. The casual observer would not recognize this reversal, for he only notices the daily westward motion of the sky. However, if he were to plot the position of a planet on a map of the sky, night after night, he would discover that planets usually move eastward among the stars but sometimes appear to stop, move westward for a number of nights, then again stop and proceed in an easterly direction (Figure 1.14). The astronomer speaks of the westerly motion of planets, among the stars, as being their *retrograde* (backward) motion. The flip pages beginning on page 114 show the retrograde motion of Jupiter, seen among the stars in the constellation Leo, in a very dynamic way. Hipparchus not only recognized this motion of a planet but made detailed observations and kept accurate records of planetary positions—records that were invaluable in later years.

In A.D. 140 Ptolemy (Claudius Ptolemeus), the last great astronomer of the Alexandrian school, gathered together all the accumulated astronomical knowledge of the ancient world and published his *Almagest*. This series of 13 volumes reflected the thinking of such men as Aristotle, Pythagoras, and Hipparchus, in combination with a few of Ptolemy's own ideas, and this combined picture of the universe is called the *Ptolemaic system*. Remember that Aristotle viewed the earth as standing still in space at the center of the cosmos. Pythagoras had proposed a spherical earth, and Hipparchus had introduced the concept of the epicycle, as applied to the sun. However, the observation which demanded Ptolemy's

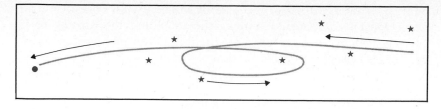

Figure 1.14 *The apparent retrograde motion of a planet (as seen by an observer facing south).*

main attention, and that of his immediate predecessors, was the fact that planets sometimes appear to *retrograde* (back up) against the background of stars. In an attempt to preserve the concept of uniform circular motion, thought to be an absolute necessity, Ptolemy assigned each planet its own epicycle, the center of which moved along the larger orbit called the deferent (Figure 1.15). As the sun moved about the earth, so the outer planets (Mars, Jupiter, and Saturn) moved on their respective epicycles. This motion produced two desired effects: the planets appeared to retrograde at the proper time, and they sometimes appeared closer to the earth and therefore brighter at that time. The right-hand flip pages beginning on page 25 show Ptolemy's model of planetary motion.

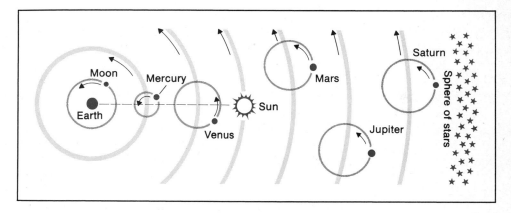

Figure 1.15 *The Ptolemaic system.*

It should be particularly noted that Mercury and Venus move so that the centers of their respective epicycles always lie on an imaginary line joining the earth and the sun. This is consistent with the fact that Mercury is always seen within 28° of the sun. From our point of view on earth, Mercury would appear to flit back and forth on either side of the sun. Likewise Venus, while it has a much slower apparent motion among the stars than Mercury, also moves from one side of the sun to the other, and never more than 47° from it. The epicycle of the sun has been omitted in the drawing for the sake of simplicity. Of course, the sun never appears to retrograde among the stars, and the turning of its epicycle had to be synchronized with its motion on the deferent to explain its lack of retrograde motion.

How Good Was Ptolemy's Model?

We should now ask ourselves some interesting questions: Just how good was the Ptolemaic system? How well did it work as a model of the universe? Did it seem to agree with the observations of Ptolemy's time? Could one use this model to predict the location of any planet at some prescribed time in the future? To the latter two questions we can answer, "Yes, within fairly good accuracy," and to that degree the Ptolemaic system was a good model. There were no observations at that time which seemed to invalidate the model. As time passed, however, it became evident that the accuracy with which the position of a planet could be predicted was not good enough. Small corrections were frequently necessary, and these multiplied until the model became quite cumbersome.

Despite these difficulties Ptolemy's was destined to remain the principal model of the universe for more than 1300 years. How did a theory based upon so many erroneous ideas stand unchallenged for so long? Ptolemy's model was one of the last to emerge from the classic age of philosophy and natural science. Successor to the Greek power, the once mighty Roman Empire had already begun its gradual decline in Ptolemy's time. By the middle of the fifth century Rome had fallen prey to barbarian hordes from the north—the Vandals, Visigoths, and other tribes. The plundering that ensued destroyed much of the classical culture, the fine works of art, and the literature of Europe. These were the "Dark Ages" of Europe; however, when the city of Alexandria fell to Arab invaders in A.D. 642, the culture and "science" which characterized that city were preserved. Ptolemy's *Almagest* was translated into Arabic and found numerous readers among scholars in Bagdad, a new center of learning. Arabs not only made significant contributions in mathematics and other fields, but their respect for knowledge led them to compile the scientific works of many cultures. With later expansionism, this knowledge flowed across North Africa into Spain in what is termed the Moorish influence.

In Spain, a new awakening may be traced to Alphonso X, king of Castile (northern Spain), who in 1222 commissioned the preparation of new tables describing the motion of the sun, moon, and planets. Within these Alphonsine Tables we see the Greek influence transmitted through Ptolemy's *Almagest*; however, certain improvements had been made by both Arab and Jewish astronomers, working in the observatory of Toledo, Spain. (Study of the errors that remained in the Alphonsine Tables was destined to be a prime motivating force for the major discoveries of the sixteenth century.)

In the thirteenth century St. Thomas Aquinas, a prominent Dominican educator, philosopher, and theologian, had succeeded in elevating the Aristotelian model, which placed the earth at the center of the universe, to the level of religious dogma. From then on, any person who taught a contrary view might be tried before the Inquisition as a heretic and, if found guilty, be put to death. It is within this atmosphere that we witness the work of Nicolaus Copernicus, Tycho Brahe, Johannes Kepler, and Galileo Galilei. As late as 1600 Giordano Bruno, an opponent of Aristotelianism and champion of the new Copernican cosmology, was burned at the stake by the Inquisition.

Renewed interest in science had been evident in Europe as early as the fourteenth century, but that interest really blossomed in the fifteenth and sixteenth centuries, especially in the universities of Italy and Germany. With the successful voyages of exploration around Africa to India and East Asia, discovery of the Americas, and circumnavigation of the globe (in 1522), there was renewed interest in astronomy and mathematics as aids to ocean mariners. The astronomical-philosophical classics of ancient Greece were examined in their original form, and the study of these works produced a flood of reactions. Many of the concepts of Aristotle, Philolaus, Aristarchus, and Ptolemy, together with the writings of the Arab world, were now scrutinized very critically. The time was ripe for the emergence of a great Polish astronomer, Nicolaus Copernicus (1473–1543).

Nicolaus Copernicus, Founder of Modern Astronomy

Under the influence of a learned uncle who had adopted him at the age of 10, Nicolaus Copernicus attended the University of Cracow to study the classics, philosophy, theology, law, medicine, and mathematics (Figure 1.16). Continuing his education in Italy, he acquired a doctoral degree in canon law and also became competent in medical practices of the time, but his contact with a noted astronomer at the University of Bologna sparked a lasting interest in astronomy. In 1500 he lectured in that subject at the Vatican in Rome. His training in the Greek language allowed him to pursue a first-hand study of the classics. These interests, together with the errors he had found in the Alphonsine Tables, spurred him to spend some 30 years in the development of a model of the solar system (Figure 1.17), published in a six-volume work called *De Revolutionibus Orbium Coelestium*. Although the manuscript of this great work was virtually completed in 1530, Copernicus delayed publication until the final years of his life because of political and religious considerations. The final volume only reached him at his deathbed.

The essence of his scheme is expressed by his own words:

At rest in the middle of everything is the sun. For in this most beautiful temple, who would put this lamp in another or better

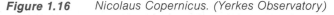

position than from which it can illuminate the whole thing at the same time? Thus, indeed, as though seated on a royal throne, the sun governs the family of planets revolving around it.

Copernicus had thus displaced the earth from its favored position at the center of the universe and had relegated it to the position of one of several planets which orbit the sun. He might well have incurred the wrath of the Inquisition for such a heretical idea, if he had not died within a few days of the publication of his theory.

Copernicus believed that the earth both rotated on its axis and revolved around the sun, as did all the other planets. He viewed the stars as being very distant and fixed in place. It was their great distance, he argued, that explained the lack of any stellar parallax (apparent shifting

Figure 1.16   *Nicolaus Copernicus. (Yerkes Observatory)*

in the position of nearby stars as the earth moves in its orbit around the sun). Copernicus explained that the daily "rising" and "setting" of all objects in the sky could easily be explained by the daily rotation of the earth on its axis. The apparent yearly motion of the sun among the stars could be explained by the revolution of the earth around the sun. But what about the retrograde motion of the planets? Could Copernicus explain why planets sometimes appear to move backward among the stars?

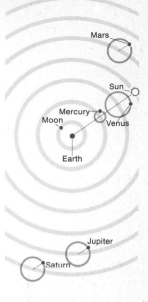

The right-hand flip pages, starting on page 115, illustrate the motions of the planets in the Copernican model. As you flip these pages, you will see that Mercury and Venus periodically pass the earth in its orbit, and likewise the earth passes the outer planets of Mars, Jupiter, and Saturn periodically in their orbits. This phenomenon of passing produces the apparent retrograde motion of the planets. Consider the earth and Mars as seen in Figure 1.18. Note that as the earth moves from positions A through K, Mars also moves through the corresponding positions on its orbit; however, the earth moves faster. At any one position the stars with which Mars appears to align vary. Moving from A to E, Mars seems to move eastward among the stars, but while moving from E to G it seems to

Figure 1.17   *The Copernican system, as shown in* De Revolutionibus Orbium Coelestium, *1566. (Yerkes Observatory)*

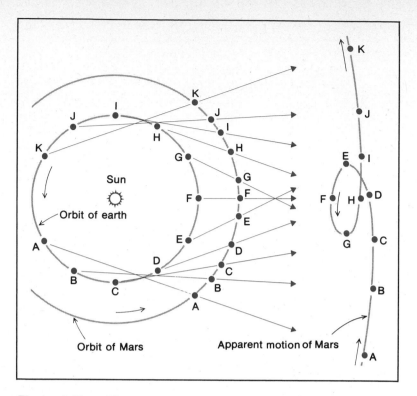

Figure 1.18 *The apparent retrograde motion of Mars (shown at the right).*

move westward. Finally, moving from G to H, it appears to move eastward again. A similar explanation applies to the apparent retrograde motion of the other planets.

It is also easy to see from the flip pages that planets are much closer to the earth on some occasions and farther from it at others. This explains their apparent changes in brightness. While Copernicus no longer needed the epicycle to explain the retrograde motion of planets, he retained the idea for other reasons. He believed that all objects move in a uniform way (at the same rate) on circular orbits or on combinations of circular orbits. Yet he saw planets sometimes moving rapidly against the background of stars and sometimes slowly; he thought that through a set of epicycles, which were assigned the proper uniform motion, the complex motion of the planet could be explained. Some of his earlier works contained epicycles upon epicycles. The model of Copernicus appears to explain the observations of the time, but the model of Ptolemy did so as well. Although there was not yet any substantial evidence as to which model most closely represented reality, the publication of the Copernican theory created a furor which was to last more than 200 years. The daring of Copernicus encouraged others to devise better means for observation so that his theory might be put to the test. Typical of this new spirit of observation was the work of Tycho Brahe.

Tycho Brahe, the Diligent Observer

Tycho Brahe (1546–1601), a Danish nobleman (Figure 1.19), was trained as an astronomer and, under the sponsorship of the king of Denmark, he devoted himself to a systematic observational approach to the subject. He was granted the tiny Baltic island of Hven (Figure 1.20), where he established an observatory and constructed instruments capable of measuring angles between stars quite accurately (Figure 1.21). Tycho made thousands of observations of star positions, averaging his readings to produce a catalogue of approximately 800 stars. The positions of many stars were accurate to within $\frac{1}{100}$ of a degree. Tycho's catalogue was far more accurate than that of Hipparchus or Ptolemy, and it stood as a reliable reference source for more than a century. Tycho also kept precise records of the changing positions of planets, believing that only through an understanding of the orderliness and unity of the universe might man better order his own affairs. In a lecture at the University of Copenhagen in 1574 Tycho said:

> To deny the forces and influence of the stars is to undervalue firstly the divine wisdom and providence and moreover to contradict evident experience. For what could be thought more unjust and foolish about God than that He should have made this large and admirable scenery of the skies and so many brilliant stars to no use or purpose—whereas no man makes even his least work without a certain aim.

Figure 1.19   *Tycho Brahe, the observer.* (*Yerkes Observatory*)

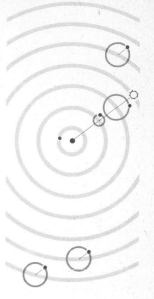

Figure 1.20 *Uraniborg, Tycho Brahe's observatory. (Yerkes Observatory)*

Figure 1.21 *Tycho Brahe at Uraniborg: (a) Tycho observing with his great mural quadrant; (b) sextant used to measure angles between celestial objects. (Rare Book Division, The New York Public Library, Astor, Lenox and Tilden Foundations)*

(a)   (b)

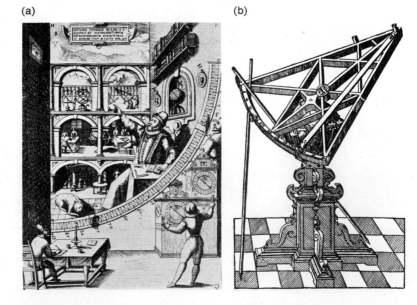

Tycho's labors epitomized an important aspect of the scientist's work that is often overlooked: that of keeping absolutely honest and careful records, which are often valuable regardless of the correctness of conclusions drawn at the time. Tycho's observational records were his greatest contribution to astronomy, for they were to assist those who followed him to present a clearer picture of the solar system. He could not accept Copernicus's premise that the earth is in motion, so he devised a very interesting scheme that pictured Venus and Mercury as circling the sun but had the sun and the other planets orbiting around the earth (Figure 1.22).

In his later years Tycho fell out of favor with the new king of Denmark and lost his island sanctuary. He sought and found the support of Rudolf II, the scholarly Holy Roman emperor, and built a new observatory near Prague, in Bohemia. Here the master astronomer took as his

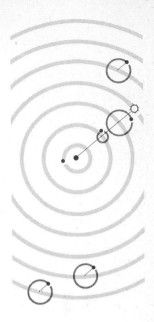

Figure 1.22   *Hypothetical planetary system devised by Tycho Brahe in 1578: here Venus and Mercury circle the sun, the sun and moon circle the earth, and Mars, Jupiter, and Saturn move in epicycles, with their deferents centered on the earth.*

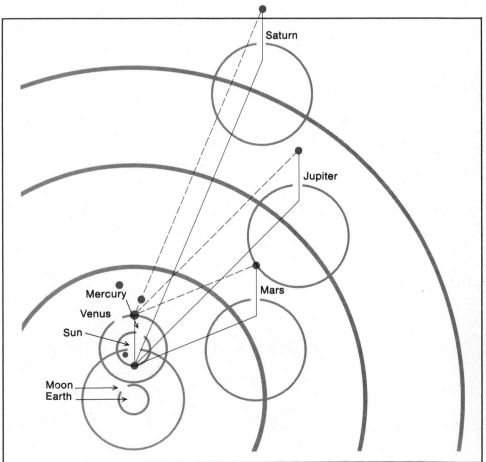

aide a young German philosopher and mathematician, Johannes Kepler (1571–1630), who upon Tycho's death fell heir to his voluminous observational records.

Johannes Kepler, the Inventive Mathematician

Kepler was educated at Tübingen University and taught at the University of Graz in Austria (Figure 1.23). When he was relieved of this professorship for religious reasons, he received an invitation from Tycho Brahe to become his assistant. Joining Tycho, Kepler soon acquired a keen respect for the high degree of accuracy with which Tycho observed the positions of stars and planets.

Kepler immediately recognized that Mars did not move against the background of stars at a constant rate but moved faster during one part of its orbit and then slowed down during another part. This fact alone must have suggested to him that the planet did not move around the sun in a circular orbit as Copernicus had proposed, for if the orbit were circular, a constant rate of motion would have been expected. Kepler attempted to fit various orbital shapes to the observed motion of Mars, shapes which would provide for the variation in speeds he had witnessed. He eventually concluded that the best fit was that of an *ellipse*.

Figure 1.23   *Johannes Kepler. (Yerkes Observatory)*

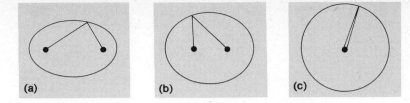

Figure 1.24   *Drawing ellipses of different eccentricities.*

A device for drawing ellipses can be fashioned using two thumb-tacks and a string of fixed length. By attaching each end of the string to a thumbtack and by stretching the string taut with a pencil, it is possible to draw an ellipse (Figure 1.24). By varying the spacing of the thumbtacks, one may draw ellipses of varying eccentricity. Figure 1.25 shows a number of terms associated with an ellipse. The two points (thumbtacks) are called the *foci* (plural of *focus*), and a line drawn through these points and extending to the curve itself is called the *major axis* of the ellipse. One-half that distance is called the *semimajor axis,* and this is equivalent to the average distance (radius) from one focus to any point on the ellipse. In Table 1.1 the semimajor axis for each planet is given, thus expressing its average distance from the sun. The *eccentricity* of an ellipse is found by dividing the distance between the foci by the length of the major axis. A circle has an eccentricity of zero, and the eccentricities

Figure 1.25   *Axis and average radius of an ellipse.*

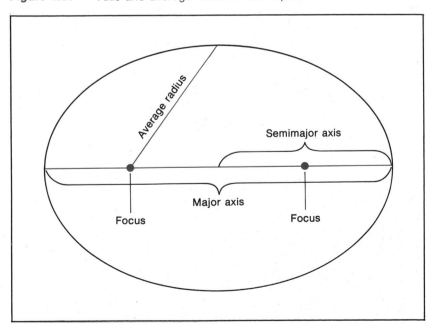

Table 1.1   Planetary Data

PLANET	SEMIMAJOR AXIS (r), A.U.	r^3	PERIOD (p), YEARS	p^2
Mercury	0.39	0.058	0.24	0.058
Venus	0.72	0.378	0.62	0.378
Earth	1.00	1.000	1.00	1.000
Mars	1.52	3.54	1.88	3.54
Jupiter	5.20	140.7	11.86	140.8
Saturn	9.54	867.7	29.46	867.9

of the planetary orbits range from 0.007 (Venus) to 0.25 (Pluto), with that of the earth being 0.017, not very much different from a circle.

Kepler found that he could best fit the observed motions of Mars to an elliptical orbit with the sun at one focus, the other focus being only an imaginary point in space. He generalized this idea to include all planets, and the statement of it is now known as Kepler's first law of planetary motion:

> Each planet moves around the sun in an orbit whose shape is that of an ellipse, with the sun at one focal point.

Although this law is called Kepler's first, this numbering may not represent the order of its discovery. That which is called his second law was actually more obvious—the fact that a planet moves more slowly when far from the sun and more rapidly when nearer the sun. This lack of uniformity in motion suggests something other than a circular orbit. Kepler's second law is stated:

> A straight line joining the planet and the sun sweeps out equal areas in space in equal intervals of time.

As an illustration of Kepler's second law (Figure 1.26), let us suppose that Mars moves from points A to B in a month. Some time later it also

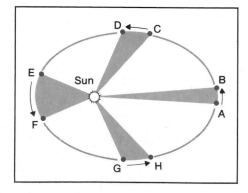

Figure 1.26 *Illustration of Kepler's law of equal areas in equal time.*

moves from C to D in a month, but now it is closer to the sun, and so the distance from C to D must be greater in order for Mars to sweep out an equal area in the same time. To cover this greater distance, Mars must move faster. Likewise, when the planet travels from E to F, still closer to the sun, it must move still faster to sweep out an equal area in a month; however, as it moves from G to H it is moving more slowly again, and as it passes from A to B it has returned to the speed that it had at the start of this illustration. Thus Kepler's second law provides a mathematical model whereby the speed of a planet may be computed for any given position on its orbit.

To recognize a third basic relationship required several years of effort on Kepler's part. Of course, it was obvious to him that the farther a planet was from the sun, the longer it took to complete one revolution around the sun. But Kepler sought to find the exact mathematical relationship between its period of revolution and its average distance from the sun (the semimajor axis of its orbit), if such a relationship did indeed exist. Although the actual distances between the sun and the planets were then not known in miles, it was possible to calculate the ratio between a planet's distance from the sun and the earth's distance from the sun. For example, Kepler knew that the semimajor axis of Mars's orbit was equal to approximately 1.5 times the semimajor axis of the earth's. If we let the semimajor axis of the earth's orbit be called 1 A.U. (astronomical unit), then the semimajor axis of Mars's orbit can be specified as 1.5 A.U. Table 1.1 lists the semimajor axes for each planet, as measured in astronomical units. This table also shows the period of revolution for each planet, measured in years. Note that for any planet, if the semimajor axis (r) is cubed and the period of revolution (p) is squared, the two numbers produced are the same, with only minor discrepancies for Jupiter and Saturn. This fact may be expressed by the simple equation $p^2 = r^3$, where p is measured in years and r is measured in astronomical units. Should we choose to measure the period of the planet in days or the distance in miles, then we would not expect $p^2 = r^3$; however, we can allow for this possibility by writing $p^2 = kr^3$, where k is a constant whose value is determined by the units of measure used. Still another equation may be written to specify Kepler's third law:

$$\frac{(p_1)^2}{(p_2)^2} = \frac{(r_1)^3}{(r_2)^3}$$

which may be stated as

The squares of the periods of any two planets have the same ratio as the cubes of their semimajor axes.

Assuming that this relationship would hold true for any planet in our solar system, let us suppose an imaginary planet X has a period of 8 years. Can we find its average distance from the sun in astronomical units? Let the period and semimajor axis of planet X replace p_1 and r_1 in the equation, and let the period and semimajor axis of the earth replace p_2 and r_2. For the earth, we know that $p_2 = 1$ year and $r_2 = 1$ A.U. (see Table 1.1). The period of planet X has been given as 8 years, so $p_1 = 8$. If we make these replacements in the equation and solve for r_1, we have

$$\frac{(p_1)^2}{(p_2)^2} = \frac{(r_1)^3}{(r_2)^3}$$

$$\frac{(8)^2}{(1)^2} = \frac{(r_1)^3}{(1)^3}$$

$$\frac{64}{1} = \frac{(r_1)^3}{1}$$

$$4 = r_1$$

Therefore the imaginary planet X would have a semimajor axis of 4 A.U.

A warning. All of Kepler's laws are highly idealized, for they neglect the fact that each planet influences every other, so that the orbits are never smooth ellipses nor are the areas swept out exactly the same. Furthermore, Kepler's third law is limited to a restricted set of conditions, namely, to very small (low-mass) objects going around very large (high-mass) objects, such as planets around the sun, or moons around a large planet. Nonetheless, Kepler's discoveries were remarkable approximations that were later confirmed and generalized by Sir Isaac Newton, whose astronomical knowledge and mathematical techniques were more advanced.

Although Kepler was a mystic, his work served as the foundation for the removal of astronomy from the realm of mysticism and for the establishment of its cause-and-effect nature. He sensed that some unknown force, emanating from the sun, kept the planets moving in their orbits. He was on the verge of discovering gravity. We know that Kepler communicated with Galileo, a man in whom his thoughts were to find fruition.

Galileo Galilei, Father of Experimental Science

The seventeenth century was one of discovery, experimentation, and invention. The invention that had the largest impact on astronomy was the telescope, of which the exact history is unclear. The Dutch spectacle maker Hans Lippershey is usually credited with having combined several lenses to produce an enlarged image of a distant object. Although he produced his instrument in 1608, Lippershey apparently did not fully grasp its enormous potential. The following year in Padua, Italy, Galileo Galilei (1564–1642) heard of this invention and, without any detailed knowledge of how the task had been accomplished, set out to produce a telescope of his own (Figure 1.27). Galileo's first success resulted in an instrument that enlarged objects three times, a *three-power* telescope. Later refinements resulted in a 30-power instrument. These instruments he put into immediate use. By directing them skyward, Galileo soon discovered four moons circling the planet Jupiter, each moving around the planet in its own particular period (Figure 1.28). This sight reminded Galileo of the solar system itself, as Copernicus had described it, and it surely demonstrated that the earth is not the only center of revolution in the universe.

Using the telescope, Galileo also found that what had appeared as a fuzzy patch in the sky, a nebula, could now be seen as a group of separate stars. In fact, in all regions of the sky the instrument revealed many stars that were too faint to be seen with the naked eye. Concerning a familiar constellation, Galileo wrote:

> I had determined to depict the entire constellation of Orion, but I was overwhelmed by the vast quantity of stars and by want of time, and so I have deferred attempting this to another occasion, for there are adjacent to, or scattered among, the old stars more than five hundred new stars.

Galileo also observed that the planet Venus goes through phases (see Figure 1.29), sometimes appearing as a thin crescent and at other times as a disk almost fully lighted. This was a startling observation; it provided evidence, for the first time, that the Ptolemaic system could not represent

Figure 1.27   *Galileo Galilei. (Yerkes Observatory)*

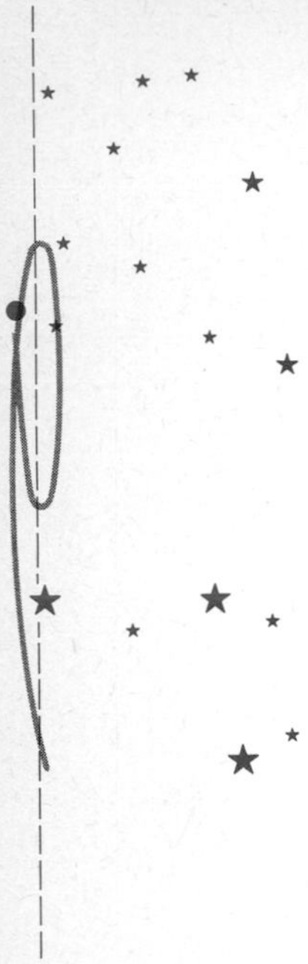

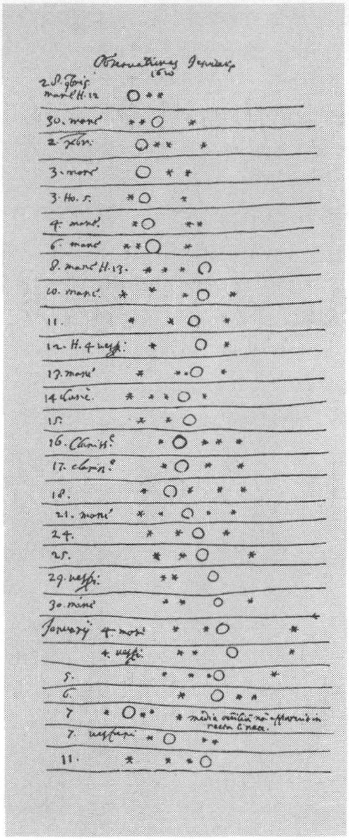

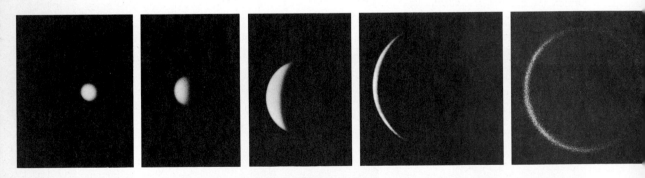

Figure 1.28 Galileo's drawings showing the motions of the moons of Jupiter. (Yerkes Observatory)

Figure 1.29 The phases of Venus. (Lowell Observatory)

reality. Remember that in the Ptolemaic system the center of Venus's epicycle must always remain on an imaginary line joining the earth and the sun. Under this restriction, the planet Venus could never appear any fuller than a mere crescent (see Figure 1.30). Galileo observed that Venus sometimes appears almost full. Could this fact be better explained by the Copernican system? Yes, as you can see in Figure 1.31. Venus may be seen in all possible phases, except when it is closely aligned with the sun.

The revelation that Venus goes through phases dealt the death blow to the Ptolemaic system. Confident in his new-found proof, Galileo tried to convince the leadership of the Roman Catholic church that its interpretation of the Holy Scriptures was inconsistent with observed facts. He failed in this attempt. In fact, the church hierarchy pronounced his findings to be false and heretical and forbade anyone to teach them. So blind are the eyes of men when they do not want to see!

Galileo also observed sunspots and identified them as being on the surface of the sun itself. By charting their movement, he determined the rotation period of the sun (Figure 1.32). To the philosophers of that time, however, any "imperfection" on a celestial body was unthinkable; and so his report was again ridiculed.

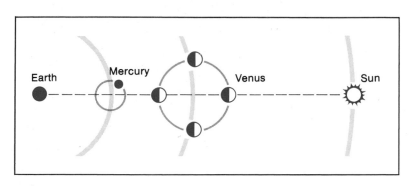

Figure 1.30   *The phases of Venus—Ptolemaic system.*

Figure 1.31   *The phases of Venus—Copernican system.*

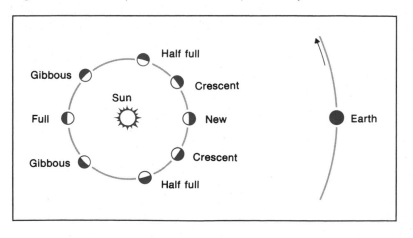

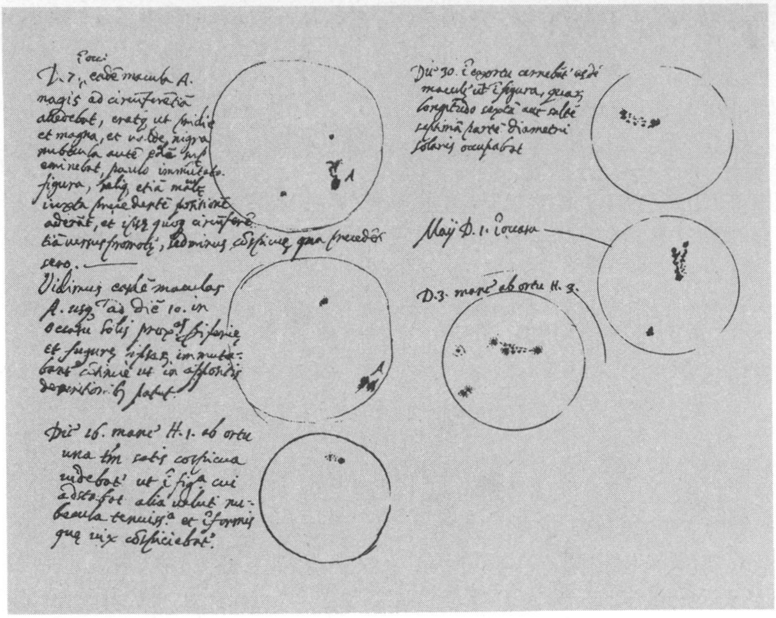

Figure 1.32 *Galileo's drawings of sunspots. (Yerkes Observatory)*

Despite all these obstructions, Galileo opened up a new era of scientific experimentation. For example, he wondered about Aristotle's claim that when two objects of different weight are dropped from a high place, the heavier object hits the ground first. According to legend, Galileo climbed the Leaning Tower of Pisa (in the city of his birth) and from the top dropped two round stones, one of which was much heavier than the other. To the amazement of the crowd that gathered, the stones fell side by side—hitting the ground together. In order to measure the effect of gravity more accurately, Galileo caused a ball to roll down an inclined ramp and timed its acceleration. He caused an object to slide along different surfaces and observed that when friction was least, the object would travel farther. He may have come to the realization that an object which is moving through space, where no friction exists, will travel onward at its given speed forever; this concept is now credited to Sir Isaac Newton. That great British scientist surely was recognizing the work of Galileo, as well as of Kepler, Tycho, and Copernicus, when he said, "If I have been able to see farther than other men, it is because I was standing on the shoulders of giants."

Isaac Newton, the Young Genius

Newton (1642–1727) entered Cambridge at 19. Four years later, in 1665, the year of the great London plague, the university was closed during the epidemic and he returned home (Figure 1.33). At this time he

Figure 1.33 *Sir Isaac Newton.*
(Yerkes Observatory)

made lasting contributions in three areas: in mathematics, by inventing the calculus; in optics, by demonstrating that ordinary white light is composed of various colors that can be separated upon refraction; and in astronomy, by precisely formulating the idea of gravitation and the laws of motion that govern bodies that are in each other's gravitational field.

The idea of the mutual attraction of objects had earlier been suggested by Copernicus in his explanation of the earth's spherical shape. Newton recognized this force as being responsible for the tides, for the fact that objects fall toward the center of the earth when released, and for the fact that the planets remain in orbit around the sun. Realizing that circular motion is not the natural motion of an object, he stated his first law—the law of inertia:

> *Every body perseveres in its state of rest, or of uniform motion in a right (straight) line, unless it is compelled to change that state by force impressed thereon.*

This says that a body will, if it is at rest, remain at rest unless acted upon by an external force. It also says that a body will, if it is moving, continue to move in a straight line forever—unless acted upon by an external force. It follows that the natural motion of an object moving in space is not circular but in a straight line (assuming that no other force is acting upon that object).

Newton's second law says:

> *The alteration of motion is ever proportional to the motive force impressed, and is made in the direction of the right (straight) line in which it is impressed.*

He explained this law further by saying:

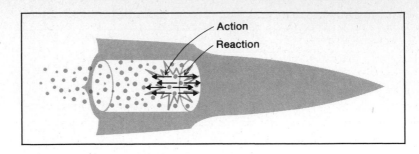

Figure 1.34   *Newton's third law in action.*

> *If any force generates a motion, a double force will generate a double motion, a triple force triple the motion.*

This second law also expresses the idea that if a force is applied that opposes a motion, the object will slow down, but if the force is applied partially or entirely in line with the motion, the object will speed up. Newton's second law also embodies the idea that the amount by which an object is speeded up or slowed down by a given force depends on the mass of the object. For a given force (f), the greater the mass (m), the less the acceleration (a): $F = ma$.

Newton further stated, in his third law:

> *To every action there is always opposed an equal reaction; or the mutual actions of two bodies upon each other are always equal, and directed to contrary parts.*

To illustrate this law, imagine jumping from the stern of a boat into the water. While your body is propelled in one direction, the boat moves in the other direction. This law explains the force that propels a rocketship (Figure 1.34). As particles of gas are expelled from the rear of the rocket, the equal but opposite reactive force propels the ship forward. The action and reaction forces always act on different bodies (*action*—force on gas; *reaction*—force on rocketship).

Let us apply Newton's laws to the planets and their orbits around the sun. His first law says that the natural tendency is for the planet to move in a straight line, thus "flying off" along the tangent (A) to the circle (see Figure 1.35). However, the force of gravitation between the planet and the sun is acting in a direction (B) toward the sun. The effect of this force of gravity is such that the planet continually falls away from its natural path (A) by an amount which just maintains it in an elliptical orbit.

Kepler had already described the motions of the planets, based on what he had observed. Now it was possible for Newton to deduce these same motions from his universal laws of gravitation. He realized that the total mass of the objects in question also entered into the relationship, as shown in the following equation:

$$(m_{sun} + m_{planet})p^2 = kr^3$$

Since the mass of the sun (m_{sun}) is so much greater than the mass of any one of its planets, we may say that $(m_{sun} + m_{planet})$ is approximately the same as the mass of the sun alone; therefore

$$(m_{sun})p^2 = kr^3$$

$$(m_{sun}) = \frac{kr^3}{p^2}$$

This shows that the mass of the sun can be found with only the knowledge of the period of any planet and its average distance from the sun. Furthermore, Newton's law of planetary motion reduces to that of Kepler when we permit the approximation

$$m_{sun} + m_{planet} = m_{sun}$$

$$(m_{sun})p^2 = kr^3$$

$$p^2 = \frac{kr^3}{m_{sun}}$$

Let k/m_{sun} be set equal to a new constant K; then

$$p^2 = Kr^3$$

Thus Newton was able to confirm Kepler's laws in essence and to refine them in detail. Using the refined statements, it is possible not only to accurately describe the motions of natural bodies but also to predict to a high degree of accuracy the orbits of today's artificial satellites.

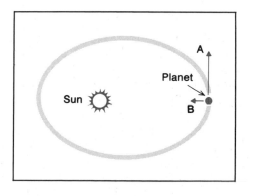

Figure 1.35   *The orbit of a planet.*

Questions

1. List several facts that indicate that humans made astronomical observations before 1000 B.C.

2. What are some ways in which early man lived more harmoniously with nature as a result of observing cycles in nature?

3. What object in the sky appears to have determined the length of a month?

4. Of what practical value were astronomical observations in Egypt?

5. Why did the Mesopotamian calendar of twelve 30-day months fail to record the length of a year accurately?

6. What are several similarities and differences between astronomy and astrology?

7. What evidence led Pythagoras to conclude that the earth was shaped like a ball?

8. While Philolaus's model of the planetary system did not represent reality, it did contribute an important true concept. What was that concept?

9. In what sense was Aristarchus's model many years ahead of his time?

10. List the principal points of Aristotle's picture of the sun, earth, planets, and universe.

11. By what scheme did Hipparchus explain that the sun was not always the same distance from the earth?

12. How did Ptolemy explain the apparent retrograde (backward) motion of the planets?

13. How can the apparent retrograde motion of the planets be explained in terms of the Copernican system?

14. Why did the incorporation of the Aristotelian model into church dogma tend to delay man's gaining a more accurate picture of the solar system?

15. How was Tycho Brahe able to make a catalogue of accurate star positions when the telescope had not yet been invented?

16. What do you think was Tycho's motivation for studying the planets and stars?

17. What is the shape of the orbits of the planets?

18. Express Kepler's second law in common language.

19. Explain the significance of Galileo's discovery that Venus goes through phases like the moon.

20. What did Newton view as the most natural tendency for a body moving in empty space?

21. Design a calendar which would make the new moon fall on the first day of each month. Would your calendar make the shortest day of the year fall at the same date (give or take one day) each year? Why?

22. If at noon of the shortest day of the year you placed a stick in the ground so that it had no shadow, how could you use that stick to tell when a year had passed?

Suggested Readings

BUSH, S., Ancient astronomers—Fact or fancy? View of Alexander Thom. *Psychology Today 10*, 100(1976).

DE VAUCOULEURS, GERARD, *Discovery of the universe*. London: Faber and Faber, 1957.

GINGERICH, OWEN, Johannes Kepler and the Rudolphine Tables. *Sky and Telescope 42* (6), 328–333 (1971).

GINGERICH, OWEN, Copernicus and Tycho. *Scientific American 229* (6), 86–101 (1973).

KRUPP, E. C. (ed.), *In search of ancient astronomies*. New York: Doubleday, 1978.

LOCKYER, J. NORMAN, *The dawn of astronomy*. Cambridge, Mass.: M.I.T. Press, 1964.

PANNEKOEK, A., *A history of astronomy*. New York: John Wiley (Interscience), 1961.

ROSEN, EDWARD, Copernicus' place in the history of astronomy. *Sky and Telescope 45* (2), 72–75 (1973).

SHAPLEY, HARLOW, *A source book in astronomy: 1900–1950*. Cambridge, Mass.: Harvard University Press, 1960.

SHAPLEY, HARLOW, and HOWARTH, HELEN E., *A source book in astronomy*. New York: McGraw-Hill, 1929.

STILLMAN, DRAKE, Galileo's discovery of the law of free fall. *Scientific American 228* (5), 84–92 (1973).

WILSON, CURTIS, How did Kepler discover his first two laws? *Scientific American 226* (3), 92–106 (1972).

TWO

METHODS OF ASTRONOMY

Unlike physicists, chemists, and geologists, astronomers cannot get their hands on much of what they study. While humans have now walked on the moon and brought home samples, and while they have placed a probe on the surface of Mars, for the most part astronomers are restricted to working with the one thing which almost every celestial object sends to them, namely, radiation. Light is the most familiar, but not the only, form of radiation received from stars and galaxies, others are radio, infrared, ultraviolet, X rays, and gamma rays. All these forms of radiation have certain aspects in common with light, so as we study the various properties of light, most of them will apply to the other forms as well. The important thing is to learn how to extract the maximum information from the radiation that reaches us.

In 1610 Galileo constructed one of the earliest telescopes (Figure 2.1), by which he could observe some of the moons of the planet Jupiter. When these moons were eclipsed by the planet, he recorded the times when they went behind the planet and then reappeared on the other side. By 1675 the Danish astronomer Ole Roemer became interested in the possibility that the motion of these moons might be so regular as to form a natural clock that could be used for purposes of navigation (Figure 2.2). First he observed the length of time between

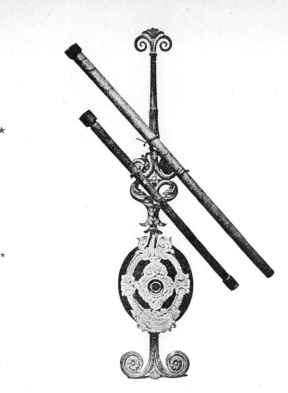

eclipses of the moon Io when the earth was at position E_0 (Figure 2.3). These time intervals seemed to be the same. As the earth moved on to positions E_1, E_2, E_3, E_4, and E_5, however, the length of time between eclipses was not the same; it became a bit longer after each eclipse. As the earth reached E_6, Roemer could not see Jupiter because of the sun's glare, but when he saw Jupiter again from position E_7 and began his timing once again, a strange thing occurred. Now the intervals of time between eclipses began to decrease, and this continued until he again returned to E_0. While Roemer apparently had failed to find a regular clock in the motions of Io, he had discovered something more important—the velocity of light (Figure 2.4). He reasoned that as the earth moves farther from Jupiter, it should take longer for the light that signals Io's eclipse to reach the earth. He was then able to determine the speed with which light travels. It took approximately 1000 sec (seconds) more time to reach the earth at position E_6 than at E_0. The best estimate for the major axis of the earth's orbit at that time was 150,000,000 miles, and so the velocity of light was computed as

$$\frac{150,000,000 \text{ miles}}{1000 \text{ sec}} = 150,000 \text{ miles/sec}$$

Today we know that the value of the major axis is closer to 186,000,000 miles, yielding

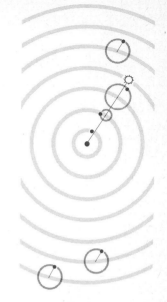

Figure 2.2   *Roemer's telescope. (Yerkes Observatory)*

Figure 2.3   *Roemer's determination of the velocity of light.*

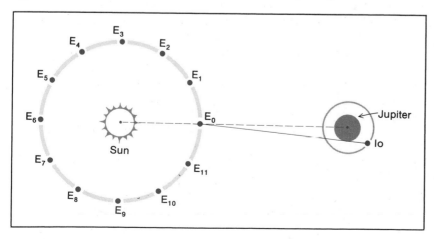

$$\frac{186{,}000{,}000 \text{ miles}}{1000 \text{ sec}} = 186{,}000 \text{ miles/sec}$$

To appreciate how fast that really is, suppose you had the privilege of climbing into a jet which had unlimited speed capabilities. Flying over the equator of the earth, you attained a speed of 186,000 miles/sec. At that speed it would be possible to make more than seven complete trips around the earth in 1 sec. If you can comprehend such speed, it will help you to relate to the size of the universe in later sections of the book, because the unit of length used to measure the distances to stars and galaxies depends directly on the speed of light. If light travels 186,000 miles in a second, and there are 60 sec in a minute, then light travels $186{,}000 \times 60$ miles in a minute. Since there are 60 minutes in an hour, light travels $186{,}000 \times 60 \times 60$ miles in an hour, $186{,}000 \times 60 \times 60 \times 24$ miles in a day, and $186{,}000 \times 60 \times 60 \times 24 \times 365\frac{1}{4}$ miles in a year. By multiplying this latter set of numbers, we have the distance light travels in a year—a *light-year*. Perform this arithmetic yourself and your answer will be just under 6 trillion (6,000,000,000,000) miles/year. The closest star, α (Alpha) Centauri, is 4.3 light-years away: thus its distance from the earth is $4.3 \times 6{,}000{,}000{,}000{,}000$ miles, or roughly 26 trillion miles.

Figure 2.4   *Ole Roemer, discoverer of the speed of light. (Yerkes Observatory)*

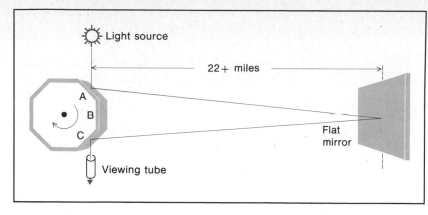

Figure 2.5   *Michelson's rotating mirror experiment.*

In 1924 Albert A. Michelson, an American physicist, developed an experiment by which he could more accurately determine the velocity of light. In order to determine velocity he had to know two things: the distance traveled and the time required. He set up part of his apparatus on Mt. Wilson, in California, and part on Mt. San Antonio, approximately 22 miles away. It is not an easy task to measure the distance from one mountain peak to another; however, by carefully using standard surveying techniques, a Coastal Geodetic Survey team was able to measure the distance accurately to a fraction of an inch. The device whereby he might measure the time for light to travel from one mountain top to the other and return consisted of a rotating mirror set on a motor which would turn it at any desired speed. A bright source of light was directed onto mirror face A, one of eight mirrors on the eight-sided wheel (Figure 2.5), and mirror face A reflected the light to a flat mirror on Mt. San Antonio. When the wheel was standing still, the flat mirror returned the light to mirror face C and into the viewing tube. Once the wheel was rotating, the speed was adjusted so that the flash of light sent by face A would be caught by face B, which had turned to the position that face C had occupied by the time the flash traveled to Mt. San Antonio and back—a distance of 43.978 miles. If the wheel turned at the rate of 529.37 rps (revolutions per second), the computation would look like this:

529.37 rps × 8 flashes/revolution

× 43.978 miles/flash = 186,250 miles/sec

Expressed in metric units, this velocity is 299,729 km/sec, or almost 300,000 km/sec. Of course this experiment was not performed in a vacuum, and so the result represents the velocity of light in air. Modern determination in a vacuum produces a velocity only slightly faster: 299,793 km/sec (186,288 miles/sec). This is the accepted value for the velocity of light in empty space. Light travels considerably slower than that in media such as water or glass.

International System of Units (SI)

While the English system of weights and measures has long been used for everyday transactions in the United States and certain other countries, we are now gradually moving toward a conversion to the International System (SI), already in use by scientists around the world. This system is essentially the same as the metric system. The fundamental unit of length is the *meter* (m), which was originally defined as one ten-millionth of the distance from the north pole to the equator, as measured on the earth's surface. A more reasonable way to determine the length of a meter is presented in Appendix 2, where the International System is more completely discussed and conversion units are given. The fundamental unit of mass is the *kilogram* (kg), and the unit of time is the *second* (sec). It will be very helpful to you if you can visualize distances in terms of meters and kilometers (km), where 1 km = 1000 m. Figure 2.6 will help you to do this.

Figure 2.6   *Relationships between units of the International System (Système International—SI) and the English system are shown in graphic form.*

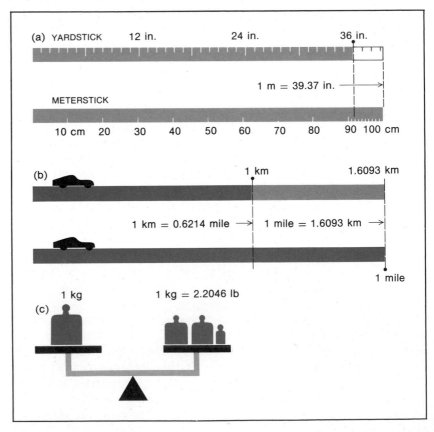

In the remainder of this chapter we will use the SI units and give the English equivalents in parentheses; in later chapters the SI units will be used exclusively. (Refer to Appendix 2 for a more detailed explanation.)

The Wave Nature of Light

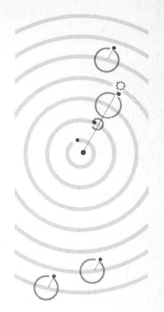

We are all familiar with the expanding wave pattern that is set up when a pebble is dropped into a quiet pond. Molecules of the water will rise and fall periodically in a very definite and predictable manner. While light may be created by the oscillation (up-and-down motion) of a charged particle, it is not transferred from one place to another by causing particles to move up and down. We know this to be true because light travels through empty space, where no particles exist. Therefore, when we speak of the wave nature of light, we are speaking of what appears to be a sequence of electrical and magnetic changes, the graphical representation of these changes appearing as a wave form.

Perhaps you have experienced the phenomenon of becoming charged electrically by scuffing your feet on a nylon rug. Picture a small object that is charged and standing still. The space around the charged particle can be thought of as a "field," a region of influence. If another charged (test) particle is brought into the field, the test particle will experience a constant force. If the test charge is like that of the fixed charge, the test particle will experience a force that repels. If the test particle is charged in a manner opposite to the fixed charge, it will experience a force of attraction. If the first particle is set into oscillation, then the test particle will experience a changing field and will respond by oscillating also.

Thus, the oscillation of one particle can be transferred to another without there being any material in between the two. This is a partial model of how light travels through empty space, but to complete the model we should recognize that whenever a changing electrical field occurs, a changing magnetic field accompanies it. To picture a changing magnetic field, hold a bar magnet in one hand and set a compass nearby. Now rotate the magnet back and forth. You are creating a changing magnetic field, which is evidenced by the action of the compass needle (Figure 2.7). Again the oscillating magnet will influence the compass even if no material exists between the two. Now we have a more complete picture of light; we envision it as an electromagnetic disturbance, produced by an oscillating charge that creates a changing electrical and a changing magnetic field simultaneously. This idea is represented graphically in Figure 2.8. Because the electric component of the wave is responsible for all optical effects, and because the magnetic component will always accompany a changing electric field, we will henceforth speak only of the electrical component.

Light is only a small portion of the electromagnetic spectrum,

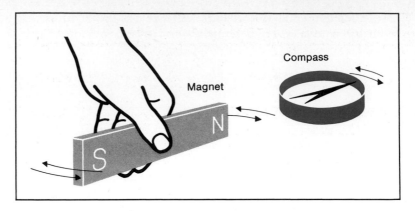

Figure 2.7   *A changing magnetic field.*

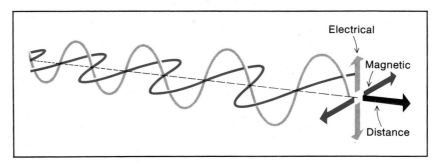

Figure 2.8   *A graphic representation of electrical and magnetic disturbances.*

which includes radio, radiant heat (infrared), ultraviolet radiation, X rays, and gamma rays. What makes these forms of radiation different from one another? It is their wavelength. We may picture one *wavelength* in a changing electrical field as the distance the disturbance travels as it goes through one complete oscillation—say, from the crest on one wave to the crest of the next (Figure 2.9). Remember that all forms of electromagnetic disturbance travel through empty space at approximately 300,000 km/sec; accordingly, if a radio wave travels 1000 m while going

Figure 2.9   *The waveform.*

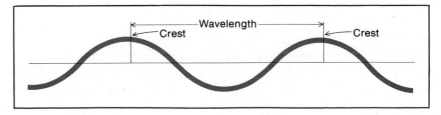

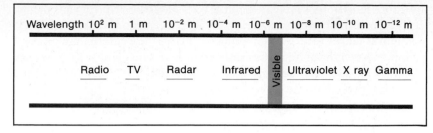

Wavelength 10^2 m 1 m 10^{-2} m 10^{-4} m 10^{-6} m 10^{-8} m 10^{-10} m 10^{-12} m

Radio TV Radar Infrared Visible Ultraviolet X ray Gamma

Figure 2.10   *The electromagnetic spectrum.*

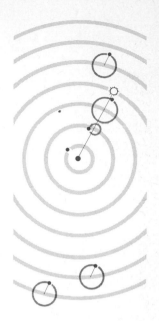

through one wave cycle, we speak of its wavelength as 1000 m. It should be apparent at this point, that if a wave travels 1000 m in one cycle and it travels 300,000,000 m in 1 sec, then it must go through 300,000,000/1000 = 300,000 cycles/sec. This is the frequency of the disturbance. While we will continue to emphasize wavelength, it will always be possible for you to obtain the frequency of the disturbance by dividing the velocity of light by the wavelength (using the same units, of course).

The wavelengths of radio signals which we receive on AM radio fall within the range of 200 to 500 m, whereas the amateur radio operator typically uses wavelengths in the range of 2 to 160 m. Radio astronomers are particularly interested in radio wavelengths between 0.01 m (1 centimeter) and 1 m. The wavelengths of these and other forms of electromagnetic radiation are shown in Figure 2.10. Note that the expression 10^{-2} m means 0.01 m, 10^{-3} m means 0.001 m, and so on. (Appendix 3 provides a full explanation of powers of 10.) As you can see, the wavelength of visible light is very short, between 0.0000004 and 0.0000007 m, with ultraviolet, X rays, and gamma rays being characterized by even shorter wavelengths. Rather than measuring such small lengths in meters, we use a much smaller unit called the *angstrom*, denoted by the symbol Å:

$$1 \text{ Å} = 10^{-10} \text{ m} = 0.0000000001 \text{ m}$$

The range of visible light, measured in angstroms, is thus 4000 to 7000 Å (7000 Å = 0.0000007 m).

Polarization of Light

Light is typically emitted by atoms oscillating in many different orientations, and we refer to such light as being unpolarized. If unpolarized light is passed through certain filters that have been manufactured by laying down many long, needlelike crystals in the same direction, then only the oscillations which are parallel to the length of the crystals will be freely transmitted—thus polarizing the light. Analogous to such a filter is a common picket gate. The long narrow spaces between the pickets will permit a wave motion to be transmitted only in one plane, and so we say that the first gate polarizes the wave. If the second gate is oriented in the same direction, then the polarized wave is transmitted [Figure 2.11(a)]; however, if the second gate is turned 90°, then the polarized wave is blocked [Figure 2.11(b)].

When sunlight strikes a flat surface, such as a lake or the hood of a car, the reflected rays tend to be horizontally polarized because the surface absorbs oscillations that are perpendicular to it but reflects oscillations parallel to it. Such reflected light is called "glare," and because it is polarized, it may be prevented from reaching the eye by use of polarized sunglasses with vertical "gates" (Figure 2.12).

The astronomer is particularly interested in the polarization of starlight, because it provides clues to the material through which the light has passed in reaching the observer. For instance, dust is thought to polarize starlight. This topic will be more fully discussed in Chapter 11.

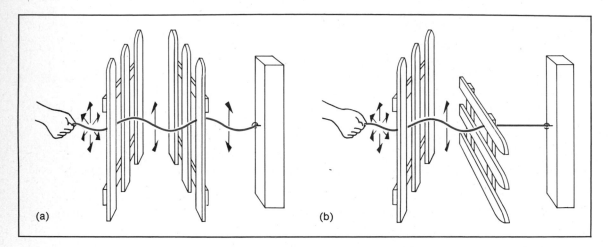

Figure 2.11   An analogy for the polarization of light: (a) both vertical gates; (b) vertical and horizontal gates.

Reflection of Light

Does light travel in a straight line? We assume that it does in space that is free of all matter; however, in our own environment light is subject to many influences, one of which is *reflection*. In order to describe the reflection phenomenon more accurately, we must first understand what is meant by the expression "a normal to the surface."

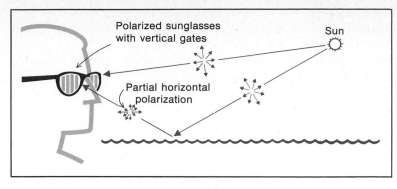

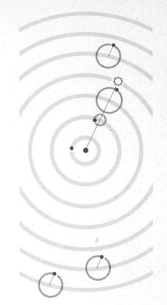

Figure 2.12 *Sunlight which is reflected from the surface of the water is partially polarized in a horizontal orientation, and the vertical "gates" of the polarized sunglasses prevent its entry into the eyes.*

The *normal* is a line perpendicular to the surface at a given point. A vertical flagpole erected on level ground is an example of a normal to a surface. The angle between the incident (incoming) ray of light and the normal is called the *angle of incidence,* and the angle between the reflected ray of light and the normal is called the *angle of reflection.* The law of reflection may thus be stated:

The angle of reflection is equal to the angle of incidence.

This law forms the basis for design of the reflecting telescope, for it is true for every point on the surface of a mirror, even if that mirror is curved (Figure 2.13).

Figure 2.13 *The angle of reflection always equals the angle of incidence (a), even for a curved mirror (b).*

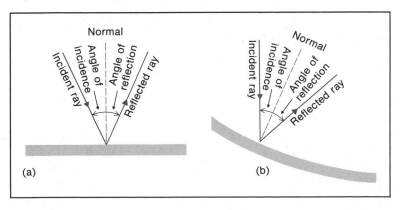

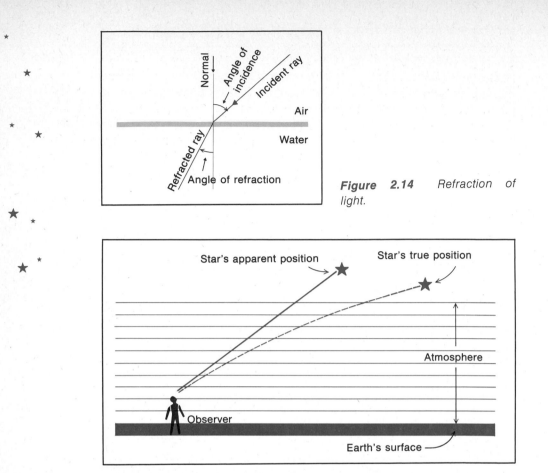

Figure 2.14 Refraction of light.

Figure 2.15 Refraction by the earth's atmosphere.

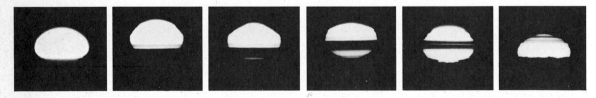

Figure 2.16 Photographs of the setting sun, showing effects of atmospheric refraction. (Lick Observatory)

Refraction of Light

Another way in which light is influenced by its environment is by *refraction.* We have shown that light travels at 299,793 km/sec (186,288 miles/sec) in a vacuum. It travels at about 224,000 km/sec (140,000 miles/sec) in water, and at still different velocities in glass or other transparent materials. The refraction (bending) of light results from this fact—that light travels at different velocities in different materials. For light of a given wavelength, the ratio of its velocity in a vacuum to its velocity in a given medium is called the *index of refraction* for that medium. In Figure 2.14 you can see that the angle the refracted ray makes with the normal is called the *angle of refraction*. As the light passes from air into water, the angle of refraction is smaller than the angle of incidence; the reason is that the velocity of light is less in water than in air. This illustrates the basic relationship between the angle that a ray makes with the normal and the velocity of light in that medium. Perhaps you have noticed that a spoon, when placed in a glass of water and viewed from an angle, appears to be bent. This illusion is due to refraction. A less obvious result of refraction occurs in the atmosphere of the earth. As light from a star enters the atmosphere, it passes through "layers" of increasing density and thus is refracted. Because the observer judges the star's position by the direction in which the light enters his eye, he sees the star as if it were higher in the sky than its true position (Figure 2.15).

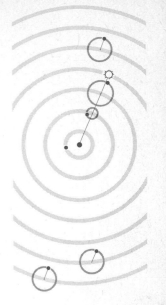

The effects of atmospheric refraction are quite obvious when the sun is seen setting over the ocean: as it approaches the horizon, its bottom part is refracted more than its top part, and so the sun appears flattened (Figure 2.16).

Dispersion by Refraction

A phenomenon that occurs simultaneously with refraction is that of *dispersion*. White light is composed of all colors, and each color represents a different wavelength. Since the angle of refraction also depends upon the wavelength of the light, each color is bent at a slightly different angle when refracted by a transparent medium. Among visible colors blue light has the shortest wavelength and is bent more than red light. By use of the prism, white light may be separated into the full range of colors, an array called the *spectrum* [Figure 2.17(a)]. Reflection, refraction, and dispersion take place in raindrops, producing a rainbow [Figure 2.17(b)]. See Color Plate 1.

Figure 2.17   *Dispersion of white light by (a) a prism and (b) a raindrop.*

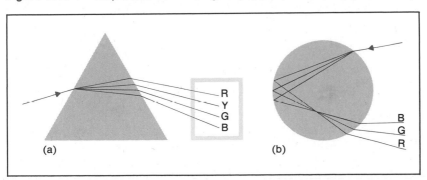

As we shall see, the phenomenon of dispersion creates a serious problem in the simple refracting telescope, yet it also serves as the basis for one of the most useful tools of astronomy, the spectrograph.

Diffraction of Light

As light passes the edge of an object or through a small opening, it spreads out in all directions as though the edge were a new source of light. This phenomenon, called diffraction, is illustrated in water waves in Figure 2.18(a). Diffraction may be useful in certain applications but is somewhat detrimental in astronomical instruments, for as light enters the telescope it must travel past numerous edges. A telescope generally has a round opening at the front and braces to support parts of the telescope, as in the case of the secondary mirror of a reflector (see Figure 2.36, page 68). Each of these parts acts as a new source of light. Figure 2.19

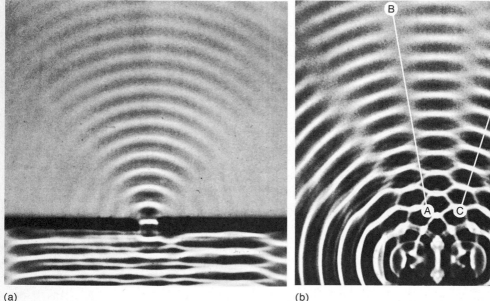

(a)   (b)

Figure 2.18   (a) Diffraction in water waves. (b) Interference in water waves. Destructive interference occurs along line AB; constructive interference, along line CD. (Educational Development Corp.)

reveals the pattern of diffraction spikes (cross shape) seen on bright star images. Diffraction prevents any telescope from forming perfect images of stars.

Figure 2.19 *Diffraction spikes are produced by the structure which supports the secondary mirror of the telescope. (Hale Observatories)*

Interference of Light

When two (or more) small openings act as new sources of waves, a pattern of interference is seen, as in Figure 2.18 (b). Along the line AB no wave motion is evident, which indicates that the waves from the two sources have canceled each other. This is called *destructive interference.* We visualize this as occurring when the *crest* (high point) of one wave meets the *trough* (low point) of the other, as in Figure 2.20.

On the other hand, along line CD in Figure 2.18 (b) the wave action is distinct. This is a consequence of *constructive interference,* when the two waves interact with crest meeting crest and trough meeting trough, as shown in Figure 2.21.

Figure 2.20 *The crest of one wave meets the trough of another wave, resulting in destructive interference.*

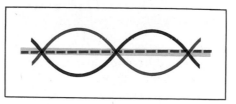

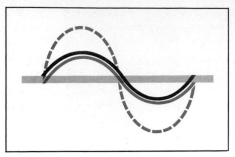

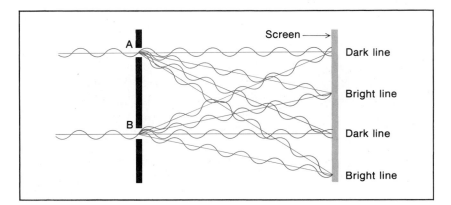

Figure 2.21 *The crest of one wave meets the crest of another wave, resulting in constructive interference.*

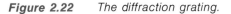

Figure 2.22 *The diffraction grating.*

Figure 2.23 *The interference of light. The light bands represent constructive interference; the dark bands, destructive interference. (Don Jenkins)*

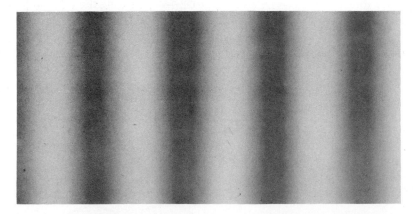

Both diffraction and interference are easily seen in water waves. What evidence do we have that similar phenomena occur in light? Suppose that many fine lines are ruled (scratched) on a sheet of glass, in effect providing many edges. Such a device is called a *diffraction grating*. When the light from a single source passes these edges, they act as many new sources. Consider the light that passes two adjacent edges, lines A and B, in Figure 2.22. The wave nature of light tells us that at certain angles the light waves from adjacent lines will interfere with each other destructively, thereby producing dark bands, and that at other angles the light waves will interfere constructively, producing bright bands on the screen. The spacing between the bright and dark regions will depend on the spacing of the lines and on the wavelength of the light that is being considered (Figure 2.23). Furthermore, could these bright bands be seen in color, the full spectrum would be evident in each band, because different colors (different wavelengths) interfere constructively at slightly different places on the screen. Thus, the diffraction grating becomes an important device whereby the light of a given source may be dispersed into its spectrum. Often the diffraction grating is used in place of a prism to disperse light.

The Dual Nature of Light

We have noted several experiments demonstrating that light behaves like waves. Some experiments require another kind of explanation. Such an experiment may be performed using an electroscope with a clean zinc plate attached to its knob. An electroscope is a device that is capable of detecting the presence of charged particles. When there is no charge on the zinc plate, the foil leaves hang downward together, but if the zinc plate becomes charged, that charge is conducted to the foil leaves and they experience a like charge, causing them to be repelled from each other.

Suppose that a negative charge (a surplus of electrons) is placed on the plate, causing the foil leaves to separate, and then ultraviolet light is allowed to fall onto the plate. The leaves will be seen to fall back together, indicating that electrons have been knocked free of the plate, leaving it neutralized. To explain this phenomenon, called the *photoelectric effect* (Figure 2.24), imagine that light behaves like particles, called *photons*, which are capable of knocking electrons free because they represent

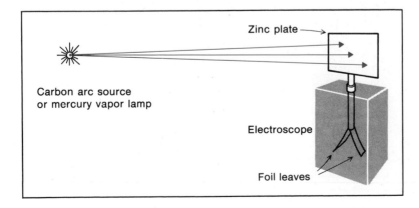

Carbon arc source or mercury vapor lamp

Zinc plate

Electroscope

Foil leaves

Figure 2.24
The photoelectric effect.

energy; but unlike ordinary particles, photons cannot be placed on a scale and weighed—that is to say, they have zero rest mass. Their energy is inversely proportional to their wavelength: thus electromagnetic energy that is characterized by short wavelengths, such as ultraviolet light, carries more energy per photon than does blue light. Blue light of any desired intensity can be directed onto the plate of the electroscope and no electrons will be freed, but when ultraviolet light falls onto the plate, electrons are ejected instantaneously. Evidently the ultraviolet photon carries enough energy to free the electron, whereas no collection of blue light photons could accomplish the task. If light behaved only like a wave (with its energy distributed uniformly over a wave front), and not as particles, then the photoelectric effect could not be explained.

Could it be possible that all matter exhibits a dual nature—that of a wave and that of a particle? The following experiment suggests that it does. When electrons (particles) are shot through the very small opening of a crystal lattice, a diffraction pattern is formed [Figure 2.25(a)] that closely resembles the diffraction pattern created when X rays (usually thought of as a wave phenomenon) are passed through the small openings within the structure of aluminum foil [Figure 2.25(b)].

Figure 2.25   *(a) An electron diffraction pattern created by passing a beam of electrons through a crystal lattice, in this case beryllium. (b) The diffraction pattern created by X rays directed through polycrystalline aluminum. (a, RCA Laboratories, Princeton, New Jersey; b, courtesy of Mrs. M. H. Read, Bell Telephone Laboratories, Murray Hill, New Jersey)*

(a)

(b)

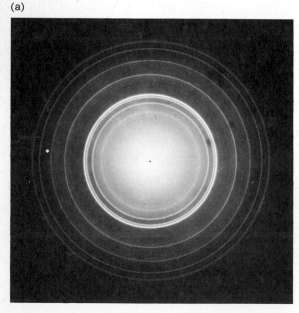

Telescope Design

When Galileo viewed the sky with his telescope, he realized that he could see stars which were too faint to be seen with the naked eye. This fact illustrates the real purpose for assembling either lenses or mirrors to make a telescope—that is, to gather light over a large surface and concentrate its energy into a small area so as to produce an image which is brighter than the object appears to the naked eye. The effects of refraction were known before the time of Galileo, but mere refraction, say, into water or a flat piece of glass, would not concentrate the light energy at one point; Galileo's important contribution was the discovery that when he gave a piece of glass the proper curvature, light rays from a distant star that passed through the glass near its edge would be bent (refracted) more than those which passed near its center, causing the rays of light from a given source to bend toward a focal point (Figure 2.26). The distance between the lens and the focal point is called the *focal length* of the lens.

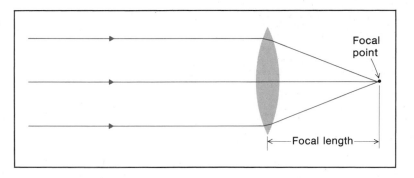

Figure 2.26   *A simple convex lens.*

Figure 2.27   *Formation of the extended image.*

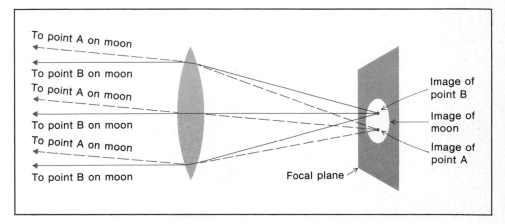

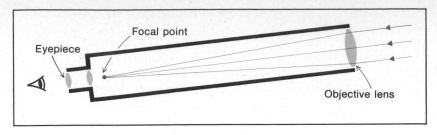

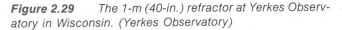

Figure 2.28   *The refracting telescope.*

Were we to view an object like the moon, it would present many points from which light travels to our telescope. Each point on the object is brought to a focus at a different point, and the collection of all such focal points forms the image of the moon on the *focal plane* of the telescope (Figure 2.27).

Figure 2.29   *The 1-m (40-in.) refractor at Yerkes Observatory in Wisconsin. (Yerkes Observatory)*

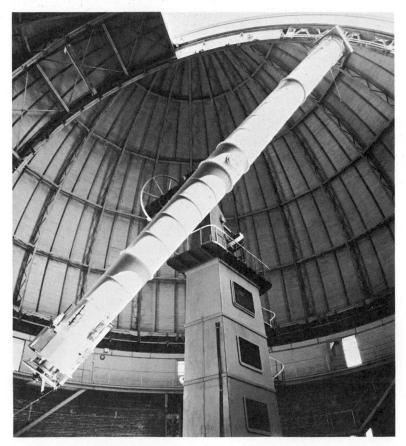

Today, a piece of film can be placed at the focal plane and a picture of the moon may be made in which every point is said to be in focus. To use such a telescope for visual observations, it is necessary to place an eyepiece just behind the focal plane (Figure 2.28). The eyepiece serves as a magnifying glass, making the image of the moon appear larger. For a given focal length, the larger the objective (first) lens is made, the more light will be collected and the brighter the image will be. In an attempt to see fainter objects, opticians have constructed several large objective lenses, the largest of which is approximately 1 m (40 in.) in diameter and is located at Yerkes Observatory in Wisconsin (Figure 2.29).

Limitations in Refractors

Refractor telescopes are limited in size because large objective lenses are difficult to support without sagging, which changes their shape. The glass from which the objective is ground must be of high quality, free of bubbles and other imperfections. Still further limitations are imposed by the design of the lens itself.

Chromatic aberration. Since white light is composed of all colors and blue light is refracted more than red light, blue light is brought to a focus ahead of red light; similarly, all colors between red and blue are brought to slightly different focal points by a single lens, resulting in the formation of a blurred image. This produces a color defect called *chromatic aberration* (Figure 2.30). This defect may be corrected by the addition of a second lens, which, because of its shape and a slightly different index of refraction, will deflect the rays and bring all colors back to the same focal point, producing a sharp image. A compound lens of this type is called an *achromatic lens* (Figure 2.31). This discussion also applies to cameras. An inexpensive camera generally has only a single-element lens, hence will suffer from chromatic aberration. The more expensive, multiple-element lens will produce a sharper image, and its achromatic nature is usually stated in writing on the frame of the lens itself.

Figure 2.30   Chromatic aberration (see text).

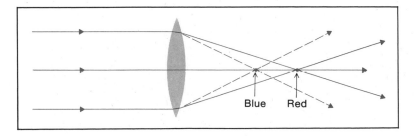

Blue   Red

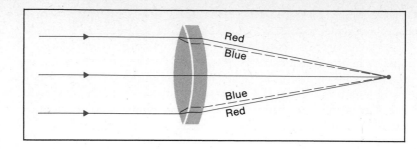

Figure 2.31   *The achromatic lens (see text).*

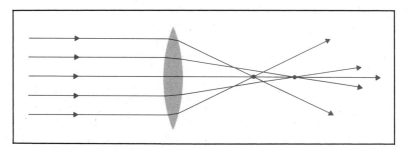

Figure 2.32   *Spherical aberraton (see text).*

Spherical aberration.   While it is much easier to grind a lens of spherical shape, a simple lens of this design suffers a defect called *spherical aberration* (Figure 2.32): the rays of light from a distant object are not brought to the same focal point. Those rays passing through the lens near its outer edge are brought to a focus closer than those passing nearer the center. This defect may also be corrected by the addition of a second element in the lens system.

The Reflecting Telescope

Sir Isaac Newton was the first to fully recognize that whenever light is refracted, it is also dispersed, causing the phenomenon of chromatic aberration in refractors. He also recognized that light is not dispersed when it is reflected, hence that a telescope depending only upon reflection will not be plagued by chromatic aberration. Obviously a flat mirror would not concentrate the light. Newton may have tried a spherical mirror—only to find that it possessed spherical aberration, because rays of light (from a star) which struck the outside portions of the mirror did not reflect to the same focal point as those which struck the mirror toward its center (Figure 2.33). Newton deduced that if he were to give the mirror a parabolic shape by hollowing out its center (Figure 2.34), he could cause all rays from a given star to converge on a point, the focal

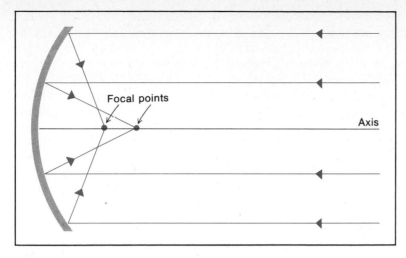

Figure 2.33 *Spherical aberration in a mirror.*

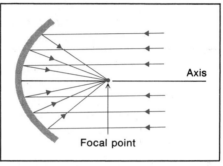

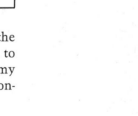

Figure 2.34 *The parabolic reflector.*

point. As you can see in Figure 2.35(c), the parabola is one of the distinctive curves which may be obtained by cutting a cone parrallel to one edge. Other curves which are useful in our discussion of astronomy may also be obtained by cutting the cone in various ways, as demonstrated in Figure 2.35(a, b, c, d).

Figure 2.35 *Sections of a cone: (a) circle; (b) ellipse; (c) parabola; (d) hyperbola.*

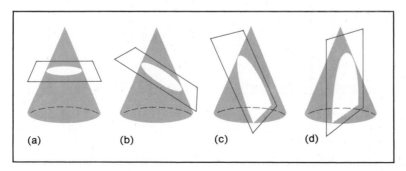

Reflecting telescopes are usually of one of the following designs, differing primarily in the placement of the eyepiece (Figure 2.36). The *prime focus* is used only on larger instruments like the 5-m (200-in.) Hale telescope at Mt. Palomar Observatory, since placement of the observer's head at that position on a small instrument would obscure all incoming rays of light. The *Newtonian focus* was the original design used by Sir Isaac Newton and is still one of the most popular among amateur astronomers. The *Cassegrain focus* allows the light to make on additional trip through the tube. This fact, together with the curvature of the secondary mirror, allows the design of a long effective focal length in a short tube, thus increasing the portability of the instrument. The *Coudé focus* utilizes a third mirror which directs the light down the polar axis of the telescope into a fixed eyepiece or spectrograph. Because this focal point does not move, even when the telescope is driven to compensate for the earth's rotation, the heavy equipment which characterizes spectroscopic studies can be placed in the Coudé focus room. Most modern telescopes, like the 4-m Mayall, at Kitt Peak (Figure 2.37) utilize the prime focus, the Cassegrain focus, or the Coudé focus in the same instrument by moving the proper secondary mirror(s) into place. The focal length of the Palomar 5-meter is 16.8-m (55 ft), thus the prime focus position is 16.8 m from the mirror. At this position the astronomer moves with the telescope while making long exposures on film. An astronomer may also ride at the Cassegrain focus while the telescope is in motion (Figure 2.38). Figure 2.39 shows the largest telescope in the world, the 6-m reflector operated by the Soviet Union.

Figure 2.36   *Four reflecting telescopes, utilizing different focal points: (a) prime focus; (b) Newtonian focus; (c) Cassegrain focus; (d) Coudé focus.*

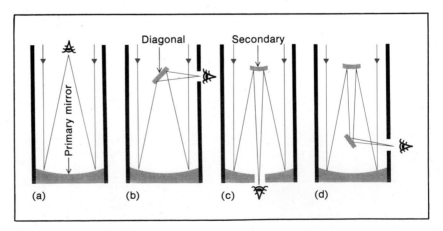

Figure 2.37   The 4-m Nicholas U. Mayall telescope at Kitt Peak National Observatory in Tucson, Arizona. (Kitt Peak National Observatory)

Multiple-Mirror Telescopes

Astronomers have been motivated to build larger and larger reflecting telescopes for two reasons; first, when more light is collected, dimmer (and hence more distant) objects may be seen; second, larger mirrors produce sharper images, as you will see in forthcoming sections. On the other hand, the cost of a mirror is proportional to its diameter cubed. That is, if a 10-in. cost $200, then a 30-in. (three times as large) would cost 3^3 or 27 times as much, or $5,400. This relationship will demonstrate the fiscal motivation to assemble six 72-in. mirrors in a

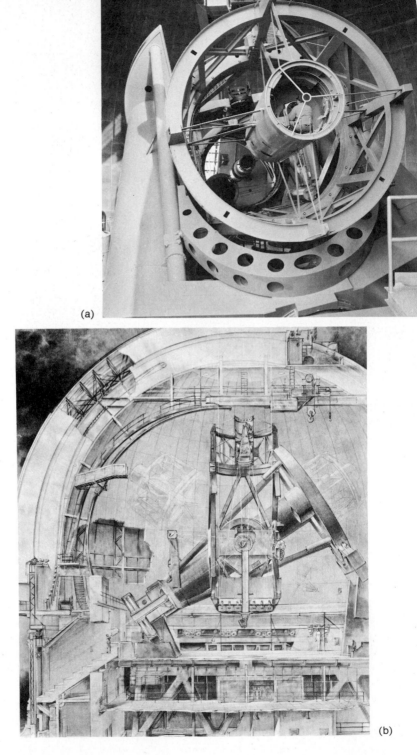

Figure 2.38 *(a) Looking down the tube of the 200-in. (5.1-m) Palomar telescope, past the observer at the prime focus position, to the mirror at the far end. (b) A drawing of the Palomar telescope. (Hale Observatories)*

(a)

(b)

Figure 2.39 *The largest telescope in the world, a 6-meter reflector, operated by the USSR Academy of Sciences near Zelenchukskaya in the southern part of Russia. (TASS Fotokhronika)*

single structure, as was done at Mount Hopkins near Tucson, Arizona (see Figure 2.40). Obviously, problems exist that are not encountered in the conventional design, such as making seven mirrors of identical focal length (the seventh is a guide telescope; see the center opening in Figure 2.40) and directing each mirror to the same focus precisely. On the other hand the effective light-gathering power will equal that of a 176-in. telescope.

The Schmidt Camera

We speak of the Schmidt design as a camera because it is used exclusively for photographic work and has no provision for visual observation. All guiding is done through an auxiliary scope mounted on the exterior. The principles of refraction and reflection are both employed in the Schmidt camera. Light enters the system through a weak lens called a corrector plate, where it is refracted slightly, and then the rays are reflected by the spherical primary mirror to the focal point. A piece of film, 35.6 × 35.6 cm (14 × 14 in.), exposed at this point will record a wide-angle view of the sky. The 1.22-m (48-in.) Schmidt camera at Mt. Palomar records a square portion of the sky 6° on a side, covering 36 square degrees in a single exposure. The 5-m (200-in.) telescope covers only about one square degree per exposure. The Schmidt camera is very

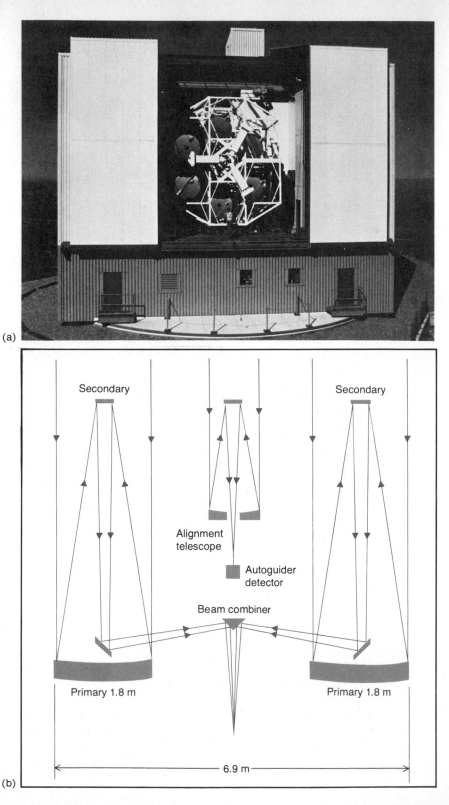

Figure 2.40 (a) The Multiple-Mirror telescope on Mt. Hopkins in Arizona. The object of this design is to provide the light-gathering power and the resolution of a very large telescope in a smaller package and at less cost. (b) This drawing shows how the light of two of the six telescopes is combined to form a single image. The central scope is used to guide the entire array. (Smithsonian Astrophysical Observatory and Steward Observatory, University of Arizona)

(a)

Secondary   Secondary

Alignment
telescope

Autoguider
detector

Beam combiner

Primary 1.8 m   Primary 1.8 m

6.9 m

(b)

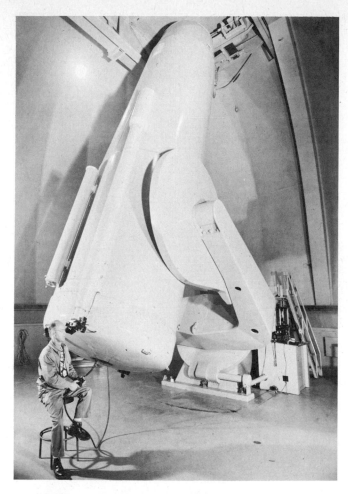

Figure 2.41   *The 48-in. (1.2-m) Schmidt telescope of the Palomar Observatory. (Hale Observatories)*

well suited to a survey of the visible sky, a program that required 7 years (over 1800 photographs) to complete (Figure 2.41).

Schmidt Cassegrain

One of the most popular amateur telescope designs is a variation on the Schmidt (see Figure 2.42). Its popularity stems partially from its compact design. For instance, a telescope with an effective focal length of 2 m can be built in a tube 0.5 m long. The angle at which the rays of light converge from the secondary determines the effective focal length of this design. Note that four surfaces (either lens or mirror surfaces) must be ground to a very close tolerance if this design is to perform properly.

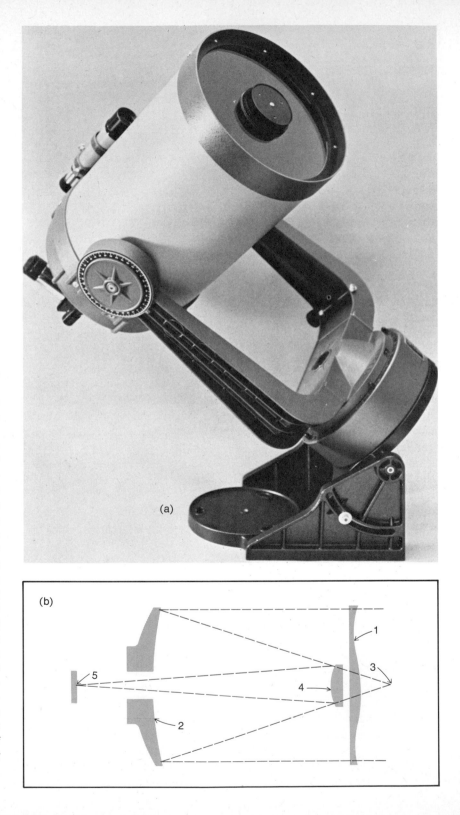

Figure 2.42 *(a) The Schmidt Cassegrain design, providing a relatively long focal length in a very portable (short tube) telescope. (b) Optical diagram of the Celestron Schmidt Cassegrain lens system: The light enters the system through a thin aspheric corrector lens (1); it is then reflected by a large spherical primary mirror (2) toward the prime focus (3). The light from the primary is intercepted by the convex secondary mirror (4) and reflected back through a hole in the primary mirror to the Cassegrain focus (5). The effect of folding the optical path back with a convex secondary mirror also increases the effective focal length by a factor of three to seven times that of the primary alone. (Celestron International)*

(a)

(b)

Telescope Performance

The factors that ultimately affect the size, brightness, and quality of an image formed in a telescope include the aperture (diameter of lens or mirror), the focal length, and the quality of the objective lens or mirror and of the eyepiece used. Of first importance is the quality of the lens or mirror, by which basically we mean its shape. We have already stressed that there is a correct shape for a lens or a mirror, so the degree to which this shape is achieved determines the quality of the telescope. If the mirror is ground so that its surface is like that of a parabola with no variations larger than one-tenth of the wavelength of blue light [no more than 0.00000004 m, or 40 nanometers (nm)], then its quality is specified as being "tenth-wave," considered to be very adequate for an amateur telescope. Some mirrors have a superior "twentieth-wave" quality, with no deviations greater than one-twentieth the wavelength of blue light.

Before we can specify the factors which determine size, brightness, and sharpness of the image, we must realize that the astronomer views two classes of objects: those which have an apparent size, such as the sun, moon, planets, clusters, nebulae, and galaxies, called *extended objects*, and those which have no apparent size, such as stars, called *point sources*. While the earth's atmosphere sometimes blurs images to produce an apparent size in stars, no amount of magnification will actually produce a real size in a star image. Stars are simply too far away to be seen as disks. As an exception to this general rule, a few stars are both large enough and close enough to be resolved into a disk. Such a star is Betelgeuse in the constellation of Orion.

Size. For a given extended object, the size of the image formed at the prime focus depends upon the focal length of the lens or mirror. The longer the focal length, the larger the image formed. If f is the focal length of the lens or mirror expressed in inches, then the size of the image (s), expressed in inches, that will be produced by an object 1° in diameter is given by the formula $s = 0.01744f$. The scale of a typical amateur telescope—for example, one of 152-cm (60 in.) focal length—would be

$$s = 0.01744 \times 152 \text{ cm} = 2.67 \text{ cm/degree}$$

Using the full moon as an object that has a diameter of 0.5°, we would expect its image to be one-half the figure computed above—that is, 1.33 cm (0.5 in.) in diameter. This would be the size of the image that could be recorded on film without the aid of an eyepiece. A telescope of twice the focal length would produce an image twice as large on a piece of film, without the aid of an eyepiece. Thus the longer focal lengths are particularly useful in observation of the planets.

Brightness. In the case of a point source (a star), the *brightness* of the image depends only upon the aperture (diameter) of the mirror or lens. The larger the aperture, the brighter the image. For an extended source, the brightness of the image produced by various telescopes of equal focal length depends upon the aperture of the mirror or lens. The brightness of the image increases with the square of the aperture, meaning that if the diameter of the mirror or lens is doubled, the total light concentrated in the image will be multipled by four. On the other hand, if the aperture of several telescopes is the same and they differ only in focal length, the one with the longer focal length will produce the dimmer image. This is an inverse relationship. The combined effect of aperture and focal length upon the brightness of the extended image is expressed by the formula

$$B = C\left(\frac{a}{f}\right)^2$$

where B represents brightness, a represents aperture, f represents the focal length of the telescope, and C is a constant. The ratio of focal length to aperture (f/a) is called the *focal ratio* or *f-stop,* as when used in connection with a camera. A focal ratio of $f/10$ means that the focal length of

Figure 2.43   *Region of the Orion Nebula, showing the effect of aperture and time exposure on the formation of star images: (a) small aperture and/or short exposure time; (b) large aperture and/or long exposure time. (Lick Observatory)*

(a)   (b)

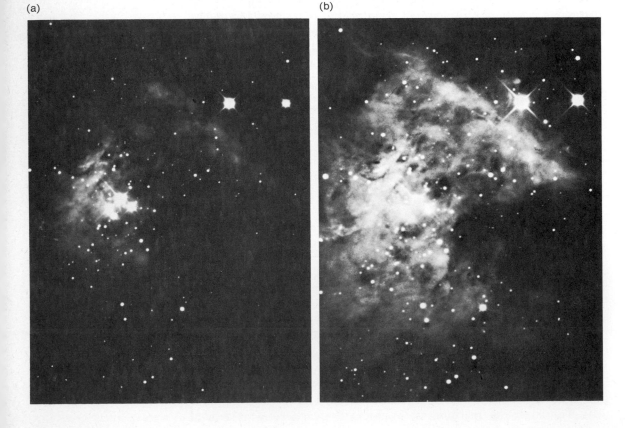

the lens or mirror is 10 times the aperture. A telescope in which the focal ratio is $f/5$ will provide images approximately four times as bright as with the $f/10$ ratio.

We have been referring to the brightness of the image as viewed by the human eye. The effective brightness may be greatly increased by photographic and/or electronic devices. Imagine an object so dim that it cannot be seen in the telescope. If the light from that object is allowed to fall on a sensitive photographic plate for a prolonged period, an image will be "built up" on the plate itself. When developed, the plate will reveal the presence of that object. Dim objects that normally require hours to expose on film may now be intensified by means of an electronic device that senses their presence and builds the photographic image at a much faster rate. Thus the effective aperture of the telescope has been increased (Figure 2.43).

Resolution. The *resolution* of a telescope measures its ability to show detail in an image—that is, to separate objects that appear to be close together. The resolving power of a telescope is expressed in terms of the smallest angle between two stars that can be distinguished as separate objects. The resolving power of an optical telescope depends primarily upon its aperture and is expressed by the formula $a = 11.58/d$, measured in seconds of arc, where d is the diameter of the lens or mirror measured in centimeters. A telescope with a 15.24-cm (6-in.) lens can separate star images that are at least 0.76 second of arc apart. (In angular measures, $1° = 60$ minutes and 1 minute $= 60$ seconds.) A 25.4-cm (10-in.) telescope can separate stars that are at least 0.5 second of arc apart. Figure 2.44 indicates how two images might appear in telescopes of the indicated aperture, under ideal atmospheric conditions. But that's just the problem—ideal atmospheric conditions seldom, if ever, exist.

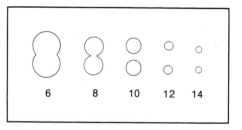

Figure 2.44 *Resolution of star images using different apertures (shown in inches).*

Atmospheric Turbulence: A Resolution Spoiler

Ground-based telescopes have always been plagued by the fact that the light from a star (or other source) must pass through the earth's atmosphere, an atmosphere which is in constant motion. In moving from the vacuum of outer space to the increased density of the atmosphere, individual rays of light experience a random effect in relation to their reduction of velocity, producing a random effect in relation to their refraction. When all of the rays of light from a single point source are brought to a focus by the lens or mirror of a telescope, the individual rays arrive at the focal point with their waves out of step with one another and they tend to partially destroy each other. Recall the discussion on interference on page 59. The random nature of the wave interference causes the star to appear to twinkle and it also spoils the resolution, spreading the light which would normally fall into a very small dot into a larger disk of light (see the brighter stars in Figure 2.43). Under such atmospheric conditions, a 10-inch telescope may not be able to separate binary stars as shown in Figure 2.44. Only with the advent of the high-speed computer have astronomers begun to compensate for this atmospheric effect. The device used is called the "rubber" mirror. Several mirrors are mounted individually so as to permit minute corrections to be made in the shape of the composite mirror. These corrections can be made in a time scale measured in microseconds. Detectors constantly are sensing the light which arrives at the focus, from each component mirror, and by rapid trial-and-error corrections, the computer directs each mirror to formulate the smallest possible image. The fluctuations due to atmospheric turbulence may occur at the rate of 100 per second, so you can see that the high speed of the computer is essential to this approach.

Seeing

The term "seeing" applies directly to the concept we have been discussing. If astronomers say the "seeing" is good tonight, they mean that the atmosphere is unusually calm (not as turbulent as usual) and star images are small—binaries are easier to separate. "Poor seeing" would, of course, mean the opposite. You can judge the seeing conditions yourself by noticing the degree to which the stars appear to twinkle.

Magnification

While the size of an extended image, formed at the prime focus, is determined by the focal length of the objective lens or mirror, the *power* (magnification) of the telescope is determined by both the focal length of the objective and the focal length of the eyepiece used. This is expressed in the following equation:

$$\text{power} = \frac{\text{focal length of the objective}}{\text{focal length of the eyepiece}} = \frac{f_o}{f_e}$$

Suppose that, for a given telescope, $f_o = 2$ m, and we choose an eyepiece of 10 mm (millimeters) focal length (f_e). We must first express the focal length of the objective in millimeters:

$$2\,\text{m} = 2000\,\text{mm}$$

The effective power of this combination is then found:

$$\text{power} = \frac{f_o}{f_e} = \frac{2000\,\text{mm}}{10\,\text{mm}} = 200$$

When we use this telescope with the 10-mm eyepiece, an extended object will appear 200 times as large as with the naked eye. Frequently, the useful power of a telescope is limited by the object being viewed and/or the atmospheric conditions. If an eyepiece with a focal length of 20 mm is chosen, the power will be 100:

$$\text{power} = \frac{f_o}{f_e} = \frac{2000}{20} = 100$$

As a general rule, the useful limit of power for any telescope is 20 times its aperture in centimeters.

The Spectrograph

Spectroscopy is one of the most important areas in astronomy. Its operations depend upon the same phenomenon that was shown to be the cause of chromatic aberration in the simple refractor: the fact that light of different colors is bent by differing amounts when refracted by glass or some other transparent medium. Thus white light, which is composed of all colors, may be separated into these various colors by refraction, thereby producing a full spectrum of colors (Figure 2.45; see also Color Plate 1).

Light from a star is passed through a narrow slit, then through a collimating lens, which bends the rays so that they will be parallel as they pass through the prism. The light, which has been broken into its various colors, is then focused onto a photographic plate.

Figure 2.45   *The spectrograph.*

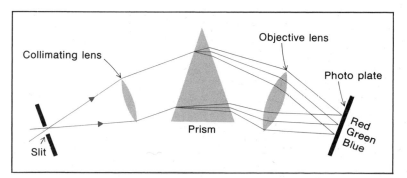

The picture of the spectrum, called a *spectrogram*, consists of a series of adjacent images of the slit, each representing a slightly different wavelength. When sunlight is viewed through such an instrument, it will be noticed that dark lines appear at certain places in the spectrum. While at first these were not recognized as being significant, in 1814 the German optician Joseph Fraunhofer recorded the position of several hundred of these dark lines, which are still referred to as *Fraunhofer lines*. In 1859 Gustav Kirchhoff, a German physicist, discovered that this same phenomenon could be produced in the laboratory by passing white light through various gases and then through the spectrograph. He particularly noticed two close lines in the yellow portion of the spectrum and found that he could produce these same dark lines by passing white light through sodium vapors. To date more than 30,000 *absorption* (dark) *lines* have been found in the visible portion of the solar spectrum, and more than half of these have been identified with elements known on earth (Figure 2.46).

Figure 2.46 *The solar spectrum, with certain lines identified with known elements. (Hale Observatories)*

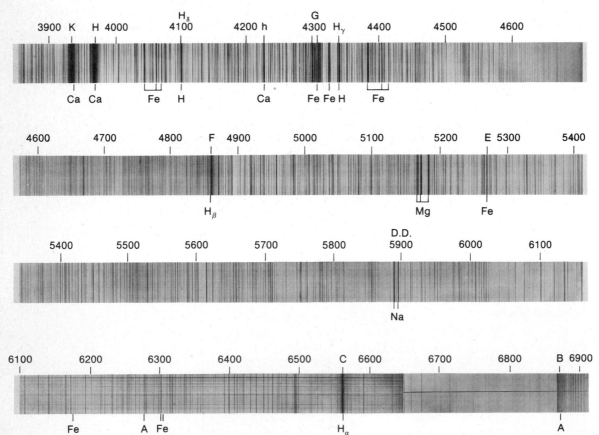

Kirchhoff also observed that a *glowing* (excited) gas produced a spectrum consisting of a series of bright lines on a dark background. As a result of this observation he stated the following three basic laws:

Continuous spectrum. A heated liquid or solid, or a gas under high pressure, emits light of all wavelengths, producing a continuous spectrum consisting of all colors.

Bright-line spectrum. A low-pressure gas that has been excited (say, by electrical current or heat) produces a bright-line spectrum, consisting of only certain colors.

Absorption spectrum. If light of a continuous nature is passed through a gas, the gas may absorb certain wavelengths, producing an absorption spectrum consisting of dark lines on a continuous background.

One might suspect that the wavelengths absorbed by the gas would be reemitted as bright lines and simply fill in the dark lines immediately. It is true that they are reemitted, but they are reemitted in all directions, and only a small fraction of their energy continues in the original direction of the light ray, and so the lines appear dark.

The Atom Signs Its Name

Suppose we place some hydrogen gas in a tube from which the air has been removed. Then we pass an electric current through this low-pressure gas by means of electrodes sealed at either end of the tube. Light will be radiated by the gas, and if that light is passed through a spectroscope, only four dominant bright lines in its spectrum are visible (Figure 2.45). Each line represents one wavelength: the red line has a wavelength of 6563 Å; the blue line, a wavelength of 4861 Å; one violet line, of 4341 Å; and the other violet line, of 4102 Å. But why are these four lines produced in the hydrogen atom?

We visualize the atom, the simplest form of a given element, as being composed of a central nucleus around which electrons move, not in a completely predictable manner but at least in a fashion such that we may predict their most probable levels of energy. These energy levels are represented in Figure 2.47 in rather oversimplified terms as circles surrounding the nucleus. Circle 1 represents the lowest energy the electron may possess, circle 2 represents the second energy level, and so on. In this model of the hydrogen atom, proposed by the prominent Danish physicist Niels Bohr (1885–1962), no intermediate levels are possible. The single electron of a hydrogen atom would normally be at the lowest energy level; if an electric current were passed through the tube of gas, however, the electrons of some of the atoms would receive energy from the electric current, permitting them to move to a higher

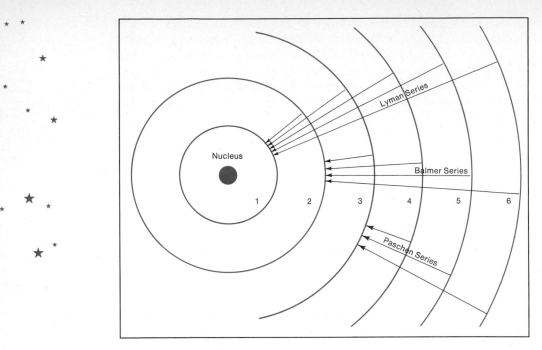

Figure 2.47   *Electron transitions in the hydrogen atom.*

energy level. An electron would not remain at that higher level but would return to the lowest energy level. In the process of making such a downward transition, the electron may stop temporarily at any level, releasing energy as it makes the downward transition. In other words, the atom has absorbed energy initially to make the upward transition, and now it emits energy as it makes a downward transition. If the electron stops at energy level 2, even temporarily, it will produce visible light, but only those colors of visible light which correspond to the amount of energy the electron lost in making a specific downward transition. In other words, corresponding to each possible change in energy level is a definite amount of energy and a definite wavelength.

In the hydrogen atom, when the electron makes a transition from the third to the second energy level (a "3-to-2" transition), it produces the red line (6563 Å), called the *hydrogen-alpha* (H_α) *line*. A "4-to-2" transition produces the blue line (4861 Å), called the *hydrogen-beta* (H_β)*line*; a "5-to-2" transition produces the first violet line (4341 Å), called the *hydrogen-gamma* (H_γ) *line*; and a "6-to-2" transition produces the second violet line (4102 Å), called the *hydrogen-delta* (H_δ) *line*. Other transitions from still higher energy levels that stop temporarily at level 2 produce numerous faint violet lines near the end of the visible spectrum. Only these specific lines are possible for the neutral hydrogen atom at low pressure; no partial transitions are possible between energy levels. For example, the electron cannot move from level 4 to $2\frac{1}{2}$ or from level 5 to $3\frac{1}{4}$, for only whole-number transitions are possible. The series of visible lines produced by the hydrogen atom is called the Balmer series. Of course, some downward transitions stop temporarily at energy level 3, and these produce a series of lines in the infrared portion of the spectrum called the Paschen series. Transitions to the lowest level produce lines in the ultraviolet portion, and these lines are called the Lyman series.

Whereas downward transitions *emit* light of definite wavelengths, upward transitions *absorb* light of certain wavelengths. Suppose we caused white light, which is a mixture of all visible wavelengths, to pass through hydrogen gas at low pressure. The photons having wavelengths corresponding to transitions from level 2 to 3 (or from 2 to 4, or from 2 to 5, and so on), would be absorbed, causing such upward transitions in the atoms. We would see the same dominant four lines of hydrogen, but now they would be dark lines on the full rainbow background of white light. The spectrum of white light is called a continuous spectrum because it contains all colors, but when certain colors (wavelengths) are removed by absorption, it is called an *absorption spectrum* (also called a *dark-line spectrum*). Since the lines occur in an absorption spectrum at the same position as the bright lines of an emission spectrum, either type is equally useful to the astronomer. Both types occur in stars, nebulae, galaxies, and the like.

Now suppose we look at the spectrum of helium, which consists of an entirely different set of lines from that of hydrogen. While the model of the helium atom or of other heavier elements is more complex, it is evident from experimentation that each kind of atom has its own characteristic set of spectral lines. Thus the atom "signs its name" in its spectrum. One of the great values of spectroscopy to the astronomer now becomes clear. He can identify the atoms which compose a distant object by recording its spectrum and comparing that to the spectra of different elements produced in the laboratory. For example, some of the lines of calcium (Ca), iron (Fe), hydrogen (H), magnesium (Mg), and argon (A) have been identified in the solar spectrum (see Figure 2.46, page 80).

Doppler Shift

The spectrogram can also be used to tell something about the motion of an object which is emitting light, owing to a phenomenon called the *Doppler effect*. The Doppler effect is a change in wavelength due to the motion of the source or of the observer or both (Figure 2.48). Imagine that a fire engine is approaching you with its siren operating at a certain note (emitting a certain wavelength of sound). The firemen riding in the truck hear that wavelength, but you hear a higher-pitched note (shorter wavelength) than they do as the engine approaches you. Why? Because each time the siren emits a new wave crest it is closer to you, and therefore the wave crests occur closer together from your point of view, producing a shortened wavelength. Then as the engine passes and recedes from you, the note you hear is lower (longer in wavelength).

This occurs because each time the siren emits a new wave crest of sound it is farther from you and the wave crests appear to be farther apart from your point of view. Furthermore, the faster the engine is moving, the greater the change in pitch.

The Doppler effect also occurs in electromagnetic phenomena. The wavelength of a given line in the spectrum of a star will be lengthened if the star is moving away from the observer, causing the line to appear shifted in position toward the red end of the spectrum. If the star is moving toward the observer, a given line will be shifted in position toward the blue end of the spectrum. The amount by which the wavelength is changed ($\Delta\lambda$) depends only on the relative velocity of the star with respect to the earth, along the observer's line of sight. If c represents the velocity of light and λ represents the laboratory wavelength of the given line, then the change in wavelength due to motion is given by

$$\Delta\lambda = \frac{v}{c}\lambda$$

This formula, when solved for v, indicates the method by which the *radial velocity* of a star may be determined:

$$v = c\frac{\Delta\lambda}{\lambda}$$

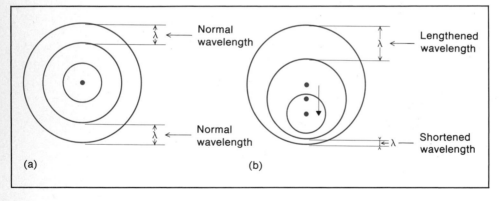

Figure 2.48   *The Doppler effect: (a) stationary source; (b) moving source.*

Here v, a radial velocity, is that part of the relative velocity which is in the direction of the observer's line of sight.

The first line in the hydrogen spectrum normally has a wavelength of 6563 Å. If in the spectrum of a star this line is observed to have a wavelength of 6565 Å, we may calculate the radial velocity of the star as follows:

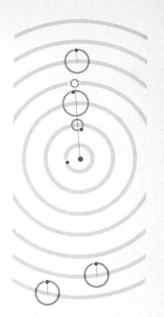

$$\Delta\lambda = 6565 \text{ Å} - 6563 \text{ Å} = 2 \text{ Å}$$

$$v = c\,\frac{\Delta\lambda}{\lambda} = (300{,}000 \text{ km/sec})\,\frac{2 \text{ Å}}{6563 \text{ Å}}$$

$$v = 91.4 \text{ km/sec}$$

Since the wavelength of the line from the stellar source appears longer than that of the laboratory source, we know that the star is moving away from us. Thus a shift toward the red end of the spectrum—a "red shift"—indicates that the star is receding from us, whereas a shift toward the blue end of the spectrum—a "blue shift"—would indicate its approach toward the earth. It should be observed that the Doppler shift tells us nothing about the distance to the star, only its radial velocity.

A Very Important Generalization

To this point we have developed concepts around the phenomenon called light, but light is only a very small portion of the electromagnetic spectrum, which includes radio, infrared, ultraviolet, X rays, and gamma rays. These are just different names for various aspects of the same general phenomenon. All forms travel through empty space in a straight line but may be reflected or refracted by the proper materials. They all can be collected or concentrated in some fashion, even as light is concentrated in an optical telescope. Objects in space emit certain wavelengths better than others, creating a spectrum in radio, television, radar, infrared, ultraviolet, X rays, and gamma rays (Figure 2.49), and that spectrum reveals Doppler shifts, even as in the visible spectrum.

Why, then, has the astronomer concentrated so much on visual observations and utilized but a small portion of the spectrum? There are two very logical reasons. Until the advent of radio receivers, humans' only natural "receiver" through which they might perceive the universe was their eye—and their eye only "tunes in" the visible spectrum. With the perfection of the radio telescope, astronomers gained another portion of the spectrum: radio wavelengths between 10 and 0.01 m. This gave

Figure 2.49   *The electromagnetic spectrum.*

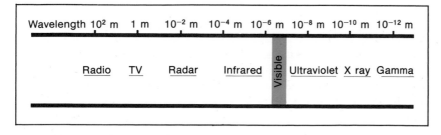

them two narrow portions of the spectrum in which to work, but what of the remaining portion? With the exception of a small portion in the infrared part of the spectrum, the earth's atmosphere reflects or absorbs most of the energy, leaving only these three small "windows" through which astronomers may observe the universe. They began to conquer the atmospheric problem by sending balloons high into the atmosphere or rockets above it. Then the real breakthrough came when scientists were able to place satellites in orbit around the earth. On those satellites they can now install telescopes, some with people aboard (as in the case of Skylab or the Apollo series) but most without (as in the case of OAO-C and others which will be mentioned later). This capability has opened up the entire spectrum for study, and an explosion of data has taken place in what might be called the new astronomy: infrared astronomy (already begun from earth-based observatories), ultraviolet astronomy, X-ray astronomy, gamma-ray astronomy, and cosmic-ray astronomy.

In order to take advantage of the broadened window, astronomers are developing new tools and techniques at a rapid pace. With the information gathered, they are beginning to unravel some of the mysteries of the universe. We will consider each of these new techniques briefly. Applications will be treated in pertinent areas of the text.

Radio Astronomy

The discovery that radio energies are emitted by objects in space came quite unexpectedly when in 1931 Karl Jansky of the Bell Telephone Laboratories noticed a certain kind of interference in his sensitive receiver. This interference occurred a few minutes earlier each day. He recalled that any given star or galaxy appears to rise 4 min earlier each night, according to our clocks, and he concluded rightly that the source of this radio energy was outside the earth.

Now, scientists reasoned, if a device could be constructed to "collect" a very weak radio signal and concentrate its energy at a focal point, and then a sensitive amplifier could strengthen that signal without mixing in its own noise, the astronomer would have a fine tool with which to expand his knowledge of the universe. In view of the similarity in properties of all electromagnetic disturbances, it seemed logical to follow the parabolic design of the optical mirror in planning construction of a *radio telescope*. Owing to the longer wavelengths of radio energy, radio telescopes can be built of steel and simply covered with perforated metal sheets. It is thus possible to construct very large models, such as the 64-m (210-ft) movable antenna at Goldstone, California (Figure 2.50), or the 305-m (1000-ft) fixed antenna at Arecibo, Puerto Rico. At Arecibo, in a mountainous area, a natural depression was bulldozed into the shape of a parabolic basin. This basin was lined with perforated metal sheets and a focal-point collector was placed on cables overhead (see Figure 2.51). This radio telescope is not fixed as to what it "sees," for, as the earth rotates,

Figure 2.50 *The 210-ft (64-m) Goldstone antenna, located near Barstow, California. This instrument is used primarily to track and communicate with deep space probes. (JPL–NASA)*

the antenna scans a path through the sky every day. Furthermore, the focal collector overhead can be moved north or south, which has the effect of pointing the telescope north or south about 20° from the zenith.

Some objects emit radio energy so efficiently that they can be sensed by a radio telescope when they are too dim to be seen in an optical telescope. One such class of objects have been dubbed *quasars* (quasistellar radio sources) by radio astronomers. Some sources of radio energy have such large red shifts in their spectra as to suggest that they are receding from us at speeds approaching that of light. These may be the most distant objects in the universe. A radio telescope can be used day and night, on cloudy or clear days, and it can penetrate many regions of our galaxy which were previously hidden from us by clouds of dust through which optical telescopes could not see. Also, whereas the optical astronomer is plagued by a background of light radiation that obscures the fainter stars, the radio astronomer has virtually no background radiation to worry about, although he is sometimes plagued by interference caused by automobile ignition or radio stations. The radio astronomer's biggest problem is how to amplify the weak radio signal coming from a distant source without introducing man-made noise that will obscure the signal, which itself has the form of noise. To solve this problem, the scientist uses a device (called a *maser*) submerged in liquid helium, which cools it to approximately $-270°C$ (almost absolute zero). In the maser the weak radio signal triggers a series of downward electron transitions in atoms that are kept in a state of continuous excitation, thus amplifying the original radio signal (Figure 2.52).

Radio telescopes in general have poor resolution because the wavelengths with which they must deal are much longer than the wavelengths of light. The length of a typical radio wave is 3 cm, compared to 0.00006 cm for a typical light wave. The formula that expresses the resolution of any telescope is given by

$$\alpha = 2.1 \times 10^5 \frac{\lambda}{d} \text{ seconds of arc}$$

The wavelength of the wave received is denoted by the Greek letter λ

Figure 2.51   *(a) The 1000-ft (304.8-m) Arecibo antenna, in Puerto Rico. The focal point for the telescope is suspended by cable from the surrounding*

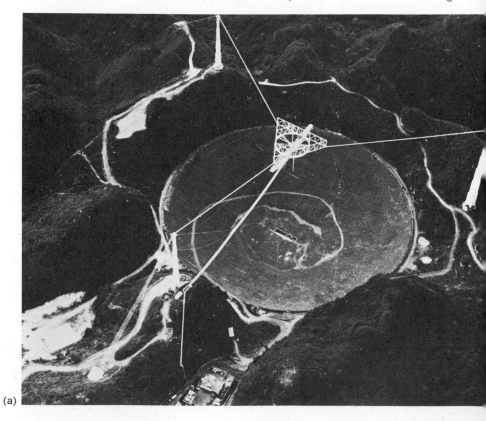

(a)

(lambda), the telescope diameter by d. For a telescope 30,480 cm (1000 ft) in diameter receiving a 3-cm signal, the resolution is

$$\alpha = 2.1 \times 10^5 \times \frac{3 \text{ cm}}{30,480 \text{ cm}} \text{ seconds of arc}$$

$$\alpha = 21 \text{ seconds of arc}$$

Under the specified conditions, it is impossible to distinguish between two sources of radio emission that are separated by less than 21 seconds of arc. In order to separate two such sources, the same region of the sky is photographed with a large optical telescope, and if the radio sources have optical (visual) counterparts, it may be possible to separate them.

The advent of the radio telescope has led to the discovery of many distant radio sources, some of which are too faint to be seen. We shall return to the subject of quasars in Chapter 14.

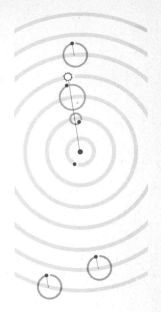

elevations and is movable for purposes of scanning the sky. (b) A recent updating of the Arecibo antenna included a new mesh covering for the dish itself. The resulting accuracy with which the surface is shaped provides added resolution. In this view you are looking upward through the mesh to the collector point. (The Arecibo Observatory is part of the National Astronomy & Ionosphere Center which is operated by Cornell University under a contract with the National Science Foundation.)

(b)

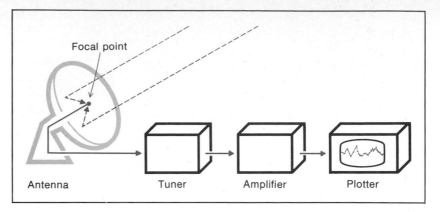

Focal point

Antenna Tuner Amplifier Plotter

Figure 2.52 *Diagram showing the essential components of a radio telescope: the antenna collects the radio energy and concentrates it at the focal point; the tuner selects the wavelength desired; and the amplifier builds up the signal, which is then plotted as a permanent record.*

Figure 2.53 *Artist's conception of the Very Large Array (VLA) radio telescope system which was completed in 1980. The Array is located in the Plains of Saint Augustin, in central New Mexico. This array of instruments permits a detailed mapping of radio sources with a resolution equal to or better than any optical telescope and better than any other radio telescope system. (National Radio Astronomy Observatory)*

Radio Interferometry

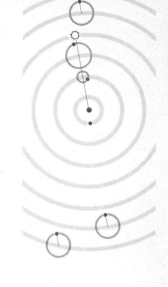

The relatively poor resolution of radio telescopes may be improved considerably by interconnecting two or more radio telescopes that have separate and variable locations (Figure 2.53). Since the radio signal from a given source will arrive at the antennas at slightly different times, the received waves will interfere with each other. From these interference patterns it is possible to locate a source much more accurately (Figure 2.54). The Very Large Array (VLA) radio telescope system (interferometer) in central New Mexico (Figure 2.53), consists of 27 movable antennae, each 85 feet in diameter and each weighing 214 tons. The radio signal from each antenna is fed into a computer, and the combined effect of these antennae produces a resolution rivaling the largest optical telescopes. In the case of extended objects, it is possible to determine both size and shape by this system.

Figure 2.54 *A graphical representation of radio sources in a region of the sky 4° in diameter, as recorded by the Cambridge One-Mile Interferometer. (From Pooley and Kenderdine,* Monthly Not. R.A.S. 139, *529 (1968); Mullard Radio Observatory, University of Cambridge)*

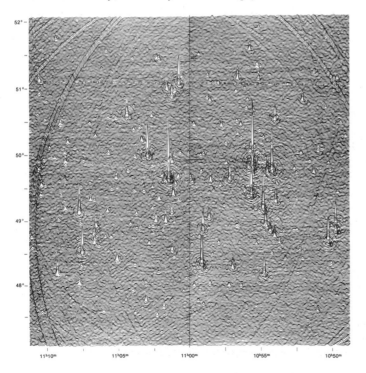

Radar Astronomy

Many radio telescopes, including those mentioned above, can be used for radar studies as well as for radio astronomy. Radar astronomy involves a two-way communication with relatively nearby objects such as the sun, the moon, and the planets. A radio signal is sent by the antenna toward an object, and then the same antenna is used to listen for the echo or reflection of that signal. Since the velocity of a radio signal is known, the length of time required by the signal to travel to a given planet and back reveals its distance. Other types of information such as the planet's velocity, its rotation, and its surface contour may also be ascertained by analyzing the Doppler shifts in the returning signal. Radar astronomy was used to penetrate the clouds of Venus, for instance, to give us our first indications of its rate and direction of rotation and its surface features.

Infrared Astronomy

The development of infrared astronomy lagged behind that of radio astronomy by approximately 20 years. The reason was that no one had perfected a detector that would record very small changes in the amount of infrared energy which fell on it and that would also discriminate as to the direction from which the radiation came. Infrared radiation falls between radio and red light in its range of wavelengths. Neither the photographic plates used by the optical astronomer nor the antennae used by the radio astronomer will record infrared radiation; however, the standard optical telescope may be employed with a suitable detector placed at the focal point.

Infrared detectors of various designs are used. The detector shown in Figure 2.55 utilizes a lead sulfide photoconductor. For each infrared photon that strikes the lead sulfide cell, a weak electrical current is created in a circuit not shown. The more infrared energy falling on the cell, the stronger the electric current generated; thus the electric current, when measured, becomes a measure of the infrared energy received. The sensitivity of such a cell is increased by cooling it to $-200°C$ with liquid helium. In order to point the device in only one direction at a given time, a small aperture admits the energy, which may then be restricted to the desired wavelengths by a filter. This apparatus is placed at the focus of an optical telescope. Its design tends to eliminate radiation from stray sources entering the cell (Figure 2.55).

One of the latest detector designs utilizes the charge-coupled diode (c.c.d.), one of the marvels of the electronic age. With an array of these very sensitive detectors, the spectrum of infrared sources can be read.

Among the very exciting objects which can be detected almost exclusively in infrared are stars in the making and stars which are dying. These objects are often surrounded by gas and dust clouds to the point of

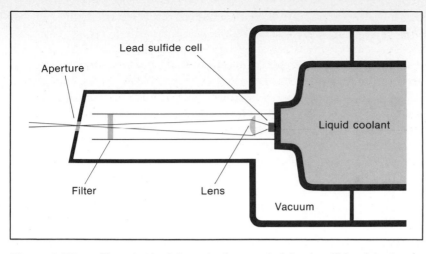

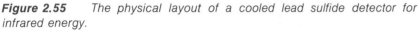

Figure 2.55 *The physical layout of a cooled lead sulfide detector for infrared energy.*

being totally obscured in visible light. The dust grains typically absorb the shorter wavelengths of light and especially of ultraviolet, reemitting that energy in the longer wavelengths of infrared. In the shells of gas and dust expelled by dying stars, astronomers find the heavier elements thought to have been manufactured in their cores—some of which take the form of organic molecules which make life in the universe possible.

Ultraviolet Astronomy

The greatest impediment to detection of radiation in the ultraviolet portion of the spectrum has been the earth's atmosphere. Success in this area has come with the use of rockets and orbiting satellites such as the

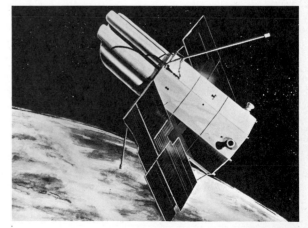

Figure 2.56 *The Copernicus satellite (OAO-C) carries a 1-m telescope designed to sense ultraviolet wavelengths and two smaller X-ray telescopes. Also illustrated here are the two solar energy panels and the knoblike star trackers near the base of the satellite. (NASA)*

Orbiting Astronomical Observatory named *Copernicus* (OAO-C, Figure 2.56). This observatory carries the largest reflecting telescope ever sent into space. Its 0.8-m mirror, spectrograph, and allied sensors are designed to detect the ultraviolet spectra of stars and of interstellar molecules with higher resolution than has been achieved previously. The satellite does not actually photograph the spectrum but reads out the photon count as it scans the source at a variety of wavelengths. A plot of such a readout is seen in Figure 2.57. The dips in this graph correspond to absorption features in the visible spectrum. Ultraviolet radiation is typically associated with very hot objects, such as stars with surface temperatures in the range of 50,000 to 100,000°K or very hot gas clouds found between the stars.

Large Space Telescope

Planned for launching in the mid-1980s is a larger orbiting telescope, dubbed simply Space Telescope (ST, Figure 2.58). Operating above the earth's atmosphere, this instrument with its 2.4-m primary mirror could represent the single greatest step forward in optical astronomy. It will provide a resolution 10 to 20 times better than that of any ground-

Figure 2.57   *A readout of a portion of the ultraviolet spectrum by OAO-C, the Copernicus satellite, showing absorption lines of ionized argon and hydrogen gas (H_2 molecules). [From Rogerson et al.,* The Astrophysical Journal (Letters) *181* (3), *L97-L102 (1973); courtesy of The American Astronomical Society and University of Chicago Press]*

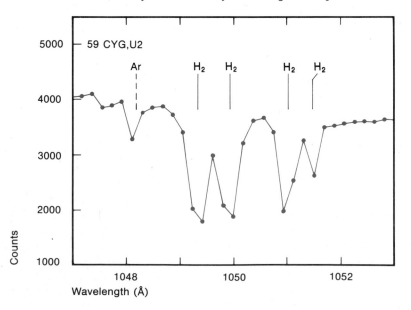

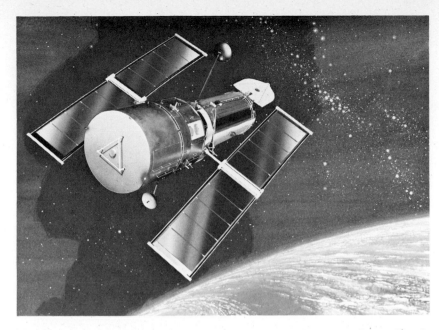

Figure 2.58   *The 96-inch Space Telescope, to be launched by Space Shuttle in 1983, will provide the clearest photos ever taken of the universe. It will respond to a wide range of wavelengths (from 1000 Å to 1 mm). (NASA)*

based telescope. Its associated detectors will "paint" an enhanced picture of the universe in the infrared, visual, and ultraviolet portion of the spectrum, yielding new information regarding faint objects, quasars, nuclei of galaxies, chemical evolution of galaxies, and tests for the models of cosmology (these topics will be treated later in this text). The ST is to be launched by the Space Shuttle, and if necessary it may be revisited or recovered by the same means, for refitting or repair.

X-ray Astronomy

The earth's atmosphere absorbs almost all X-ray radiation directed toward the earth. While this is a fortunate fact for our existence, it limits any effective examination of the universe in wavelengths between 0.1 and 100 Å. The first successful attempts to sense sources of X rays came with the firing of rockets high above the obscuring layers of our atmosphere, and 30 to 40 strong discrete sources of X rays were immediately found within our own Galaxy. The launching of a satellite exclusively designed for X-ray observation created a massive amount of additional data—the emerging X-ray "picture" of the universe. The satellite was dubbed *UHURU*, meaning "freedom" in Swahili, and was sent aloft on December 12, 1970, from a location near Kenya in honor of that country's independence (Figure 2.59). The craft spins slowly on its axis,

Figure 2.59 *Artist's view of UHURU, an X-ray satellite. [From R. Fiacconi, X-ray astronomy.* The Physics Teacher 11 *(3), 135–143 (1973)]*

Figure 2.60 *A plot of X-ray sources, based on observations by the* UHURU *satellite. The larger dots indicate the stronger sources; several are identified. (Frederick D. Seward, Lawrence Livermore Laboratory, University of California)*

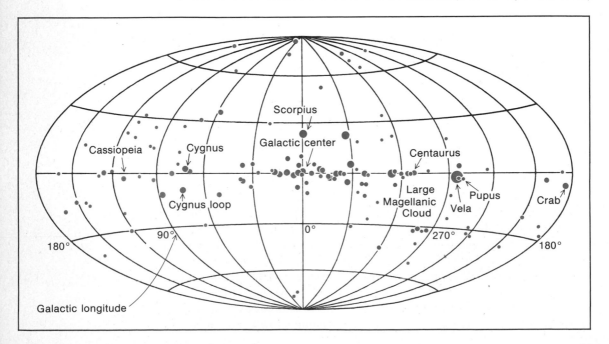

scanning the sky continually for X-ray sources, and its orientation (pointing direction) can be controlled from ground-based control centers. In the last few years, astronomers' ability to detect X-ray sources has been greatly enhanced by a series of orbiting observatories called HEAO (High Energy Astronomical Observatory). Significantly more sensitive than *UHURU*, HEAO has now detected X-ray pulsars, X-ray-emitting quasars, Seyfert galaxies, and BL Lacertae objects, in addition to the celebrated black holes. Each of these topics will be treated in later chapters. A plot of the principal X-ray sources in the Milky Way is shown in Figure 2.60.

Gamma-ray Astronomy

The latest chapter in the opening up of the entire electromagnetic spectrum to the astronomer's view involves the detection of gamma rays and particularly gamma-ray bursts. X-ray and gamma-ray photons carry more energy than any of the other portions of the spectrum, hence we may assume that the events with which they are associated represent some of the most dramatic events in the universe. Bursts of gamma rays were first detected quite unintentionally by a number of Vela satellites launched in the 1960s and early 1970s. These satellites were designed to detect atomic blasts violating the nuclear test-ban treaty—tests which would have produced gamma-ray radiation. Instead, two or three Vela satellites, orbiting at various points around the earth, sensed gamma-ray bursts almost simultaneously, indicating a source outside the earth. This and subsequent observations of gamma-ray bursts have opened a wide range of speculation as to the source of this radiation. One such source appears to coincide with the site of a black hole (Cygnus X-1), hence we may feel quite certain that gamma rays are produced whenever material falls onto a very dense object. Other possible sources will be discussed in their appropriate sections later in the text. Of one thing we may be sure: the energy poured out in one second by these events exceeds the total energy emitted by the sun in many years.

Gravitational Wave Astronomy

We have seen that surrounding every charged particle is an electrical field. If that charged particle is set into *oscillation* (vibration), a changing electrical field results. If another charged particle exists in that field, it will react by oscillating, and this oscillation permits us to detect the existence of that changing electrical field. In 1916 Albert Einstein predicted that a similar thing would happen in the gravitational field of a moving object. First picture an object—say, an uncharged steel ball—placed in the center of an empty room. We could then say that a certain gravitational field exists in the room because of the presence of the steel ball. We could demonstrate that fact by showing that another steel ball, when brought into the room, would experience a force of attraction toward the original ball. This force is so small in comparison to the electrical and magnetic forces with which we have been dealing that it would be difficult to measure. Suppose that the second ball were hung from a high ceiling by a steel wire so that it was very close to the first ball. The second ball would be deflected slightly toward the first, so that it did not hang vertically from the ceiling. That deflection proves there is gravitational attraction between the two steel balls; there is an unseen gravitational field, which is static (unchanging). Now suppose that the first ball were set into a rapid vibration; then the gravitational field around it would become a changing field, and the second ball would reveal that changing field by vibrating itself.

It was reasoned that if Einstein's prediction that an accelerating mass emits gravitational waves were right, then astronomers might detect the collapse or explosion of gigantic galaxies, the collision of two black holes, or other cataclysmic events by designing a detector for the resulting waves. Such a detector must be free from any other kind of influence, such as earthquakes, changes in electrical or magnetic fields, and air disturbances.

One of the first gravitational wave detectors consisted of a 3000-kg aluminum cylinder suspended from an arch support system. A very sensitive recording device had been placed on the cylinder, and the very small vibrations that gravity waves were thought to produce appeared to be within the detection capability of this instrument. The cylinder could not be isolated completely from its physical surroundings, however, and consequently it recorded man-made vibrations. In order to separate these vibrations from those which might represent gravitational waves, experimenter Joseph Weber installed a second detector 1600 km (1000 miles) away, reasoning that a disturbance originating outside the earth would be recorded almost simultaneously by both detectors, whereas local disturbances would show up only on one. In 1969 Weber detected what he believed to be the first gravitational waves ever recorded.

As is required in science, confirmation of Weber's observations is being sought by other scientists. One such effort centers around the work of William M. Fairbank and William Hamilton at Stanford University. These two men bring their study of *cryogenics* (the branch of physics dealing with the behavior of materials at very low temperatures) to bear upon the problem of isolating the detector from its surroundings in order to reduce the effect of background vibrations. The new detector (Figure 2.61) consists of a 6-ton cylinder of aluminum, encased in several containers of coolants to reduce the temperature in the innermost container to within $0.5°$ of absolute zero. At this temperature ($-272.5°C$) vibrations due to heat virtually cease. What is more important, the cylinder can be made to "float" in the middle of the innermost cylinder without touching anything. This is possible because when certain metals are supercooled, an electrical current, once started, will flow through them forever. The cylinder can thus be made to float on

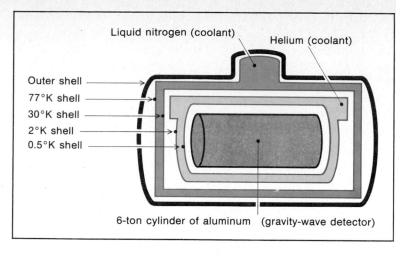

Figure 2.61   *Gravitational wave detector.*

swirls of electrical current and yet can be isolated from any electrical or magnetic disturbances from without. Two such detectors, currently being installed at Stanford University and Louisiana State University, approximately 3200 km apart, are to operate simultaneously in an attempt to confirm vibrations (gravity waves) originating in our galaxy or beyond.

Thus, using the tools we have briefly surveyed in this chapter, astronomers have increased their ability to receive data over the full spectrum of electromagnetic and particle wavelengths, and they have used these data to project a richer and clearer picture of the universe than has ever been conceived before.

Questions

1. Perform the necessary arithmetic to determine the length of a light year in kilometers, based on the fact that light travels at the speed of approximately 300,000 km/sec; then convert this distance to miles.

2. In any experiment to determine the speed of light, what two factors must be known?

3. Albert A. Michelson used a rotating mirror to measure one of the two essential factors mentioned in Question 2. Which factor did he measure with the turning mirror?

4. Other than visible light, in what ways do electromagnetic disturbances exhibit themselves?

5. Why is sound not included among electromagnetic disturbances?

6. The telescope mirror is curved so as to bring light rays from a given source to the same focal point. What basic law or principle is utilized in the design of this curvature?

7. Why do stars near the horizon appear higher in the sky than their true position?

8. The earliest telescopes utilized the principle of _____ (refraction, reflection).

9. List four examples of extended sources.

10. List three defects which may exist in the refracting telescope if it is not correctly designed.

11. Which telescope design is best suited for a sky survey?

12. Discuss the advantage of using film to photograph dim objects over merely observing them with the naked eye.

13. What is the primary reason that radio telescopes have relatively poor resolution even though they are larger than optical telescopes?

14. The functioning of a spectrograph depends upon the fact that whenever light is refracted, it is also _____.

15. The primary spectrum of the sun, as observed from the earth, is a _____ (dark-line, bright-line, continuous) spectrum.

16. A bright-line spectrum is produced by _____ (upward, downward) transitions of the electron.

17. What is indicated by a very large red shift in the spectral lines of a distant galaxy?

18. Find the size of the image of a full moon (0.5° in diameter) photographed at the prime focus of the 5-m (200-in.) telescope; focal length is 16.8 m. (Answer: 14.6 cm.)

19. What advantage does a radio telescope have over its optical counterpart?

20. What advantage does a telescope aboard an orbiting observatory have over a ground-based telescope?

21. Why should the infrared sky look different than the visible sky?

22. What kinds of events are expected to produce gravitational waves?

23. If a given star is moving through space at the same speed and direction as the sun, we do not expect to see a Doppler shift in its spectrum; but because the earth is moving around the sun

at a velocity of 30 km/sec, the earth's motion creates a periodic blue shift when it approaches the star and a red shift when it recedes from the star. Find the maximum Doppler shift produced by the earth's motion.

24. Given two telescopes having the same focal length, the one with an aperture of 25 cm collects _____ times as much light as the one with an aperture of 12.5 cm.

Suggested Readings

ASIMOV, ISAAC, *Eyes on the universe: A history of the telescope.* Boston: Houghton Mifflin Co., 1975.

GIACCONI, RICCARDO, X-ray astronomy. *The Physics Teacher 11* (3), 135–143 (1973).

GINGERICH, OWEN (ed.), *Frontiers of astronomy* (Introduction). San Francisco: W. H. Freeman & Company Publishers, 1970.

HEALY, D., Make your own astrophotographs. *Popular Photography 81,* 110–111 (1977).

HOWARD, NEALE E., *The telescope handbook and star atlas.* New York: Thomas Y. Crowell, 1967.

KELLERMAN, K. I., Intercontinental radio astronomy. *Scientific American 226* (2), 72–83 (1972).

LOGAN, JONOTHAN L., Gravitational waves. *The Physics Teacher 26* (3), 44–52 (1973).

MEYER-ARENDT, JURGEN R., *Introduction to classical and modern optics.* Englewood Cliffs, N.J.: Prentice-Hall, 1972.

NEUGEBAUER, G., and BECKLIN, ERIC E., The brightest infrared sources. *Scientific American 228* (4), 28–40 (1973).

SCHAWLOW, ARTHUR L. (ed.), *Lasers and light* (Introduction). San Francisco: W. H. Freeman & Company Publishers, 1969.

SNOW, T. P., JR., Ultraviolet spectroscopy with Copernicus (OAO-C). *Sky and Telescope 54,* 371–374 (1977).

WALLIS, B. D. and PROVIN, R. W., On the road to better astronomical photographs. *Sky and Telescope 53,* 314–318, 399–405, 484–491 (1977).

WELLS, R. A., The "first" Newtonian. *Sky and Telescope 42* (6), 342–344 (1971).

WIEGAND, CLYDE E., Exotic atoms. *Scientific American 227* (5), 102–110 (1972).

THREE

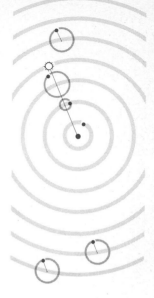

THE EARTH

Although study of the earth is often considered to be the exclusive domain of the geologist, the astronomer is also very much concerned about our planet's origin and properties, such as size, mass, density, layering, atmosphere, magnetic field, and motion. From his knowledge of the earth, he is better prepared to ask significant questions concerning the other planets and their moons.

Pythagoras in the sixth century B.C. had already recognized the shape of the earth as being spherical. Another Greek astronomer, Eratosthenes, four centuries later went a step further when he observed that the rays of the noon sun illuminated the bottom of a vertical well in Syene (now Aswan), Egypt, indicating that the sun was directly overhead. On the same day in Alexandria, 800 km to the north of Syene, the rays of the sun at noon made an angle of 7.2° with a vertical post (Figure 3.1).

Assuming the earth to be spherical, Eratosthenes ventured to compute its circumference by the following line of reasoning. The angle formed between the sun's rays and a vertical post at Alexandria must be the same as the angle formed by these lines extended to the center of the earth. This is an application of the geometric theorem that if two parallel lines are cut by a transversal (a sloping line), the corresponding angles are equal. If an angle of 7.2° at the center of the

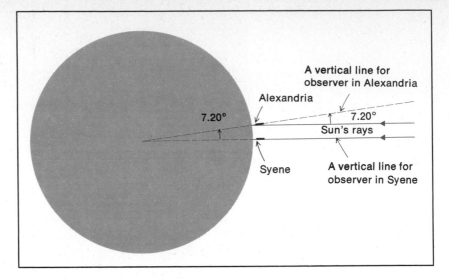

Figure 3.1   *The method by which Eratosthenes measured the size of the earth.*

earth spans a distance of 800 km on the surface of the earth and 50 such angles will fit in the circle ($360°/7.2° = 50$), then the total distance around the earth is equal to 50×800 km—that is, 40,000 km. This is very nearly the value accepted today. From the circumference, it is possible to find the diameter, remembering that $C = \pi d$ and $d = C/\pi$:

$$d = \frac{40,000 \text{ km}}{3.1416} = 12,732 \text{ km}$$

In 1687 Sir Isaac Newton suggested that the earth was not a perfect sphere, for its rotation tends to flatten it. Today we are able to measure that flattening and find the radius of the earth measured to the north or south pole to be approximately 22.5 km shorter than the radius of the earth. This flattening amounts to only about one-third of 1 percent.

Volume, Mass, and Density

The concepts of volume, mass, and density are fundamental to all of science. It will be profitable to visualize each one before applying it to the earth, planets, or other objects. Picture a brick of dimensions 12 cm × 6 cm × 4 cm. How can we find its volume, its mass, and its density? In order to answer, we must have an accurate definition of each term.

Volume is a measure of the space an object occupies. What an object is made of has no bearing on its volume. The volume of a rectangular solid (a box) is simply: $V = $ length × width × height. Therefore

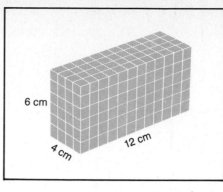

Figure 3.2 *Measuring the volume of a rectangular solid.*

the volume of the brick is $12 \text{ cm} \times 6 \text{ cm} \times 4 \text{ cm} = 168 \text{ cm}^3$ (cubic centimeters). Our unit of volume is the cubic centimeter—a cube which is 1 cm $\times$ 1 cm $\times$ 1 cm. The formula ($v = l \times w \times h$) merely counts the number of cubic centimeters that would occupy the same space as the brick (Figure 3.2).

The *rest mass* of an object is a measure of how much material it contains—a factor which is independent of its volume. We define the unit of mass to be the *gram*—the amount of material in 1 cm^3 of pure water at 4°C. How can the mass of an object be determined? A simple balance will be sufficient for an object of reasonable size. Place the object on one side and sufficient standardized masses on the other side until a balance is achieved (Figure 3.3).

Density combines the concepts of volume and mass: it is a measure of the amount of material packed into a given space:

$$\text{density} = \frac{\text{mass}}{\text{volume}}$$

Figure 3.3 *Determining the mass of the brick.*

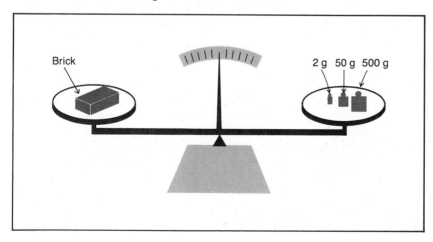

The density of the brick is

$$D = \frac{M}{V} = \frac{552 \text{ g}}{168 \text{ cm}^3} = 3.29 \text{ g/cm}^3 \text{ (grams per cubic centimeter)}$$

Volume of the Earth

Because the earth is so nearly spherical, we may use a formula which has been derived especially for a sphere:

$$V = \tfrac{4}{3} \pi r^3 \quad (r = \text{radius})$$

Using a modern value for the radius of the earth, 6375 km, we find the earth's volume to be

$$V = \tfrac{4}{3}(3.1416)\ (6375 \text{ km})^3$$

$$V = 1.085 \times 10^{12} \text{ km}^3 \text{ (cubic kilometers)}$$

The cubic kilometer is not a practical unit for use in the laboratory or for everyday situations, hence the volume might better be expressed in cubic centimeters, which we used in measuring the volume of the brick (6375 km = 637,500,000 cm):

$$V = \tfrac{4}{3}(3.1416)\ (637,500,000 \text{ cm})^3$$

$$V = 1.08 \times 10^{27} \text{ cm}^3 \text{ (cubic centimeters)}$$

Mass of the Earth

We cannot place the earth on one side of a balance to determine its mass, hence we must devise an indirect method. We will use the law of gravity as stated by Sir Isaac Newton (Figure 3.4):

Figure 3.4   *The mutual attraction between two objects is called gravity and depends upon the mass of each object, together with distance between their centers.*

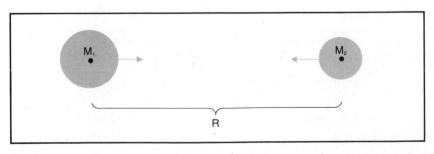

$$F = G\frac{M_1 \times M_2}{R^2}$$

where F = force due to gravity
          G = gravitational constant
          M_1 = mass of object 1
          M_2 = mass of object 2
          R = distance between centers of mass

Newton would have expressed the law of gravity as follows:

The mutually attractive force between two objects is directly proportional to the product of their masses and inversely proportional to the distance between their centers.

As shown in Figure 3.5, when an object "A" is located on the surface of the earth, its weight is actually the force (F) due to gravity:

$$F = G\frac{M_A M_E}{R^2}$$

where G = gravitational constant
          M_A = mass of object "A"
          M_E = mass of earth
          R = radius of the earth

Suppose that we place equal masses A and A' on opposite pans of a long-beam balance [see Figure 3.6(a)]. The force which the earth exerts on A is stated above. The force on A' would have a similar form:

$$F = G\frac{M_{A'} M_E}{R^2}$$

These forces are obviously equal, because the mass of A is equal to the mass of A', and so the system is in balance.

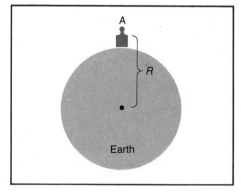

Figure 3.5   *The weight of an object (A) is the force due to gravity.*

A

R

Earth

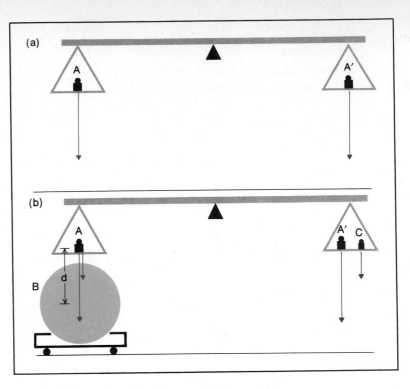

Figure 3.6   *Measuring the mass of the earth.*

Now we roll a large mass B under one pan but not touching it. Mass B exerts an additional force on mass A due to gravitational attraction in the amount of GM_AM_B/d^2, where d is the separation of centers of A and B [see Figure 3.6(b)]. In order to balance this force, a small mass C is added to the other pan, thus adding the force GM_EM_C/R^2. If a balanced condition is achieved in this way, then the force of gravity between objects A and B must just equal the force of gravity between object C and the earth (E).

$$G\frac{M_AM_B}{d^2} = G\frac{M_EM_C}{R^2}$$

Solving this equation for M_E we get

$$M_E = \frac{M_A \times M_B \times R^2}{M_C \times d^2}$$

Since each member on the right is a known quantity in the experiment, we can determine the mass of the earth (M_E) when such an experiment is performed. The answer is: $M_E = 5.98 \times 10^{27}$ g.

Density of the Earth

We have already defined density as a measure of how much material is packed into a given space:

$$\text{density} = \frac{\text{mass}}{\text{volume}} = \frac{\text{the amount of material}}{\text{the amount of space}}$$

Using the values derived for mass (M_E) and volume (V_E) of the earth, we may find its average density by

$$D_E = \frac{M_E}{V_E} = \frac{5.98 \times 10^{27}\ g}{1.08 \times 10^{27}\ cm^3} = 5.5\ g/cm^3$$

How this density compares to the density of various common materials is shown in Table 3.1.

Material such as ice, the density of which is less than 1 g/cm³, will float in water. The average density of Saturn is 0.687 g/cm³, less than that of water, whereas Mars has an average density of 3.82 g/cm³, more like that of the earth. Thus the density of an object tells us something of its physical nature. Saturn is more like a ball of gas, having a thick atmosphere, perhaps with a liquid center, whereas Mars has only a very thin atmosphere and is characterized by a crusty, solid surface.

The average surface material of the earth has a density of only 2.7 g/cm³, whereas the average density for the entire earth is 5.5 g/cm³.

Table 3.1   *Densities of Some Common Materials*

MATERIAL	DENSITY (g/cm³)
Air	0.00129
Oxygen	0.00143
Cork (average)	0.25
Gasoline	0.69
Ice	0.91
Water	1.00
Sea water	1.02
Blood	1.04
Aluminum	2.70
Granite (average)	2.70
Iron	7.86
Copper	8.89
Lead	11.35
Mercury	13.59
Gold	19.27
Platinum	21.37

What must the interior of the earth be like in order to make up for this difference? The only logical answer is that the interior must be more dense. The density of the central portion of the earth is thought to range as high as 15 g/cm^3 (Figure 3.7). We generally expect increasing density in any body toward its center because of the force of gravity. All parts of a body experience a mutual attraction that has the effect of pulling these parts toward the center. Material near the center experiences increased pressure that tends to force it into a small space, thus increasing its density. Furthermore, elements such as iron and nickel, which are naturally more dense, experience a greater force toward the center. When the earth was in a molten state and material was free to distribute itself according to the forces upon it, the more dense material tended to move toward the center of the earth.

How can we verify this change in density with increasing depth? We have studied the fact that light (a wavelike disturbance) slows down in certain materials such as glass or water and is thereby refracted. A similar phenomenon happens to earthquake waves as they travel through layers of increasing density within the earth. By timing the arrival of earth-quake waves at various points over the surface of the earth, researchers can detect distinct layers within the earth to a fair degree of accuracy. The density appears to change quite abruptly at the boundaries of these layers, as indicated in Figure 3.8. Earthquakes send out two kinds of waves: those with wavelike motion perpendicular to the line of travel (*transverse*) and those with wavelike motion parallel to the line of travel (*compressional*); the transverse waves will not pass through a liquid. Because transverse waves do not pass through the outer core of the earth, that layer is thought to be molten (a very significant fact in regard to the earth's magnetic field, which we will discuss later).

Crust of the Earth

As indicated in Figure 3.7, the crust of the earth is relatively thin, and, because it is less dense than the mantle, the crust floats on the

Figure 3.7   *Layers of the earth.*

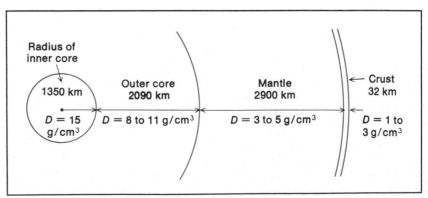

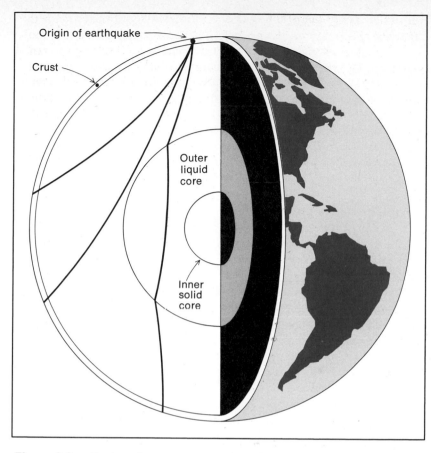

Figure 3.8 *Earthquake waves travel at different velocities in various layers of the earth. The refraction which results is an indicator of the density of these layers.*

mantle much as ice floats in water. On this floating crust the arrangement of continents is not static or permanent but rather has been changing for at least 200 million years. Today six major plates are identified as shown in Figure 3.9. Along certain plate boundaries, compressional forces are producing mountain-building, volcanic, and earthquake activity. Maps that plot the centers of earthquake activity virtually outline the crustal plates. Almost the entire Pacific coast of North and South America lies along such a boundary.

Other boundaries mark plates which are separating. Such a boundary is indicated as the Mid-Atlantic Ridge (see Figures 3.9 and 3.10). As plates separate, new material must well up to fill in the gap that is created. This action has been confirmed, since samples taken from the Mid-Atlantic ridge are very young. Note also the scars that have been left in the bottom of the Atlantic Ocean by the spreading of the sea floor (see Figure 3.10). Studies of similar geologic structure and fossil remains in the land masses of North and South America and Africa have pretty well confirmed that these regions adjoined one another 200 million years ago (see Figure 3.11).

One of the latest techniques for measuring the motions of crustal plates involves the process of radio interferometry described in Chapter 2, page 91. A radio signal is received by two different radio antenna from a distant radio source, outside our own galaxy. Because of the different path lengths, the signals arrive out of phase—they interfere in a partially destructive manner. If the two antennae are located on opposite sides of a plate boundary (a fault), then even a slight change in the relative positions of the two plates (or two sides of a fault) can be detected by a change in the interference pattern. This technique may lead to earthquake prediction based on the theory that slight dilations in the fault take place before a quake.

Age of the Earth

The rocks of the earth have a story to tell, for within them lie clues to the age and process of formation of the earth. When compared by weight, the surface material of the earth is approximately 47 percent oxygen, 28 percent silicon, 8 percent aluminum, and 5 percent iron, with lesser amounts of magnesium, calcium, potassium, and other elements.

Figure 3.9   *The major plates of the earth with arrows indicating approximate direction of motion.*

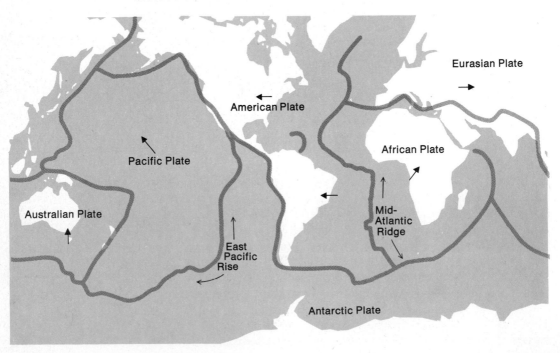

Many rocks contain traces of radioactive elements—elements which tend to break down into lighter elements at a set rate. For instance, suppose we could observe a certain quantity of uranium, say 8 g, over a period of 4.5 billion years. We would find that half of that quantity—4 g—is spontaneously transformed through several intermediate steps into lead. Now a uranium atom contains 238 neutrons and protons, whereas lead contains only 206, hence the uranium atom has rid itself of a total of 32 neutrons and protons in the process. Another 4.5 billion years would reduce the 4 g to 2 g; 4.5 billion years is thus spoken of as the *half-life* of uranium. When a rock crystallizes, the participants in this process are locked in place, and the scientist may estimate the age of the rock by determining the proportion of uranium to lead. Since lead-206

Figure 3.10   *The mid-Atlantic ridge, showing the effects of seafloor spreading as the plates drifted apart. Numbers indicate feet above or below sea level. (Photograph courtesy of Alcoa)*

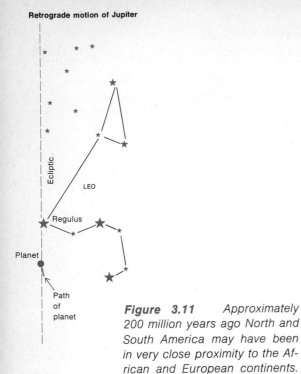

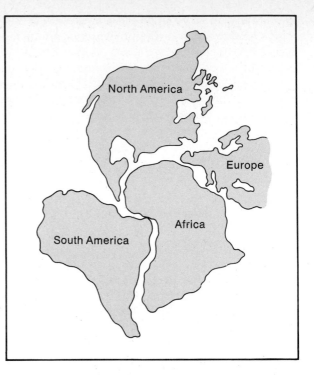

Figure 3.11 *Approximately 200 million years ago North and South America may have been in very close proximity to the African and European continents.*

derives entirely from the radioactive decay of uranium-238, we may assume that the lead-206 present has been produced since the time of crystallization. The age of the rock may be confirmed by determination of the presence of elements such as radioactive potassium, which turns into argon gas stored in the crystalline structure of the rock. We may assume that any argon gas formed before crystallization of the rock would have been lost into the air.

The oldest rocks found on the surface of the earth date back approximately 3.7 billion years to their time of crystallization. Most meteorites and certain of the oldest moon rocks date back 4.6 billion years. A very tempting generalization follows: if the oldest solid material of our solar system dates back 4.6 billion years, then the entire solar system is at least that old. This statement flows from the thought that the entire solar system with its sun, nine planets, 34 moons, and numerous asteroids, comets, and meteoroids had a common origin. Details of this theory will be presented in Chapter 5.

Atmosphere of the Earth

While the atmosphere of the earth is not divided into distinct layers, it will be helpful to think of it in this way. The first layer—the one that contacts the earth's surface—is called the *troposphere*. This layer,

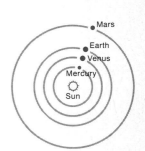

varying from 8 to 16 km in thickness, is the location of all weather disturbances. Three-fourths of the total atmosphere is "packed" into this layer by gravity, creating a normal sea-level pressure of 1 atm (1 atmosphere, equivalent to the pressure exerted by a column of mercury 76 cm high at 0°C, or about 15 lb/in.^{2}). This atmospheric pressure is exerted in all directions equally at any given point and so tends to be equalized over a body. We do not normally sense this pressure, but if we dive into a swimming pool to a depth of 2.5 to 3 m, we sense the additional pressure in our ears, owing to the overlying layer of water in addition to the normal atmospheric pressure. The troposphere is composed of 78 percent nitrogen and 21 percent oxygen, with water vapor, carbon dioxide, neon, and argon making up most of the remaining 1 percent, by volume. Traces of other elements are also found in the atmosphere. The troposphere is characterized by an average temperature that decreases from about +60°F (Fahrenheit)* at sea level to −60°F near the top of the layer.

The next layer, the *stratosphere*, extends from about 16 km to a height of 50 km and is characterized by a temperature increasing from −60°F to +32°F. Within this layer there exists a special form of oxygen called *ozone*. Three atoms of oxygen combine to form a molecule of ozone. It has a rather pungent odor that is sometimes noticeable during a lightning storm. The ozone layer absorbs ultraviolet light from the sun, thus protecting us from these dangerous rays. Although it performs this essential function, the ozone layer is nevertheless a handicap to the study of the ultraviolet portion of the spectrum. Observations in this region of the spectrum must be carried out from above the ozone layer with the aid of a balloon or rocket, or from an observatory orbiting in space. All of these methods have been utilized.

Above the stratosphere is the *mesosphere*, extending from about 50 km to a height of 90 km and characterized by decreasing temperatures with increasing height (+32°F to −130°F).

Above the mesosphere is the *ionosphere*, a series of layers of *ionized* gases. An *ion* is an atom that has lost or gained one or more electrons, hence is a *charged* atom (Figure 3.12). Ionization is caused by solar radiation, and therefore the degree to which the layers are formed depends largely on activities on the sun. These layers range in elevation between 80 and 320 km; they serve to reflect radio waves that have wavelengths longer than 15 m (Figure 3.13). This function is essential to radio communication, allowing the waves to travel around the curvature of the earth by reflection from the ionospheric layers. Wavelengths shorter than 15 m generally penetrate the ionosphere from the earth or from space; therefore this layer does not have the effect of preventing the study of very short wavelength radio energies in the universe. This is often referred to as the radio "window" into space.

The fact that the earth's atmosphere contains ozone, water vapor,

---

*Plus and minus signs are used with degrees of temperature to designate degrees above 0° and below 0°, respectively.

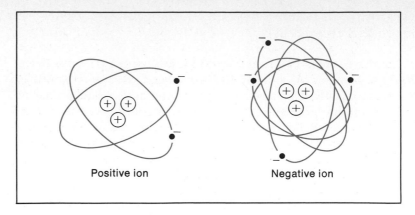

| Positive ion | Negative ion |

Figure 3.12 *Ionized atoms (neutrons not shown).*

and carbon dioxide causes it to act like a greenhouse. The earth receives many forms of energy from the sun, including some ultraviolet light, visible light, and infrared light. The crust of the earth absorbs this energy and then reemits it, primarily in the form of infrared. The atmosphere tends to absorb this form of radiation quite efficiently, and so the earth retains a large amount of heat; in fact, the earth's surface temperature would probably be 45°C cooler were it not for this "greenhouse" phenomenon.

The earth's atmosphere often puts on a real "show" called the *northern lights* (*aurora borealis*). As charged particles, which continually outflow from the sun, approach the earth, they encounter its magnetic field. This field exerts a force on the particles, directing them in a circular motion around the earth until they reach the north or south polar region. Here they collide with atoms of the earth's atmosphere, ionizing some of them. When ionized atoms recombine with free electrons, downward transitions occur and light is produced. Such a display of light near the south pole is called the *southern lights* (*aurora australis*).

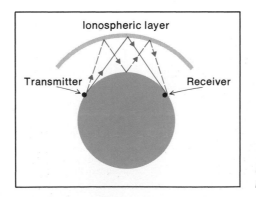

Figure 3.13 *The ionosphere reflects certain radio waves.*

Why Is the Sky Blue?

Because of their very small size, the gas molecules that compose the earth's atmosphere tend to vibrate with a wavelength corresponding to blue light more efficiently than with that of any other color. *Scattering* refers to the process whereby the molecule absorbs certain colors of light and then reemits that light in all directions. If we viewed a collection of molecules directly in line with the sun, we would see the usual color of the sun; however, if that same collection of molecules were viewed from the side, we would see the light which that collection had scattered (predominantly blue light; Figure 3.14). The red sunset is another result of scattering. As the sun appears to move toward the horizon in the late afternoon, its rays must penetrate an increasing amount of atmosphere. Since molecules in the atmosphere scatter blue light best, the colors that penetrate the atmosphere along our line of sight are predominantly red. The phenomena we call twilight and dawn also result from the scattering of light by molecules in the atmosphere.

The moon lacks any atmosphere, and its sky could best be described as black; when the sun "goes down" at any location on the moon, complete darkness immediately comes on, with a severe drop in temperature.

The Earth's Magnetic Field

It is very evident that the earth possesses a magnetic field. The fact that a compass needle aligns itself in virtually the same direction on

Figure 3.14   *The scattering of sunlight produces the blue sky.*

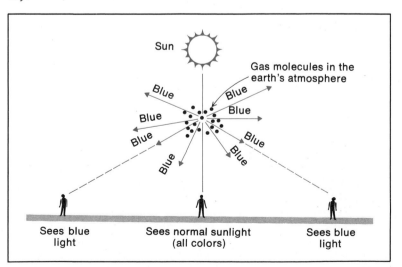

every occasion demonstrates this fact. If we traveled in a northerly direction, as indicated by the compass, we should eventually arrive at a point called the *north magnetic pole,* and yet we would find ourselves approximately 1600 km from the *geographic north pole,* which is determined by the rotation of the earth. The geographic north pole is a point on the surface of the earth that remains stationary as the earth rotates. These ideas apply equally to the southern polar region.

If we are to attempt an explanation of the cause of the earth's magnetic field, the following observations must be kept in mind. While the north magnetic pole is located in northeastern Canada at the present time, it has not always been located there but has wandered in a rather random fashion all around the north geographic pole. Furthermore, geologists have unearthed evidence in certain rocks bearing magnetic minerals that the earth's magnetic field has reversed itself many times in each of the nine epochs of the last 3.6 million years. When magnetic minerals in rocks crystallize from a liquid state, they align their magnetic components with the earth's magnetic field at the time of crystallization and then retain that magnetism until unearthed.

No fully adequate explanation has been given for sustaining the magnetic field of the earth, but we would at least rule out the probability of natural "lodestone," buried in northeastern Canada, as being responsible. If this were true, it would be impossible to explain the observed changes in location of the magnetic poles—changes measured in hundreds of miles in a period of only a few hundred years.

There is one generalized concept that provides at least a partial explanation of the earth's magnetic field. Whenever charged particles are made to flow in a circular motion, a magnetic field is generated. For instance, visualize a disk as shown in Figure 3.15. If a charge is placed on each metal disk shown and then the larger disk is rotated between the hands, a nearby compass will react by turning to align itself with the magnetic field created. If charged particles exist within the earth, they are naturally set into circular motion by the rotation of the earth. Perhaps eddy currents (little whirlpools) are set up within the liquid outer core, producing individual magnetic fields, the total field being the algebraic sum of the individual fields.

Alternately, different layers of the earth and/or different latitudes

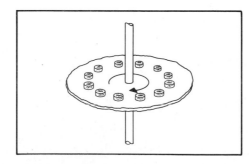

Figure 3.15 *The magnetic property of a rotating charged plate.*

may rotate at slightly different rates, producing a relative motion among the charged particles of the earth. This could also produce a magnetic field. See the explanation of the Coriolis effect on page 121. As we continue to observe the other planets with fly-by probes, orbiters, and landers, certainly we will want to compare their magnetic fields.

The magnetic field of the earth may be visualized by lines that seem to emerge from the north magnetic pole and return to the south magnetic pole. The lines are, of course, imaginary, but they indicate the direction that a compass would point if placed at a given location on a line; furthermore, a charged particle that moves into this magnetic field experiences a force that acts at right angles to the magnetic line and causes the charged particle to move in a circular path around the earth. Thus the earth seems to trap some of the rapidly moving charged particles (electrons, protons, and others) that flow from the sun. Two doughnut-shaped regions of these highly energized particles, called the *Van Allen belts*, are located at elevations of 3200 km and 16,000 km above the equator; regions over the poles are relatively free of these trapped particles (Figure 3.16).

The magnetic field of the earth is sometimes greatly disturbed by activities on the sun. Associated with a solar flare, for instance, is a

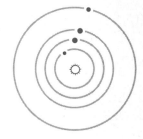

Figure 3.16 *The earth's magnetic field, distorted by the solar wind. The two Van Allen belts, somewhat doughnut-shaped, are regions of high-energy charged particles trapped by the earth's magnetic field.*

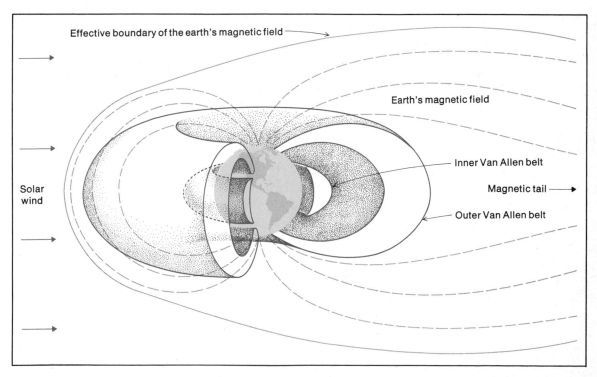

tremendous release of energy that often disrupts radio communications and produces changes in the magnetic field of the earth. Even when the sun is relatively quiet, there is a constant outflow of charged particles called the *solar wind*. This solar wind distorts the magnetic field of the earth. The records of satellites that have orbited the earth, passing in and out of the various regions of the magnetic field many times, show that this field is somewhat flattened on the side toward the sun, and that a magnetic "tail" exists on the opposite side. Figure 3.15 reveals the approximate configuration of the magnetic lines and also shows a cutaway view of the Van Allen belts.

Rotation of the Earth

Early observers saw almost all objects—sun, moon, planets, and stars—rise in the east and set in the west and concluded that they all moved around a stationary earth. This idea persisted as late as the sixteenth century. Even when theories regarding a rotating earth were introduced, no conclusive proof was available until the nineteenth century (1852), when Jean Foucault, a French physicist, demonstrated the fact of rotation. He suspended from the dome of the Pantheon in Paris a weighted pendulum that he began to swing (Figure 3.17). Only one force acts on a free-swinging pendulum, that of *gravity*. Since the force of gravity acts only toward the center of the earth, it would not produce any rotational motion, and yet the pendulum appeared to change the direction of its swing. If he started the pendulum swinging toward one entrance to the Pantheon, he would later find it swinging toward another entrance. The change of direction occurred at a given rate like the hands of a clock. It was then reasoned that the earth itself must be turning under the pendulum in order to produce the observed effect. If this experiment were performed at the north pole, the time required for one apparent rotation of the earth under the pendulum would be approximately 24 hours. As one moves from the pole toward the equator, the apparent time for one rotation of the pendulum increases, and at the latitude of Los Angeles (34° N) the period is 43 hours.

Today astronauts have witnessed the earth's rotation from their vantage point in space. This daily rotation of the earth accounts also for the apparent *diurnal* (daily) motion of the sun, the moon, the planets, and the stars. A photograph of the northern polar region of the sky, taken over a period of several hours, reveals circular star trails. In Figure 3.18 Polaris, the North Star, is seen just off center.

We know that the circumference of the earth at the equator is approximately 40,000 km. Since the earth rotates once in 24 hours, then the speed of a point on the equator is approximately 1600 km/hr (40,000 km/24 hr) or roughly 1000 mph. Farther to the north or south of the equator, say at latitude 34°, the speed of a point on the earth is closer to 1100 km/hr (700 mph). For a point at the poles the speed due to

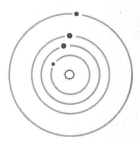

Figure 3.17   *The Foucault pendulum in the Pantheon, Paris. (Science Museum, London)*

rotation is zero. Because different latitudes have different speeds of rotation, a rocket fired northward from the equator would possess an eastward component of velocity of 1600 km/hr. As it passed northward over more slowly rotating points, the rocket would veer to the right, because its eastward component of velocity would be greater than that of the land over which it "flew." If fired southward from a region near the north pole, a rocket would still veer to the right, because its eastward velocity would be less than that of the land over which it flew (see Figure 3.19). This phenomenon is called the *Coriolis effect.* You can see that, since all moving objects experience this effect, the typical circulation of wind and ocean currents in the northern hemisphere is clockwise. By the same reasoning you can see that objects moving in the southern hemisphere experience a veering to the left, and the result is a counterclockwise pattern of currents. The very fact that the Coriolis effect can be observed constitutes another proof of the earth's rotation. The Coriolis effect may also occur within the earth, and we will draw upon this idea as a possible explanation of the earth's magnetic field.

Revolution of the Earth

Proof of the fact that the earth revolves around the sun did not come as a part of the Copernican heliocentric theory, nor as a part of Galileo's work, although they thought this idea to be correct (see Chapter 1). Rather, the proof came in the nineteenth century, when techniques were developed whereby the apparent shift of nearby stars could be detected in relation to the background of more distant stars. Were the

Figure 3.18   *Star trails in the region of the north celestial pole which result from the rotation of the earth. The trail made by the North Star, Polaris, is seen just off center. (Lick Observatory)*

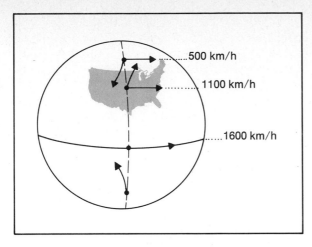

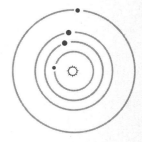

Figure 3.19   *The Coriolis effect, due to the earth's rotation.*

earth stationary, a given alignment of a nearby star and one more distant would not vary in a period of a few months. However, if the earth does in fact change its position in space, the alignment will also change. This apparent shifting of nearby stars against the background of more distant ones has been observed, and the phenomenon is called *stellar parallax*. It is a periodic kind of change, a given star apparently shifting first one way and then the other during the course of one year, hence it must be due to the fact that the earth is revolving around the sun. The change is small—only $\frac{1}{4800}$° for the nearest star. In Figure 3.20 the parallax angle (α) is greatly exaggerated.

As a consequence of the earth's revolution, the sun appears to move against the background of stars, aligning itself with different stars each day. The particular constellations through which the sun appears to move are called the signs (or "houses") of the zodiac. In Figure 3.21 the sun appears to be aligned with Pisces (the fish), but as the earth revolves about the sun, it will appear next in Aries, and so it will spend about one month moving through each of the 12 signs of the zodiac. The exact number of days required for the apparent motion of the sun through all the signs of the zodiac, returning to a given starting point, is not obvious

Figure 3.20   *Stellar parallax, due to the earth's revolution.*

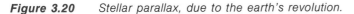

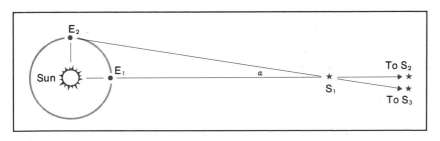

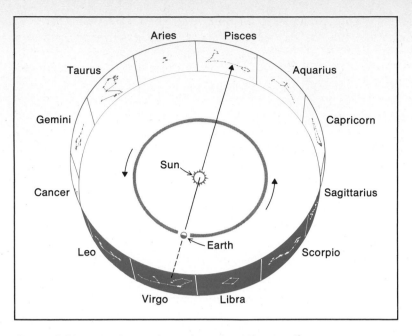

Figure 3.21 *As the earth revolves about the sun, the sun appears to align itself with each of the signs of the zodiac. The sun is shown aligned with Pisces, but in one month's time it will align with Aries, etc. In one year, the sun will again align with Pisces.*

from only one year's observation. However, when this period is observed for many years and an average is taken, the period of the earth's revolution is 365.2564 days. This is called the sidereal year—the year as measured by the stars.

The Seasons

As the earth revolves around the sun, its path defines a plane (flat surface) called the *ecliptic*. The word "ecliptic" derives from eclipse, for it is within this plane that eclipses of the sun and moon occur. If the ecliptic plane is extended until it "cuts" the sky, it will describe a circle through the signs of the zodiac, and that circle becomes the apparent path of the sun from our point of view on earth.

If the earth's axis of rotation were not tilted—that is, if it made an angle of 90° with the ecliptic plane—then the sun's rays would strike the equator of the earth directly at all times and we would not experience seasonal changes. However, the axis of the earth is tilted 23.5° away from a normal (perpendicular) to the ecliptic plane, and this tilt, together with the revolution of the earth around the sun, produces seasonal changes.

In Figure 3.22, position A shows the sun's rays striking directly over a point 23.5° north of the equator. Since direct solar radiation is the most intense, A represents summer in the northern hemisphere. Position B shows the sun's rays directly over the equator and represents autumn in the northern hemisphere. Position C shows the rays striking directly over a point 23.5° south of the equator and so represents winter in the northern hemisphere. Position C also represents summer in the southern hemisphere. Position D again shows the sun directly over the equator and represents spring in the northern hemisphere. The extreme northerly position of the sun (A) determines the Tropic of Cancer, 23.5° north of the equator; similarly, the extreme southerly sun (C) determines the Tropic of Capricorn, 23.5° south of the equator. Likewise the Arctic and Antarctic Circles are determined 23.5° from the north and south poles, respectively (Figure 3.21). Note that someone living just inside the Arctic Circle experiences continual daylight in the later part of June, which is why the Arctic has been called "land of the midnight sun." Further, this region is in continual darkness in the latter part of December.

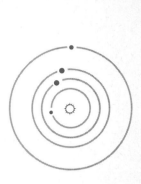

Time: How Long Is a Day?

The most fundamental unit of time to which we relate our patterns of behavior and our routine is that of a day, and yet there are several ways to measure the length of a day. Imagine a line in the sky, running

Figure 3.22   *The seasons result from the fact that the earth's equator is inclined 23.5° to its plane of orbit (the ecliptic plane).*

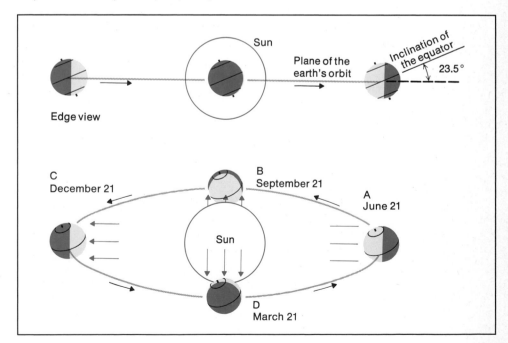

north and south through a point directly over your head. This is called your "local meridian." If you counted time from when a given star crossed your meridian until it crossed once again, you would have measured an earth rotation of exactly 360°, requiring 23 hr 56 min. This is called a *sidereal day*—a day based on a star reference. On the other hand, if you measured a day from the time the sun crossed your meridian until it crossed again, you would find that an average of 24 hr is required for such a rotation. This is called a *mean* (average) *solar day*—a day based on the sun as a reference. Figure 3.23 shows that the sun requires approximately 4 min longer to return to the observer's meridian than does the star. Why is this true?

The answer lies in the fact that while the earth is completing one rotation it also moves along its orbit around the sun approximately 1° (360°/365.25 day ≅ 1°/day); therefore, for an observer to see the sun cross his local meridian again requires one extra degree of rotation, a total of 361°. The earth rotates 361° in 24 hr, approximately 15° per hour (361°/24 ≅ 15°/hr), or 1° in 4 min; therefore, a sidereal day is 4 min shorter than a solar day (Figure 3.24). Since most of our daily routine centers about the sun's position in the sky—we eat breakfast soon after sunrise, lunch when the sun is high, and dinner soon after sunset—our clocks are built to agree with solar time (not sidereal time). As a consequence, a given star appears to rise and set about 4 min earlier each night. Thus, if a star is on your meridian at 9:00 P.M. tonight, you may expect the same star to be on your meridian at approximately 8:56 P.M. the following night. At this rate of 4 min/day, a given star rises 1 hr earlier in 15 days and approximately 2 hr earlier each month. After 1 year the star will again be on your meridian at 9:00 P.M. Likewise, night after night each constellation appears to shift a little, and gradually new constellations dominate the early evening sky. This is why we speak of some constellations as being summer constellations and others as autumn, winter, or spring constellations (see Figure 1.3, pages 8 and 9).

In determining a mean solar day, we presume that the earth is

Figure 3.23 *Observing a sidereal day.*

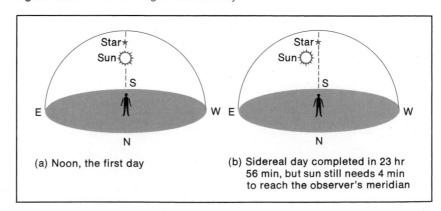

(a) Noon, the first day

(b) Sidereal day completed in 23 hr 56 min, but sun still needs 4 min to reach the observer's meridian

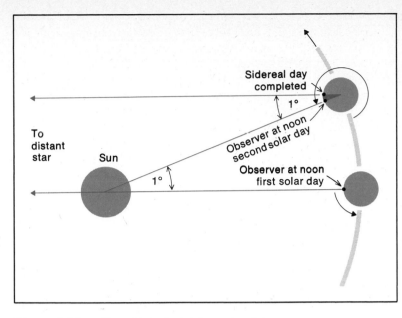

Figure 3.24   *The sidereal and the solar day.*

moving through the same angle of revolution each day. In reality, this is not the case, for the earth moves fastest when closest to the sun and slowest when farthest from the sun, as expressed by Kepler's second law (see Chapter 1). Thus the earth revolves through an angle of more than 1° per day in the winter and less than 1° per day in the summer, yet its rate of rotation is very nearly constant. For this reason, the sun may actually appear on your local meridian as much as 15 min before or after noon. The difference between true solar time and mean solar time is called the "equation of time."

The *analemma* (see Figure 3.25) found on most globes provides two pieces of information for any day of the year: (1) one can read the declination (the angle which the sun makes with the equator) on that day; (2) one can read the equation of time and correct for the difference between the "true sun" and the "mean sun" for that day. For instance, on June 21 the sun is 23.5° north of the equator, and it will cross the observer's meridian at noon (no correction is needed). On February 6 the sun is 16° south of the equator and will cross the observer's meridian 15 min after noon.

Latitude and Longitude

The location of any point on the earth's surface may be indicated by specifying two angles, one called *latitude* and the other *longitude*. How was this system devised? The north and south poles are two very special points on the earth, for the rotation (spinning) of the earth does

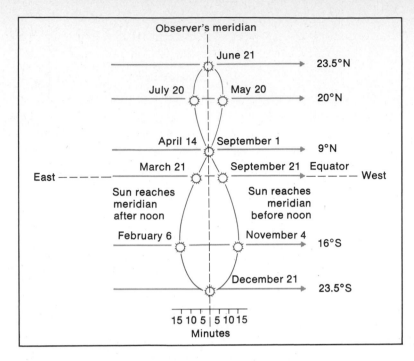

Figure 3.25   *The analemma.*

not move them. Halfway between the poles we imagine a circle called the equator. The angle which a point makes with the equator is called its latitude. In Figure 3.26, angle β (beta) is the latitude of point P. Latitude, specified as north (N) or south (S) of the equator, may be read off a chart showing a system of circles drawn parallel to the equator called *parallels of latitude.* These parallels are usually shown for each 10° of latitude.

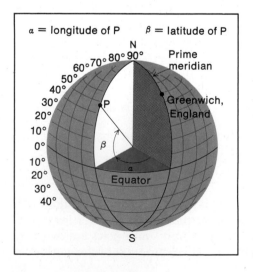

Figure 3.26   *A system of longitude and latitude.*

In measuring east or west (longitude) there was no natural starting point, and so an arbitrary choice was necessary. Based upon the large number of observations of star positions which had been made at the Royal Greenwich Observatory just outside of London, this location was selected as the starting point for measuring longitude. A line running from the north pole, through Greenwich, England, to the south pole is called the *prime meridian*. Additional meridians of longitude may be visualized through any point; however, on a globe map we usually see one meridian for each hour of earth rotation, or 24 meridians in all. Each meridian makes an angle of 15° (360°/24 = 15°) with its neighbor. The smaller angle between the prime meridian and a meridian through point P is called the longitude of P (angle α in Figure 3.26). The longitude of a point is specified as east (E) or west (W) of Greenwich, the prime meridian (Table 3.2).

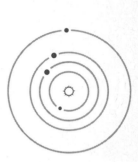

Table 3.2 *Longitude and Latitude for Several Well-known Cities*

CITY	LONGITUDE	LATITUDE
Washington, D.C.	W 77° 04′	N 38° 55′
Los Angeles	W 118° 20′	N 34° 10′
Honolulu	W 157° 45′	N 21° 22′
Rome	E 12° 32′	N 41° 50′
Rio de Janeiro	W 43° 10′	S 22° 40′

The meridians may also be thought of as hour lines, for when it is noon at Greenwich, it is 11:00 A.M. at the first meridian (15° W), 10:00 A.M. at the second meridian (30° W), and so on. Thus in Los Angeles, which is near the eighth meridian (120° W), it is 4:00 A.M., 8 hr earlier than Greenwich. Usually celestial events are given in Universal Time (Greenwich time). In order to convert to local time, subtract the number of hours equal to the number of meridians your city lies west of Greenwich—that is, for observers near Los Angeles, subtract 8 hr. Likewise, cities which lie to the east of Greenwich must add 1 hr for each meridian. These same meridians serve as the approximate center of each time zone (Figure 3.27).

The latitude of a given point in the northern hemisphere may be approximated by noting the angle that the pole star (Polaris) makes with the observer's horizon. The position of the pole star is determined by extending the axis of the earth until it penetrates the sky. If you lived at the north pole, you would expect to see Polaris approximately overhead, near your *zenith*. It would make an angle of 90° with your horizon, an angle equal to the latitude of the north pole. If you lived on the equator, you would expect to see Polaris near your northern horizon, making an angle of 0°, which is equal to the latitude of the equator. Taking an intermediate point—say, 40° north of the equator—we find that Polaris indeed makes an angle of about 40° with the horizon, another proof that the earth is basically a sphere (Figure 3.28).

Right Ascension and Declination

By a system which is very similar to longitude and latitude, astronomers specify the location of stars. Let us imagine that the stars are located on a transparent sphere, called the *celestial sphere*, at a given distance from the earth. When the earth's axis is extended until it intersects the celestial sphere, it determines the *celestial poles,* and the projection of the earth's equator onto the celestial sphere produces the *celestial equator.* In the same way, the projection of the meridians and the parallels of latitude completes the coordinate system in the sky. But again we need a starting point for our east-west measurement. This point cannot be determined by a given point on the earth; rather, it is designated as that celestial meridian which lines up with the sun on the day the sun crosses the equator on its way north. This point is called the *vernal equinox,* and it marks the position of the sun, in relation to the stars, on March 21. The projected celestial meridians are called *hour*

Figure 3.27 *Time zones. The solid lines represent standard meridians usually shown on a globe of the earth. The shaded areas show the approximate boundaries of time zones in the United States.*

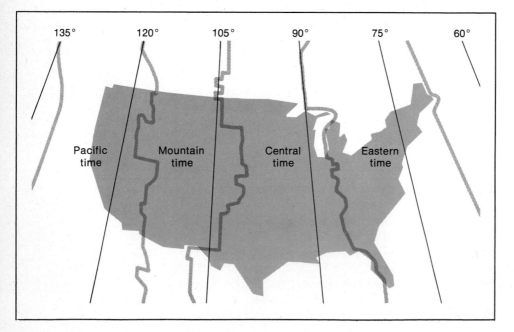

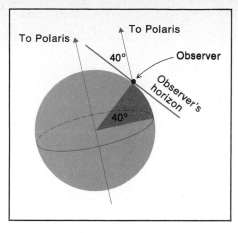

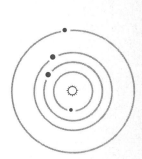

Figure 3.28 *The angle be-tween the observer's horizon and Polaris is equal to the observer's latitude.*

circles, and the hour circle that passes through the vernal equinox is called the *zero hour circle*. Successive hour circles measured to the east are called first, second, third hour circles, and so on to the twenty-third hour circle. The twenty-fourth hour circle is the same as the zero hour circle (Table 3.3). This angular measure, corresponding to the longitude on earth, is called the *right ascension* of the star and is specified like time. If a star has a right ascension of 5 hr 30 min, we know that it is halfway between the fifth and sixth hour circles. Since we have 360° divided into 24 equal hour circles, each hour circle must make an angle of 15° ($360°/24\,\mathrm{hr} = 15°/\mathrm{hr}$) with its neighbor. If a star has a right ascension of 5 hr 30 min, then it makes an angle of 82.5° with the zero hour circle ($5.5 \times 15° = 82.5°$).

Table 3.3 *Similarity Between the Terrestrial (Earth) System and the Celestial (Sky) System*

TERRESTRIAL	CELESTIAL
North and south poles	North and south celestial poles
Equator	Celestial equator

Parallels of latitude:
    Latitude (N or S)
	Declination circles:
	    Declination (+ or −)

Greenwich, England	Vernal equinox
Prime meridian	Zero hour circle

Meridians of longitude:
    Longitude (E or W)
	Hour circles:
	    Right ascension
	        (measured eastward only)

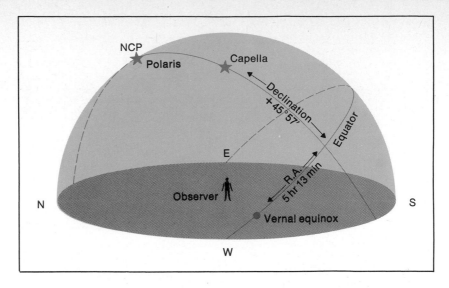

Figure 3.29 *Circles in the sky, showing the right ascension and declination of the star Capella. NCP refers to the north celestial pole (Polaris).*

The north-south measure is similar to that of latitude on the earth and is specified by the angle a star makes with the celestial equator. This angle, called the *declination* of the star, is specified as positive (+) if north of the equator and negative (−) if south of it. Therefore the position of any star may be given by two numbers. The star Capella has a right ascension (R.A.) of 5 hr 13 min and a declination (Dec.) of +45°57′ (Figure 3.29).

Apparent Motions of the Sky

While the sky seems to move overhead at the rate of one hour circle for each sidereal clock hour, we should remember that the earth's rotation produces this illusion. Since a given star appears to cross our local meridian about 4 min earlier each night, we should expect the hour circles also to shift by an equivalent amount each night. Table 3.4 presents the hour circle that will be overhead at 9:00 P.M. (local time) for various dates throughout the year.

Note that on March 21 the zero hour circle is overhead at noon, the first hour circle at 1:00 P.M. the second hour circle at 2:00 P.M., and so on. By 9:00 P.M. the ninth hour circle is overhead. Because the earth does not revolve about the sun with a constant velocity and because the axis of the earth's rotation is tilted to the ecliptic, the values given in Table 3.4 are approximated, and the time at which a given hour circle is overhead may vary as much as ±15 min.

Table 3.4 *Hour Circle Overhead at 9:00 P.M. Local Time*

DATE	HOUR CIRCLE	DATE	HOUR CIRCLE	DATE	HOUR CIRCLE
Nov. 6	0	Mar. 21	9	Aug. 6	18
Nov. 21	1	Apr. 6	10	Aug. 21	19
Dec. 6	2	Apr. 21	11	Sept. 6	20
Dec. 21	3	May 6	12	Sept. 21	21
Jan. 6	4	May 21	13	Oct. 6	22
Jan. 21	5	June 6	14	Oct. 21	23
Feb. 6	6	June 21	15	Nov. 6	24[a]
Feb. 21	7	July 6	16		
Mar. 6	8	July 21	17		

[a]Same as zero.

Altitude-Azimuth System

A coordinate system which may seem more natural to a specific observer is the altazimuth system. The observer's horizon, zenith, and the direction toward the north star become the reference points from which a star's location is specified. The *altitude* of the star is merely the angle measured along a vertical circle through the zenith, from the horizon (see Figure 3.30). The *azimuth* of the star is the angle measure from north

Figure 3.30 *The position of a star, relative to an observer, may be specified by two angles—azimuth and altitude.*

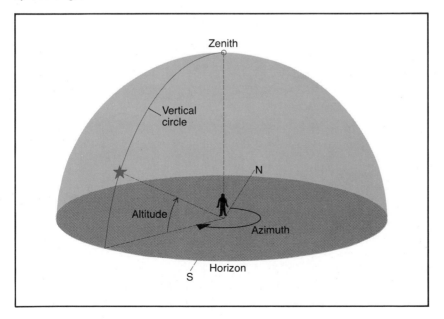

along the horizon to the vertical circle. Many modern telescopes have altazimuth mounts and must be driven in both directions in order to follow a star's diurnal motion. Only the advent of the computer-controlled drive system has made this an easy operation.

Precession of the Earth

To the more obvious motions of the earth, namely, rotation and revolution, we must add a motion that is detectable only over a long period of time. This motion, called *precession*, requires 26,000 years to complete one cycle. During this period the earth's axis will align itself with new points in the sky, points other than Polaris. We can compare this motion to that of a toy top which, while it is spinning, slowly describes a circle at the upper end of its axis (Figure 3.31). Like the toy top, the earth is bulged around its equator, and it is rotating on a tilted axis. The gravitational force which the sun and moon exert on the earth tends to straighten that tilt of axis; however, any rotating body possesses a gyroscopic effect—the tendency to maintain a given tilt of axis. Since the earth resists having its axis of rotation straightened, the gravitational energy produces the precessional motion of the earth (Figure 3.32). Because of precession, the axis of the earth, over a 26,000-year period, describes a circle in the sky with a radius of 23.5°. In the year 14,000 the bright star Vega will be within a few degrees of the north celestial pole. This precessional motion causes the vernal equinox to move westward in the sky approximately 50 seconds of arc per year, and for this reason the full name of this motion is *precession of the equinoxes*.

Since one complete cycle of precessional motion requires approximately 26,000 years, the vernal equinox remains in each of the 12 houses

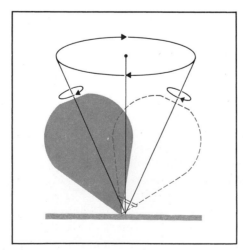

Figure 3.31   *Precession of a spinning top.*

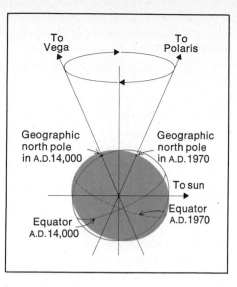

Figure 3.32 *Precession of the earth.*

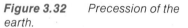

of the zodiac for just over 2000 years (refer to the *fall star map* in Appendix 12). The vernal equinox has been passing through the constellation of Pisces since around the start of the Christian era, but it has been gradually moving westward through the zodiac and will move into the constellation of Aquarius early in the twenty-first century, ushering in the "Age of Aquarius" at that time. This is a happening of special interest to astrologers; astronomers must also be aware of this gradual change, however, because any shift in the position of the vernal equinox is accompanied by a change in the right ascension and declination of all stars. The reason is that the vernal equinox defines the zero hour circle in the sky. The right ascension and declination values given in Appendixes 8 and 9 reflect the coordinates for 1950.

The Calendar

We recall from our discussion of early calendars that it was not very satisfactory to try to put exactly 12 lunar months, each of $29\frac{1}{2}$ days, into a year. The total of 354 days fell short of the seasonal year by approximately $11\frac{1}{4}$ days. Our real desire is to create a calendar which reflects the seasonal changes as accurately as possible. Because the vernal equinox marks the sun's position on the first day of spring and because the vernal equinox shifts approximately 50 seconds of arc per year, we must take this into account in determining an exact seasonal year. For the sun to return to the new equinox position, 365.2422 days are required. This period is called the *tropical* year (the year of seasons). It is shorter than the sidereal year of 365.2564 by 0.0142 days, owing to precession. How could a calendar be constructed to reflect as nearly as possible the tropical year of 365.2422 days? As early as 46 B.C. Julius Caesar decreed a superior calendar which reflected the tropical year to a fair degree of accuracy. Because he believed that the year contained $365\frac{1}{4}$ days, he added one day to each fourth year and called it a leap year. This left an error amounting to the difference between 365.2500 and 365.2422, or 0.0078 days/year. This seems insignificant, yet in a thousand years it amounts to 7.8 days. By the way, Julius Caesar was partially responsible for placing more days in one month than another. His month, July, contains 31 days. His successor, Augustus Caesar, had to have at least as many days in his month of August. He robbed February, in the process of calendar reform, leaving it with only 28 days and in leap years with 29 days.

In 1582 Pope Gregory brought further reform, first by correcting the date, which had fallen out of synchronization with the seasons by 12 days during the reign of the Julian calendar. To more closely reflect the tropical year, the Gregorian calendar, which we use today, makes the following corrections: the extra day normally placed in a leap year is deleted three times in a period of 400 years: for instance, in the century years of 1700, 1800, 1900, 2100, 2200, 2300, the extra day is deleted. In the century years of 1600, 2000, 2400 it is retained. This plan yields an average year length of 365.2425 days—very nearly the tropical-year value of 365.2422 days. The Gregorian calendar is good within a 1-day error in 3300 years.

Other Motions of the Earth

The earth, being a part of the solar system, shares the motion of the sun. The sun moves among the neighboring stars at a rate of 19.4 km/sec. The entire set of neighboring stars moves in the Galaxy of which we are a part at the rate of 242 km/sec, or about 871,000 km/hr; thus the sun and the earth also move through space at this tremendous speed. These motions are discussed in more detail in a later chapter.

Questions

1. What is the name given to the lines on the celestial sphere that most closely resemble the meridians on the earth?

2. What are the two most natural points on the earth's surface, as determined by its motion?

3. What motion of the earth is primarily responsible for creating the seasons on the earth?

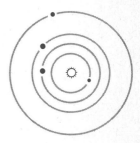

4. A spherical model of the earth (a globe) usually shows 24 meridians. How many degrees separate each meridian?

5. What is the length of time required for the earth to rotate from one meridian to the next, assuming a total of 24 meridians on the earth?

6. If the earth were tilted more than 23.5°, say, 35°, would the seasons be longer? Would the winter season be more severe? Why?

7. At the present time, the sun is in the constellation of Gemini during the month of July. In ancient times it was in Gemini during the month of May. What phenomenon has caused this change?

8. If you lived on the equator, what percentage of the sky could you see if you stayed up all night?

9. How is the vernal equinox point determined?

10. Through how many hour circles does the sun appear to move each month?

11. What property of an object is measured by each of the following: volume, mass, and density?

12. In going from the crust of the earth to its core, might we expect to find increasing or decreasing density?

13. Eratosthenes observed that the sun's rays struck one city in Egypt at a different angle than they did another city 800 km to the north at noon on a given day. How did this observation prove that the earth is not flat?

14. What is the most abundant element in the earth's atmosphere?

15. Describe the *solar wind* and discuss several effects it may have on the earth.

16. Describe an experiment by which the daily rotation of the earth is proved.

17. Describe an experiment by which the yearly revolution of the earth is proved.

18. What basic motion of the earth causes the sidereal day to differ from the solar day?

19. Twice during the year, the sun appears to move directly over

the equator of the earth. List the approximate dates of these occurrences.

20. An observer is located 34° north of the equator. Where will Polaris (the North Star) appear from his point of view?

21. Which basic motion of the earth causes different constellations to be seen on a winter evening than on a spring, summer, or fall evening?

22. What properties of the earth cause it to precess?

23. How do the stars seen by a person in North Africa (say, Casablanca, Morocco, latitude 34° N) compare with the stars seen from Mt. Wilson in California (latitude 34° N) in the evening of the same day?

24. As compared to residents of Florida, New York residents experience which of the following (several answers may be true): (a) longer summer days, (b) longer winter days, (c) same length days all year, (d) shorter summer days, (e) shorter winter days. (Note: Here the word "days" refers to the length of daylight hours.)

25. If you placed a stick in the ground at noon on the longest day of the year and then marked the tip of its shadow on the ground at precisely noon each day, what shape would the marks on the ground make in one year?

26. When it is 11:00 A.M. in New York City (approximately 75° W longitude), find the time in a city of 105° W longitude.

Suggested Readings

THE AMERICAN EPHEMERIS AND NAUTICAL ALMANAC. Washington, D.C.: Superintendent of Documents, U.S. Government Printing Office. (Annual publication.)

BOLT, BRUCE A., The fine structure of the earth's interior. *Scientific American 228* (3), 24–33 (1973).

CARRIGAN, C. R. and GUBBINS, D., The source of the earth's magnetic field. *Scientific American 240* (2), 118 (1979).

DEWEY, JOHN F., Plate tectonics. *Scientific American 226* (5), 56–68 (1972).

HALLAM, A., Continental drift and the fossil record. *Scientific American 227* (5), 56–66 (1972).

KUIPER, GERARD P. (ed.), *The earth as a planet.* Chicago: University of Chicago Press, 1954.

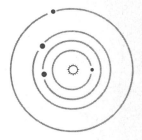

LEPP, HENRY, *Dynamic earth*. New York: McGraw-Hill, 1973.

LYNCH, DAVID K., Atmospheric halos. *Scientific American* 238 (4), 144 (1978).

MEHLIN, THEODORE G., *Astronomy and the origin of the earth*. Dubuque, Iowa: Wm. C. Brown, 1973.

MOORBATH, STEPHEN, The oldest rocks and the growth of continents. *Scientific American* 236 (3), 92 (1977).

RONA, PETER A., Plate tectonics and mineral resources. *Scientific American* 229 (1), 86–95 (1973).

FOUR

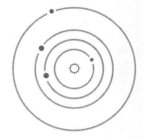

THE MOON

We often think of the motion of the earth and moon separately. We picture the earth in its annual motion around the sun and the moon in its monthly motion around the earth, but in reality the earth and moon are better thought of as a binary (dual) system which is orbiting the sun. The term *binary system* may refer to any two objects, but usually it means two objects each having an observable effect upon the other's motion due to their mutual gravitational force. The earth-moon system illustrates this very clearly, as seen in the left-hand flip pages beginning on page 182. As you flip these pages, you will see that the earth deviates from a smooth path around the sun by a significant amount, approximately 4800 km on either side of an elliptical path marked by the dashed line. The moon also orbits the sun but deviates approximately 384,000 km on either side of that same dashed path. The only point which moves along the smooth path around the sun is the *barycenter* of the system, equivalent to the point at which the earth and moon would balance if placed at either end of an imaginary rod 388,800 km long.

Since we already know the mass of the earth, this balancing idea will allow us also to determine the mass of the moon. To illustrate this concept, suppose that a father tried to seesaw with his

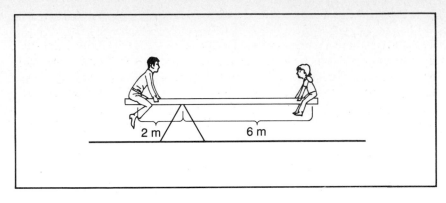

Figure 4.1   *Father and daughter balance on a teeter-totter.*

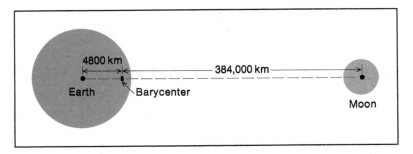

Figure 4.2   *If the earth and moon were placed at opposite ends of a rod, the system would balance at the barycenter.*

young daughter and they found that they could balance only if the daughter sat three times as far from the pivot (*fulcrum*) as the father (Figure 4.1). We would then conclude rightly that the daughter weighs only one-third as much as the father. Since the earth and moon balance at their barycenter, 4800 km from the center of the earth and 384,000 km from the center of the moon (Figure 4.2), we conclude that the moon has only $\frac{1}{80}$ the mass of the earth (4800/384,000 = $\frac{1}{80}$), or

$$\tfrac{1}{80} \times 5.98 \times 10^{27}\,g = 7.47 \times 10^{25}\,g$$

Bouncing a Laser Beam Off the Moon

We may determine the size of the moon by direct observation of its distance and the apparent angle its diameter makes with our eye. Today we merely bounce a laser beam off the moon and record the round-trip time of that signal in order to measure its distance from us. Suppose that on a certain night it requires 2.5627 sec for the signal to travel to and from the moon. We know that laser light travels at 300,000 km/sec, therefore the round-trip distance is equal to

$$2.5627 \text{ sec} \times 300{,}000 \text{ km/sec} = 768{,}800 \text{ km}$$

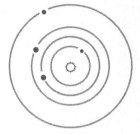

The one-way distance then is 384,400 km. Now, if on that same night the diameter of the moon appeared to make an angle of 0.518° with your eye, then we might picture these measurements as in Figure 4.3 and reason as follows: The angle of 0.518° compares to the full circle of 360° in the same way that d (a small part of the circle) compares to the entire distance around the circle (circumference); recall that the circumference C is equal to 2π (pi) multiplied by the radius of the circle ($C = 2\pi r$).

$$\frac{d}{C} = \frac{0.518°}{360°} \qquad \frac{d}{2\pi(384{,}400 \text{ km})} = \frac{0.518°}{360°} \qquad d = 3476 \text{ km}$$

We find that the moon's diameter is 3476 km, about one-quarter that of the earth, yielding a volume of 2.3×10^{25} cm³, or about $\frac{1}{50}$ the earth's volume. This is a useful comparison when we try to visualize just how large the moon really is. When both the mass of the moon and its volume are considered, its average density may be computed:

$$\text{density} = \frac{\text{mass}}{\text{volume}} = \frac{7.5 \times 10^{25} \text{ g}}{2.3 \times 10^{25} \text{ cm}^3} = 3.3 \text{ g/cm}^3$$

Many of the moon rocks have proven to be almost this dense, and before humans landed on the moon, observers had concluded that the moon was not *differentiated* (having layers of increasing density toward the center) but was homogeneous throughout. After all, a body as small as the moon would not experience pressures as great as the earth at its core. However, several seismic (moonquake) stations left on the moon by

Figure 4.3   *Finding the diameter of the moon.*

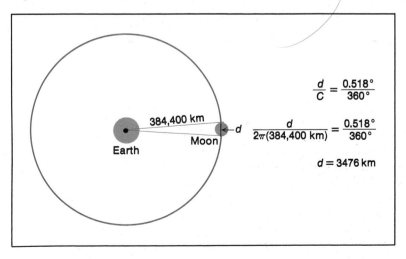

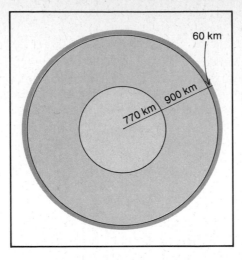

60 km

900 km

770 km

Figure 4.4
Layers of the moon.

Apollo crews reveal quite a different picture. While the typical moon-quake is rather mild, registering 1 or 2 on the Richter scale, it is possible to measure velocity changes in the seismic waves at varying depths within the moon. One such change occurs at a depth of 50 to 60 km and another at a depth of 900 to 1000 km. In fact, most moonquakes originate at these latter depths. We interpret this to mean that the moon has a crust 50 to 60 km thick, below which a solid (rigid) mantle exists, leaving a central core as illustrated in Figure 4.4. Meteor impacts on the far side of the moon yield a further clue as to the nature of the core. Such an impact generates both shear and pressure waves. However, in passing through the moon's core to the near side, where seismographs record the disturbance, the shear waves are lost—an indication of the molten nature of the core, with estimates of temperature ranging upward to 1500°C.

Another interesting method for study beneath the surface of the moon has been developed. The Apollo command module, orbiting around the moon, sends radar signals of various wavelengths toward the moon and times their return (echo). It has been found that signals of 115-cm wavelength penetrate the lunar soil approximately 20 m and signals of 13-cm wavelength penetrate as much as 200 m. Using this method, scientists have concluded that what now appear to be fairly smooth lowland regions of the moon represent valleys and basins that were filled in by molten lava at one stage in the moon's evolution. When a spacecraft flies over certain of these *maria* (lunar "seas"), it experiences an increased gravitational attraction, indicating higher concentrations of mass, called *mascons* (the circled areas in Figure 4.5). Most observers believe that the original basins, from which the maria derive, were blasted out by the impact of huge objects in the early stages of the moon's history. Whether those objects are buried in the basins, thus accounting for the mascons, is not known. It seems more likely that the basins have been filled by lava, which is more dense than the surrounding material. It is interesting to note that mascons do not exist on the far side of the moon, where large maria are not filled with dark lava. (Figure 4.14, page 154). Mascons are also known to exist on earth, where they are usually referred to as gravitational anomalies.

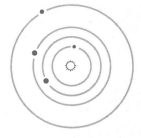

Surface Features

The most obvious distinction that can be made in the surface features of the moon is between the flat lowlands called *maria* (seas), which appear to be filled by dark lava, and the *highlands* (mountains), which appear both lighter in color and brighter than the maria. Although the lowlands are called "seas," they contain no water. Lacking an atmosphere, the moon cannot maintain liquid water on its surface. In fact, analysis of the Apollo rock samples does not show the water chemically

Figure 4.5 *Eleven of the twelve known concentrations of mass on the moon. The mascons seem to be associated with a definite type of mare. (JPL–NASA)*

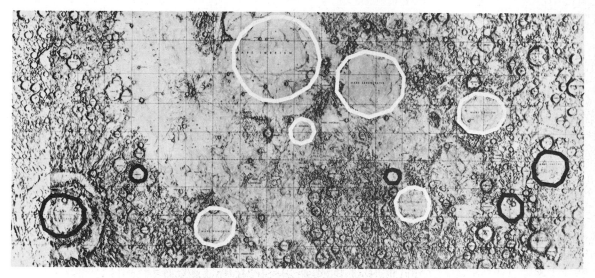

bonded in the minerals that is typical of earth rocks. The circular maria, which were identified as being associated with mascons (Figure 4.5), range in diameter up to 1130 km, and irregular maria such as Oceanus Procellarum are significantly larger.

The mountain ranges extend for hundreds of kilometers, reaching heights of over 6 km above the level of the maria. These lunar highlands bear the names of terrestrial mountain ranges, such as the Alps and the Pyrenees. Both the maria and mountain ranges are identified in Figures 4.7 and 4.9, and their photographic counterparts are shown in Figures 4.6 and 4.8. Craters represent still another major surface feature of the moon.

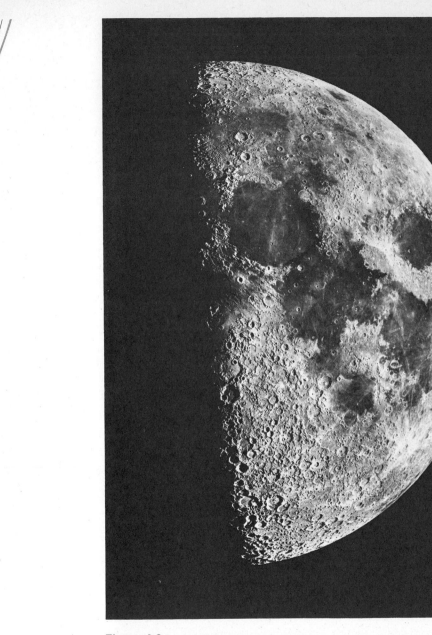

Figure 4.6 *First-quarter moon (7 days old). North is at top, as seen by the naked eye. (Lick Observatory)*

Ranging in size up to 240 km, craters are named for persons of science and politics. Note that the highlands are dominated by craters, whereas only relatively few occur within the maria, where they may have been obliterated by lava flows.

Many of the major features identified in Figures 4.5 through 4.8 can

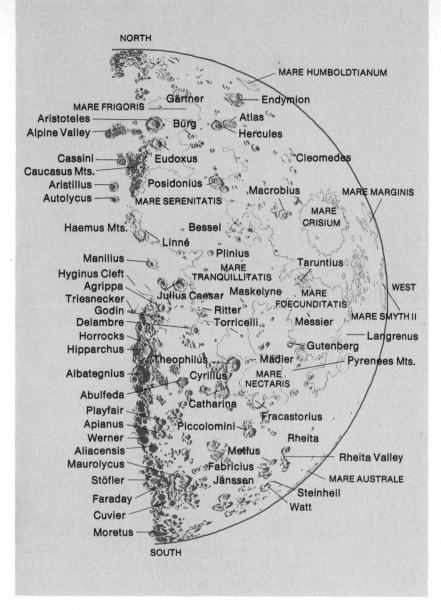

Figure 4.7 *Key to the first-quarter moon.*

be seen with a pair of binoculars or a small telescope. If binoculars are used, the moon will appear as illustrated, its first-quarter phase apparently lighted from the right. This can be called the "naked-eye" view. If a simple telescope is used, however, the image will appear upside-down and backward, right to left. A simple rotation of the figure in your book will allow for comparison. See how many features you can identify. Because

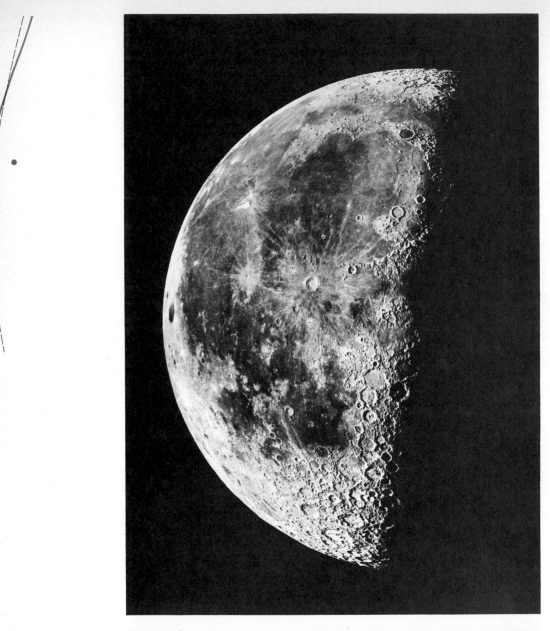

Figure 4.8   *Third-quarter moon. North is at top, as seen by the naked eye.*
(Lick Observatory)

shadows tend to clarify surface details, the best time to view the moon is in its early (waxing) phases (crescent, first quarter, and gibbous) and also in its later (waning) phases (gibbous, third quarter, and crescent) (Figure 4.20, page 159). During these phases a portion of the moon has long shadows, especially near the boundary between the lighted and dark

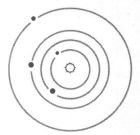

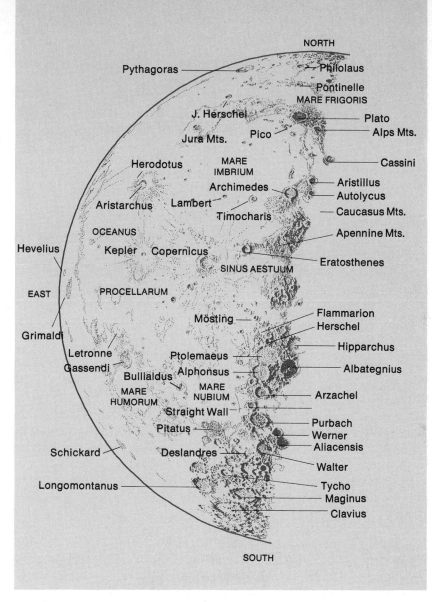

Figure 4.9   *Key to the third-quarter moon.*

portions of the moon, called the *terminator*. Each night, as the phase of the moon changes, the terminator moves to a new location, revealing added detail. By contrast, when the moon is full, no shadows exist and little detail is seen. During early or late phases, shadow lengths give very accurate clues as to heights of objects on the surface (see Figure 4.10).

Two kinds of valleys exist on the moon: the type that one would

Figure 4.10   *The Crater Theophilus. The lengths of the shadows reveal the heights of the rim of this crater to be approximately 4.2 km and the central peak to be 2.25 km. The diameter is approximately 104 km (65 miles). (Yerkes Observatory)*

expect in any mountainous region (see Figure 4.11) and the type that resembles a broad meandering river valley. The Prinz Valleys, shown in Figure 4.12, do flow downhill; however, this is not sufficient evidence to think that they were formed by water flow. Rather they may be the result of faulting (crustal movement) and slumping of the lunar soil. Typical of this later phenomena are the *rilles* which mark much of the moon's surface (see Figure 4.13 a and b).

One side of the moon is perpetually turned toward the earth, the far side never having been seen until a probe was placed in lunar orbit. Figure 4.14 reveals that the far side presents quite a different appearance than the near side. It is cratered rather uniformly and is almost devoid of dark, lava-filled maria. This fact poses a very interesting question of the origin and evolution of the moon.

Moon Rocks

The samples of the moon brought back by the Apollo astronauts have an exciting story to tell. Within their composition and age lie important clues as to the moon's history. There are three basic classifications of moon rock:

1. Basalt, an igneous rock (having once been melted) much like lava we find on earth (see Figure 4.15).

2. Anorthosite, another igneous rock, which cooled slowly beneath the surface of the moon (see Figure 4.16).

3. Breccia, a rock which formed by the continual bombardment of the moon by meteorites and micrometeorites. By such action surface material is first pulverized but on occasion is again compacted by the shock of a subsequent meteorite impact. The breccia comes closest to being like sedimentary rock on earth (see Figure 4.17).

The basalts are typical of the maria; when dated by radioactive methods, they are found to be between 3.1 and 3.8 billion years old. The anorthosites, which are typical of the highlands, are between 3.9 and 4.4 billion years old. The breccias are more typically found in the highlands, but also to a lesser degree in the maria. Their ages reflect the age of the grains from which they were formed. The oldest highland breccias grains

Figure 4.11   *Scientist-astronaut Harrison H. Schmitt standing next to a huge lunar boulder during the third Apollo 17 extravehicular activity (EVA-3) at the Taurus-Littrow landing site. (NASA)*

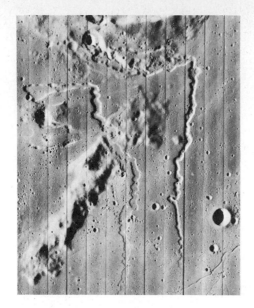

Figure 4.12 *The meandering Prinz Valleys I and II, running downhill from the Harbinger Mountain area, range in width from 300 to 2400 m. A Lunar Orbiter V photograph. (NASA)*

are between 4.3 and 4.6 billion years old. How do these facts reveal the history of the moon?

History of the Moon

Before the Apollo program and the dating of moon rocks, many observers had assumed that the moon had been cratered at a rather uniform rate since its formation. However, we now envision very definite periods of cratering and vulcanism, after which the moon was essentially frozen in time for the last 3 billion years. Let us trace these periods.

All lunar surface material shows signs of having crystallized from a molten state. Since the oldest material dates to 4.6 billion years, we presume that this dates the early formation of the moon's surface. The heat for the overall melting of the lunar surface most likely was derived from the energy of infall of the material which formed the moon initially, perhaps in its final stages of *accretion*—the gravitational gathering in of material from the surrounding region.

The first billion years of the moon's existence can be characterized by general cratering due to the impact of meteorites. A very large meteorite may have blasted out the huge basins, which were later filled with lava to form the maria as we see them today. This filling of the basins—some of a ringed nature, such as those identified with the mascons, and others of an irregular shape, such as Oceanus Procelarum—represents a second major heating of the moon. The source of heat on this occasion may have been radioactive elements in a deeper layer of the moon, somewhat like the heating which takes place in the upper mantle of the earth. This type of heating would exhibit itself in volcanic activity,

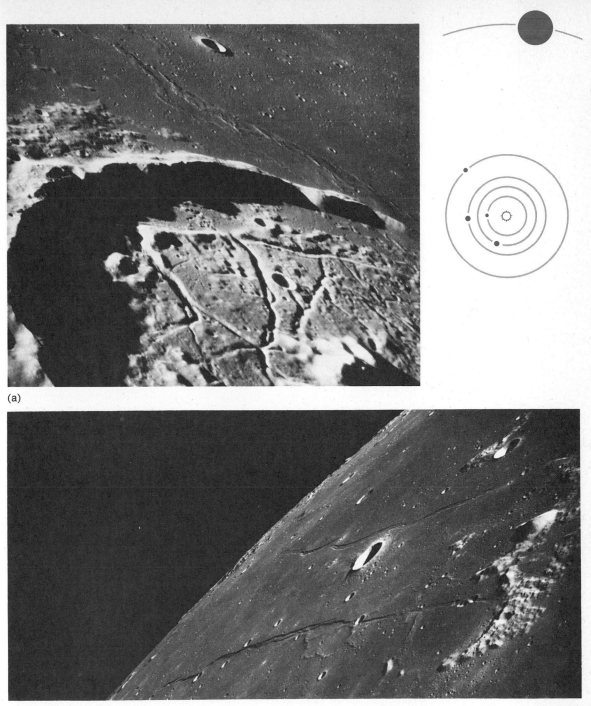

(a)

(b)

Figure 4.13   (a) *A portion of Gassendi Crater on the near side of the moon, as photographed from the Apollo 16 spacecraft in lunar orbit. This view is looking southerly into the Sea of Moisture. (b) The Sea of Tranquillity, including the Crater Cauchy and two rilles. (NASA)*

Figure 4.14 *The far side of the moon as seen by Lunar Orbiter III. This photograph was taken 1400 km above the lunar surface and shows the prominent crater Tsiolkovsky filled with dark material. This crater is about 225 km in diameter (NASA)*

flooding the low-lying areas, such as the basins, with lava; thus the maria were formed. The dating of maria basalt indicates that this flooding occurred between 3.1 and 3.8 billion years ago and could have extended over several hundred million years. The moon has been quiet since that time, with the exception of a relatively few impacts in the maria and a few young craters elsewhere, as exhibited by the bright rays which appear to emanate from the crater (see the Crater Copernicus in Figures 4.8 and 4.9).

However, one process continues to this day: that of continual bombardment of the lunar surface by micrometeorites. This action over the billions of years of the moon's existence has pulverized lunar rock to produce a layer of lunar soil. Micrometeorite impact and gravity are the two main factors of erosion on the moon. Sharp crater rims are rounded very slowly by continual bombardment, gravity making the pulverized material tumble down the slopes.

No water exists on the moon, not even in chemically bonded form within the rocks, hence we do not see the erosion factors normally seen on earth. If the rocks contained water, then the expansion and contraction (due to the freezing and thawing) which would accompany the drastic changes in temperature on the moon would tend to break the rock. Temperatures on the moon range from 94°C (200°F) directly under the sun to −128°C (−200°F) in the shaded portion. This wide variation in temperature can occur only on a body devoid of atmosphere. The earth's atmosphere provides a moderating blanket, limiting the typical variation in daytime and nighttime temperatures to 30° or 40°C. Furthermore, the phenomena we know as twilight and dawn do not occur on

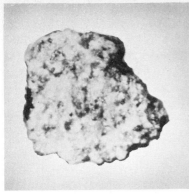

Figure 4.15   Moon rock: a coarse basalt. (Smithsonian Observatory)

Figure 4.16   A thin section of anorthosite, enlarged 22 times. (Smithsonian Astrophysical Observatory)

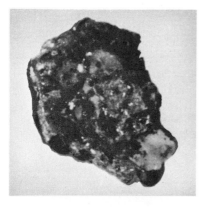

Figure 4.17   Moon rock: soil breccia. (Smithsonian Observatory)

the moon. As the sun goes down, darkness is immediate, with the possible exception of reflection from nearby peaks rising above the observer's head. Even the percentage of light reflected is very, very small, in the neighborhood of only 12 percent. The percentage of light which a body reflects is called its *albedo*, and bodies without atmospheres typically have very low albedos. By contrast a planet with a cloudy atmosphere, such as Venus, reflects about 76 percent of the light it receives.

Magnetic Field

The moon today has no detectable magnetic field, yet some of the moon rocks possess a locked-in magnetism. This would indicate that at some time, most likely more than 3 billion years ago, the moon possessed a strong magnetic field. This would be difficult to explain on the basis of its slow rotation ($27\frac{1}{3}$ days); however, the moon may have been spinning more rapidly in its early history.

Origin of the Moon

The theories concerning the moon's origin center around the following possibilities:

1. The moon was formed far from the earth and was later captured by the earth.

2. The moon was "torn" from the earth, leaving a great depression—the Pacific Ocean.

3. The moon formed out of the same nebula of gas and dust at the same time as did the sun, earth, and other planets.

Theory 1 seems unlikely. It is nearly impossible for one object to capture another unless at least one other object of the proper mass and motion is present. On the other hand, the possibility cannot be ruled out.

Table 4.1   *Percentages of Various Elements on the Crust of the Earth and the Moon*[a]

	OXYGEN (%)	SILICON (%)	IRON (%)	MAGNESIUM (%)	ALUMINUM (%)	CALCIUM AND POTASSIUM (%)
Highlands of moon	61.1	16.2	4.5	4.0	10.2	6.1
Maria of moon	60.6	16.8	1.8	5.3	6.6	4.7
Earth	61.7	21.0	1.9	1.8	6.4	3.3

[a]Percentages are given in relation to numbers of atoms, not as to weight.

Before we comment on theories 2 and 3, inspect Table 4.1, dealing with the similarities and dissimilarities in composition of the moon (highlands and maria) and of the crust of the earth. If you had to choose between theory 2 and theory 3 on the basis of this information, which would be your choice? It would seem that if the moon had been torn from the earth, the percentage composition of earth and moon should be almost identical. On the other hand, they are close enough in composition to have formed from the same nebula, allowing for some layering (differentiation) of elements as the body cooled. Most experts choose theory 3 as most plausible.

Revolution of the Moon

Owing to the earth's rotation, the moon appears to move in a westward direction during one evening's observation. Closer study will show, however, that it is moving in an eastward direction against the background of stars. You may verify this by observing the moon when it is at the western limit of a bright star or cluster, such as the Pleiades. If you observe the moon for several hours, its eastward motion will become apparent as it moves to occult (cover up) the stars one after another.

A more dramatic eastward change takes place when we observe the moon on successive nights, noting its position relative to the stars. If we continue this practice for approximately 27 days, we find that the moon returns finally to the same portion of the sky. To be exact, the moon's *sidereal period of revolution* is 27.322 days, which causes it to shift its position in the sky by an amount equal to just slightly more than 13° per day:

$$\frac{360°}{27.322 \text{ days}} \simeq 13°/\text{day}$$

We would then expect the moon to move through slightly less than one hour circle in a 24-hr period.

We can also observe that more than 27.322 days are required for the moon to complete its cycle of phases. The period from one full moon to the next is 29.531 days. This is called the *synodic period* of the moon. Suppose we started counting time when the moon is full, shown in Figures 4.18 and 4.19 as position A. We note that the moon appears near the bright star Regulus. In 27.322 days the moon will return to align itself with Regulus, but in that length of time the earth will have moved through approximately 27° to position E_2, and the moon will then be in position B. Since *full moon* occurs only when the moon is on the opposite side of the earth from the sun, it will not be full until it moves through an additional angle of approximately 29° to position C. This additional revolution requires approximately 2.209 days, hence the synodic period is 27.322 + 2.209 = 29.531 days.

In passing from one full moon to the next, the moon goes through all its successive phases, as we can see in Figure 4.20. We observe that the *new* (all-dark) *moon* occurs when the moon is between the earth and the sun and that the *first-quarter moon* occurs when the moon makes a 90° angle with the sun as viewed from earth. Midway between these two phases is the *waxing* (growing) *crescent*, the phase which most people incorrectly refer to as the "new" moon, because this phase is the first we see when the moon is starting a new sequence of phases. The next phase is that of *waxing gibbous*, when the face of the moon appears about three-quarters lighted. The *full moon* occurs when the moon is on the opposite side of the earth from the sun. Then follows a series of *waning* phases in which the visible illuminated portion appears to shrink.

Throughout all its phases, the moon presents essentially the same side toward the earth. Does this mean that the moon does not rotate on its own axis? On the contrary: in order always to present one side toward the earth, the moon must rotate once, with respect to the stars, while revolving once around the earth (Figure 4.21). This motion, referred to as *synchronous rotation*, has a period of 27.322 days; however, there are two facts that modify it. One is that while the moon's rotation is uniform, its revolution is along an elliptical orbit and hence is not uniform. These

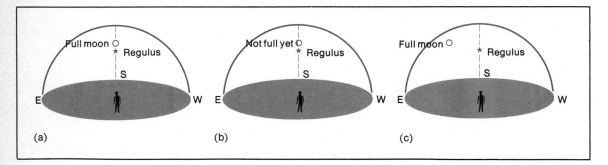

Figure 4.18   *The sidereal and the synodic month as seen in the sky. (a) A full moon, in conjunction with Regulus (start counting time here). (b) After 27.33 days—the moon, again in conjunction with Regulus, but not yet a full moon (this is a sidereal month). (c) After 29.5 days—the moon, again full, but now located approximately 30° east of Regulus (this is a synodic month).*

motions periodically allow the moon to expose small additional portions along either its western or its eastern edge (*limb*). Second, the moon's orbit is inclined 5° to that of the earth, and as it moves in that orbit we periodically view a little beyond the normal polar regions, either north or south. Over a period of several months we are able to view a total of 59 percent of the moon's surface from the earth. The various interactions of these motions are called the *librations* of the moon (Figure 4.22).

In the next section we will answer the question, "Has the moon always pointed the same face toward the earth?"

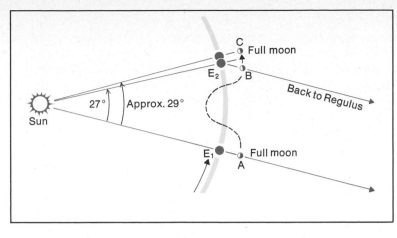

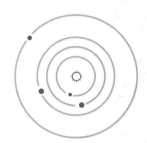

Figure 4.19 *The synodic period of the moon.*

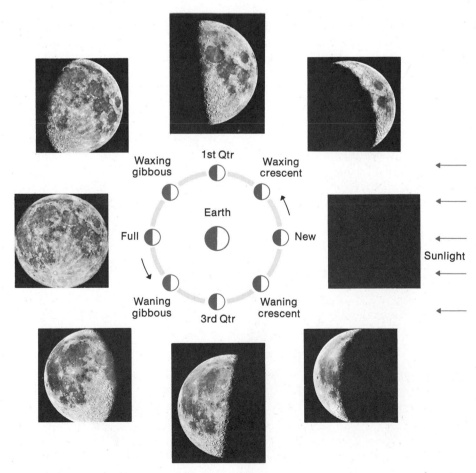

Figure 4.20 *The phases of the moon. North is at top, as seen by the naked eye. (Lick Observatory)*

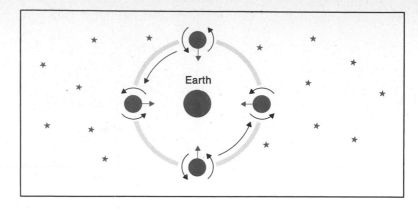

Figure 4.21 *The synchronous rotation of the moon.*

Tides Produced by the Moon

Let us conduct an imaginary experiment, observing tides of the ocean on several successive evenings when the moon is also visible overhead. On the first night the moon is on your local meridian at 8:00 P.M., and high tide occurs a short time later. On the second night the moon is on your local meridian at 8:50 P.M., and again high tide occurs a short time later. On the third night the moon passes your local meridian at 9:40 P.M.; once more high tide occurs a short time later. Were this experiment conducted for several additional nights, it would be obvious that the timing of the high tide coincides with the passage of the moon and must be due to its presence overhead. What is not so obvious is the reason why there are two high tides in a period of 24 hr 50 min. That is, why does the earth have a double bulge?

In order to explain this phenomenon, let us consider an imaginary earth—perfectly smooth, with no mountains or valleys—alone in space,

Figure 4.22 *Two photographs of the moon showing libration. (Lick Observatory)*

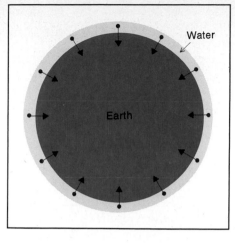

Figure 4.23 *Water molecules (dots) distributed evenly over the surface of an imaginary earth, alone in space and without any motion of its own.*

with no motion of its own. If water existed on the surface of this imaginary earth, it would be distributed evenly over the entire surface, for each molecule would experience the same force of gravity toward the center of the earth (Figure 4.23).

If this imaginary earth were then set into rotation, the layer of water would become deeper around the equator and shallower at the poles, owing to centrifugal force. However, this would only produce a constant bulge around the equator; it would not account for the fluctuating tide that we experience. If now we consider the effect of the moon, we realize that, owing to gravitation, it must exert a force on the earth and on the waters that cover it. But the force of gravitation depends on the distance between the two objects in question: the greater the separation, the less the force. Suppose that a 1-g mass is situated at each of the three positions shown in Figure 4.24: at A, in the water near the moon; at B, in the center of the earth; and at C, in the water on the side away from the moon. Because point A is closer to the moon than B, it will experience a greater force than B, thus explaining the bulge toward the moon. Likewise, point B is closer to the moon than point C, and it will experience a greater force than on C, pulling the earth away from the water on the far side.

The effect here is very similar to that of stretching the system of springs and masses illustrated in Figure 4.25. Note that not only does the

Figure 4.24 *Tides on the earth.*

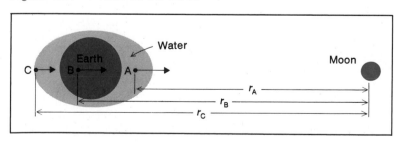

distance between M_1 and M_2 increase as the spring is stretched, but also the distance between M_2 and M_3. Imagine that the earth, a rigid body, is attached to M_2 and that we see a bulge in the earth both in the direction of the force (pull) and in the opposite direction.

Unlike the imaginary earth, our planet is not smooth but has mountains, beaches, and deep trenches under the ocean. As the earth rotates within the bulged surface water, raised portions of the land literally run into the bulge and friction results, slowing the earth's rotation. While this slowing trend extends the length of a day by only about 0.002 sec in a period of 100 years, it is the very same phenomenon that once slowed the moon's rotation until it now turns the same side toward the earth at all times. Although the moon is devoid of oceans, the earth once raised moving land tides on a rather flexible primal moon. The friction that resulted changed the moon's rotation until it coincided with its revolution. Now the cool, semirigid moon has its two earth-produced bulges permanently fixed, one pointing always toward the earth and the other always away from it.

The rate of slowing of the earth's rotation appears rather insignificant, yet in the time periods with which the astronomer deals, it is very

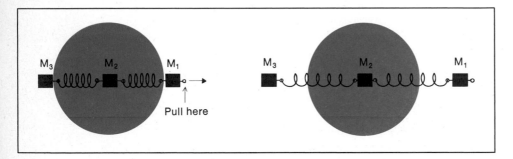

Figure 4.25   *A system of springs and masses that illustrates the effect of the moon's gravitational force in producing tides on the earth. The circle attached to the center mass represents the earth.*

significant. As the day lengthens on earth, the moon also is moving farther from the earth at the rate of 3 cm/year. Could we turn time backward, say, 500 million years, we would find the moon somewhat closer to the earth and the length of a day to be only about 21 hr. This fact is confirmed by the fossil record of certain corals, for they show 416 daily growth rings in each yearly band. Assuming that the number of hours in a year is approximately the same today as it was 500 million years ago, then a 416-day year would have approximately 21-hr days.

The sun at an average distance of 150 million kilometers (15×10^7 km) also has its tidal effect on the earth's waters, but less than one-half the magnitude of the moon's effect. When the sun and moon

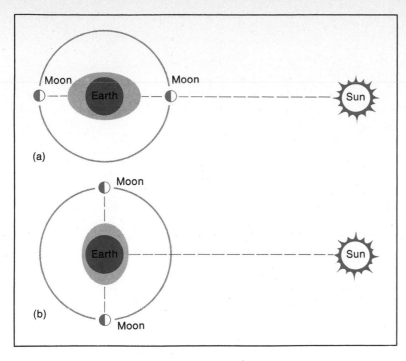

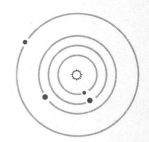

Figure 4.26 (a) *Spring tides.* (b) *Neap tides.*

tend to make the waters bulge in the same direction, they produce extremely high and low tides, called *spring tides*. As we can see in Figure 4.26(a), spring tides occur during new and full moon. When the forces of the sun and moon act at right angles to each other, the tides are less extreme and are called *neap tides*. Figure 4.26 (b) shows that neap tides occur during first- and third-quarter phases of the moon. These same forces act upon the rigid earth to produce land tides and may sometimes trigger the release of stresses in the earth, producing earthquakes.

Eclipses

Eclipses occur when one object passes within the shadow of another, and the structure of the shadow itself provides a key to our understanding of the eclipse. If the source of light were that of a star—a point source—then the boundaries of the shadow cast by an object would be simple and well defined, as indicated by Figure 4.27. In our discussion the sun is the source, however, and because it is nearby it presents an extended source. Under such conditions, the shadow that an object casts is more complex. The dense portion of the shadow is called the *umbra*. If you found yourself within this region, you would see the sun completely covered by the object, in this case the moon (Figure 4.28). The less dense portion of the shadow is called the *penumbra*, and if you were observ-

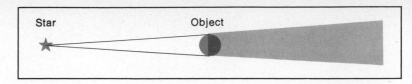

Figure 4.27   *The shadow produced by a point source.*

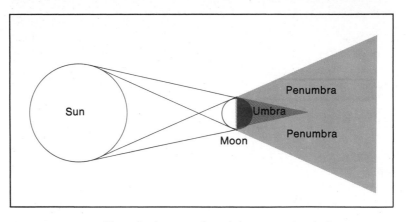

Figure 4.28   *The shadow produced by an extended source.*

ing from within this region, the moon would appear to cover only part of the sun.

Solar Eclipse

The length of the umbral portion of the moon's shadow is slightly more than 383,000 km. The distance from the moon to the earth varies between 362,000 and 406,600 km. Thus we may conclude that under certain conditions the shadow of the moon will strike the earth but cover only a very limited region of it at any one time (Figure 4.29).

Figure 4.29   *A solar eclipse. An observer within the circle of totality will see a total eclipse of the sun, while an observer in the circle of partiality will see only a partial eclipse.*

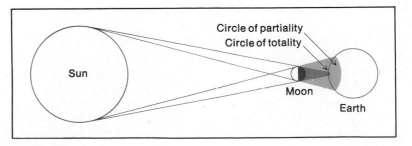

Figure 4.30 *The diamond ring effect. (Cliff Holmes)*

If you were standing within the region of the earth covered by the path of the dark umbral portion on March 7, 1970, you were able to see a truly spectacular event: total eclipse of the sun. The combined motion of the earth and the moon caused this umbral portion of the shadow to sweep out a narrow path across the face of the earth, traveling at approximately 1600 km/hr. The sun's disk was gradually covered by the disk of the moon.

For just an instant before total covering, a portion of the sun was visible shining through a depression in the moon's profile and producing the beautiful diamond-ring effect seen in Figure 4.30. When the sun was completely covered—the time of *totality*—the sky darkened, and the beautiful *corona*, or outer atmosphere of the sun, appeared (Figure 4.31).

Figure 4.31 *The corona of the sun, as seen during a total solar eclipse—March 7, 1970. (High Altitude Observatory)*

The corona of the sun is not usually visible, owing to the brilliant glare of the *photosphere*, the body of the sun itself. During the time of totality, many strange things happened: dogs howled, chickens went to roost, birds ceased their singing, and some flowers closed. Much of the surrounding landscape took on strange hues of color. The eclipse lasted only a few minutes, and then the moon moved on, again to reveal part of the sun and finally the entire disk (Figure 4.32).

The path of totality for the solar eclipse of March 7, 1970, is shown in Figure 4.33. The eclipse was witnessed by millions of people in the United States and Mexico. Those who lived within this narrow band, or who traveled to reach it, were rewarded by a spectacular sight. Almost everyone in Mexico, the United States, and Canada was in a position to see at least a partial eclipse, since all were in the penumbral portion of the moon's shadow. Television via communications satellite played an unprecedented role in bringing this spectacular event to millions of people all over the world.

In addition to the partial and total eclipse, a third situation may arise. We know that the moon is often farther from the earth than 383,000 km and that its shadow will therefore not reach the earth. If you were standing directly in line with the moon's shadow, you would see the central portion of the sun's disk covered, leaving a ring of sunlight visible around the moon's disk. This is called an *annular* or *ring* eclipse (Figure 4.34).

The paths of solar eclipses between July 1963 and May 1984 are mapped out in Figure 4.35.

Figure 4.32   *The solar eclipse of March 7, 1970. An exposure was made every 6 min. (Maurice E. Snook)*

A Warning. Never look at the sun directly, whether by naked eye or through a telescope or binoculars, for your eye may be irreparably burned by the sun's radiation. No pain accompanies this experience, so that a person who has observed the sun directly might not immediately be aware that he is being blinded. A solar eclipse may be safely viewed by holding a small piece of cardboard in which a pinhole has been made, allowing the sun's rays to pass through the pinhole and fall on a second piece of cardboard (screen). A small telescope may be substituted for the pinhole, as illustrated in Figure 4.36.

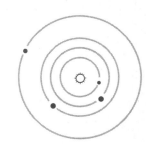

Lunar Eclipse

A total lunar eclipse occurs when the moon passes completely within the umbral portion of the earth's shadow, which extends outward approximately 1,380,000 km. At the moon's distance, the umbral portion is almost 9700 km in diameter. The moon's diameter of 3476 km allows it to fit easily within this shadow (Figure 4.37). The penumbral portion is approximately 16,000 km in diameter at the moon's distance. When a total eclipse occurs, the moon first moves into the penumbral portion of the earth's shadow with only slight dimming. As it moves on into the

Figure 4.33   *The path of totality for the solar eclipse of March 7, 1970. (Nautical Almanac Office, U.S. Naval Observatory, Washington, D.C.)*

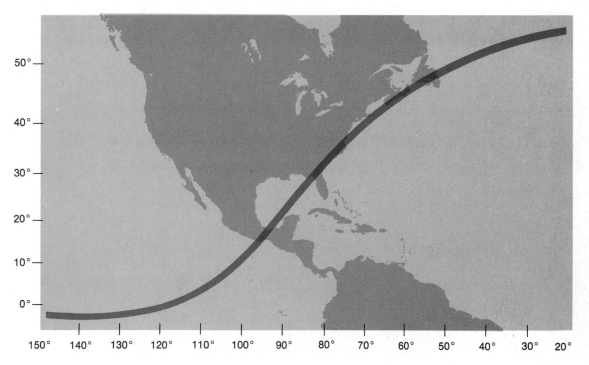
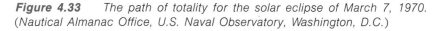

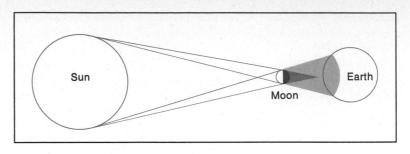

Figure 4.34   *An annular eclipse of the sun.*

Figure 4.35   *The paths of total eclipses between July 1963 and May 1984. A solid line corresponds to a total eclipse; a dashed line, to an annular one. The beginning of the path is indicated by a small white circle; the end, by a black one.* (*From J. Meeus, K. Grosjean, and W. Vanderleen,* Canon of solar eclipses. *Elmsford, N.Y.: Pergamon Press, 1966*)

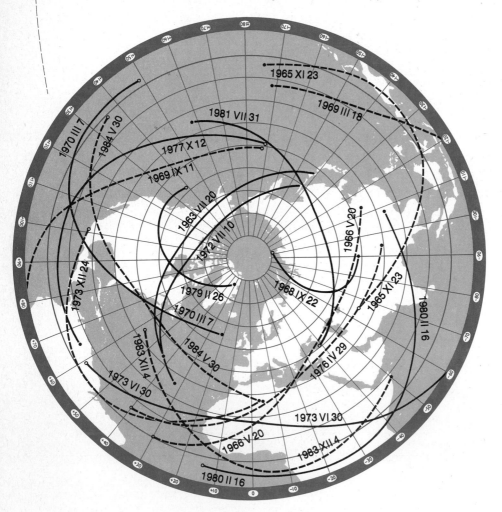

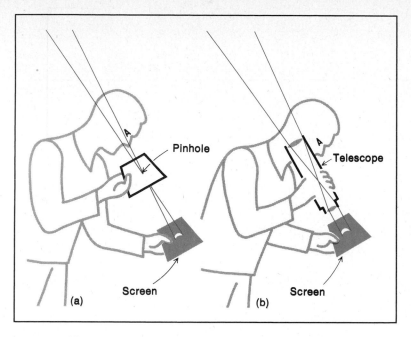

Figure 4.36 *Observation of a solar eclipse using (a) a pinhole and (b) a small telescope.*

umbral portion, the curvature of the earth's shadow may be seen on the moon (Figure 4.38). During totality the moon does not become darkened entirely but rather takes on a copper hue, because sunlight that is refracted by the earth's atmosphere still illuminates the surface of the moon.

Eclipse Seasons

It would be easy to surmise that we should have a solar eclipse at each occasion of a new moon, when the moon passes between the earth and the sun, or that we should have a lunar eclipse on the occasion of

Figure 4.37 *Lunar eclipses.*

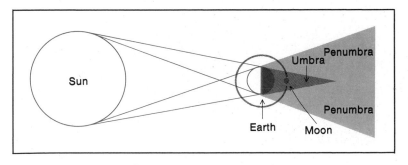

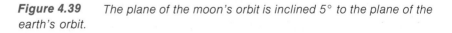

Figure 4.38 *The moon is seen moving into the umbra of the earth's shadow.*
(R. T. Dixon)

Figure 4.39 *The plane of the moon's orbit is inclined 5° to the plane of the earth's orbit.*

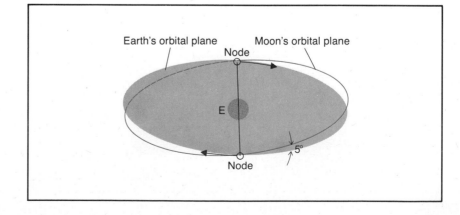

each full moon, when the earth is between the sun and the moon. This turns out not to be true, however, because the moon orbits the earth on an imaginary plane tilted 5° to the plane of the earth's orbit (Figure 4.39).

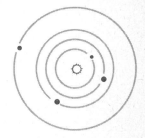

The points at which the two orbital planes intersect are called node points. Whenever the sun and moon are within a few degrees of a node point simultaneously, an eclipse occurs. If both sun and moon are near the same node, then a solar eclipse occurs, and if they are near opposite nodes, a lunar eclipse occurs. Thus we may expect eclipses to occur every six months approximately, during what may be called eclipse seasons. During any given season one would expect a solar and a lunar eclipse about 14 or 15 days apart. (See Figure 4.40.)

However, eclipse seasons do not occur exactly at six-month intervals because there is a slow regression (motion) of the nodes along the orbital plane of the earth. A given node moves through 360° in a period of 18 yr 10 days, a cycle called the Saros. Thus the circumstances which produced a given eclipse repeat themselves in that period. When both the phasing cycle of the moon and the seasonal changes are considered, yet another cycle—the *Metonic cycle*—is found with a period of 18.61 years. "Stonehengers" may have recognized this cycle with their 56 Aubrey holes, since $3 \times 18.61 = 55.83$ (almost 56) years.

A generalization. What we have learned about the earth's moon will continue to form a basis for questions about the additional moons of the solar system.

Figure 4.40 *Eclipse seasons.*

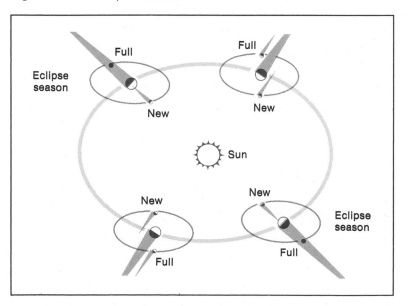

Questions

1. In which direction does the moon appear to move *against the background of stars?*

2. Why does the sidereal month (27.322 days) differ from the synodic month (29.531 days)?

3. The moon presents almost the same "face" toward the earth at all times. Therefore, which of the following is true: (a) its period of rotation is 27.322 days; (b) its period of rotation is 29.531 days; (c) it does not rotate.

4. At what time of night would you expect to see a full moon overhead on your local meridian?

5. Do coastal ports usually experience one, two, or three high tides per day?

6. The moon appears to shift among the stars: (a) approximately one hour circle (15°) per day, (b) less than one hour circle per day, (c) more than one hour circle per day. Which is true?

7. True or false: the sun and the moon "work" together to produce exceptionally high tides when the moon is new or full.

8. Where is the barycenter of the earth-moon system located?

9. How often does a point on the moon experience a cycle of changing tides?

10. Describe a mascon.

11. List six types of surface features on the moon.

12. Which major type of surface feature is most conspicuous by its absence on the far side of the moon?

13. Solar eclipses always occur during what phase of the moon?

14. An eclipse season usually lasts for about a month. How many eclipse seasons may occur in one year?

15. What part of the sun is visible during a total solar eclipse?

16. Describe two safe methods for viewing a partial eclipse of the sun.

17. Why do you not see an eclipse of the moon every month?

18. Describe the relative positions of sun, moon, and earth when the moon is full.

19. If high tide occurs at 9 A.M. on a given day, when can you expect the high tide to occur the next morning? Why?

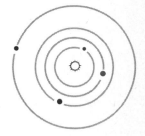

20. Why is the sun a less powerful factor than the moon in producing tides?

21. What is meant by land tides?

22. What does the term *earthshine* mean in relation to the moon?

23. If you lived on Mare Crisium (on the moon), you might expect to see a sunrise once every _____ earth days.

24. What factors of erosion on the earth are *not* found on the moon? Are there any factors of erosion which the earth and moon have in common?

Suggested Readings

ALTER, DINSMORE, *Pictorial guide to the moon.* New York: Thomas Y. Crowell, 1973.

CHERRINGTON, ERNEST H., JR., *Exploring the moon with binoculars.* New York: McGraw-Hill, 1969.

EGLINTON, GEOFFREY, MAXWELL, JAMES R., and PILLINGER, COLIN T., The carbon chemistry of the moon. *Scientific American* 227 (4), 80–90 (1972).

GOLDREICH, PETER, Tides and the earth-moon system. *Scientific American* 226 (4), 42–52 (1972).

LEVINSON, A. A. (ed.), *Proceedings of the Apollo 11 lunar science conference,* 3 vols. Elmsford, N.Y.: Pergamon Press, 1970.

PAGE, THORNTON L., The Third Lunar Science Conference: I and II. *Sky and Telescope* 43 (3), 145–150; (4), 219–222 (1972).

PAGE, THORNTON L., Notes on the Fourth Lunar Science Conference: I, II, and III. *Sky and Telescope* 45 (6), 355–358 (1973); 46 (1), 14–17 (1973); 46 (2), 88–90 (1973).

TAYLOR, STUART R., *Lunar science: A post Apollo view.* Elmsford, N.Y.: Pergamon Press, 1975.

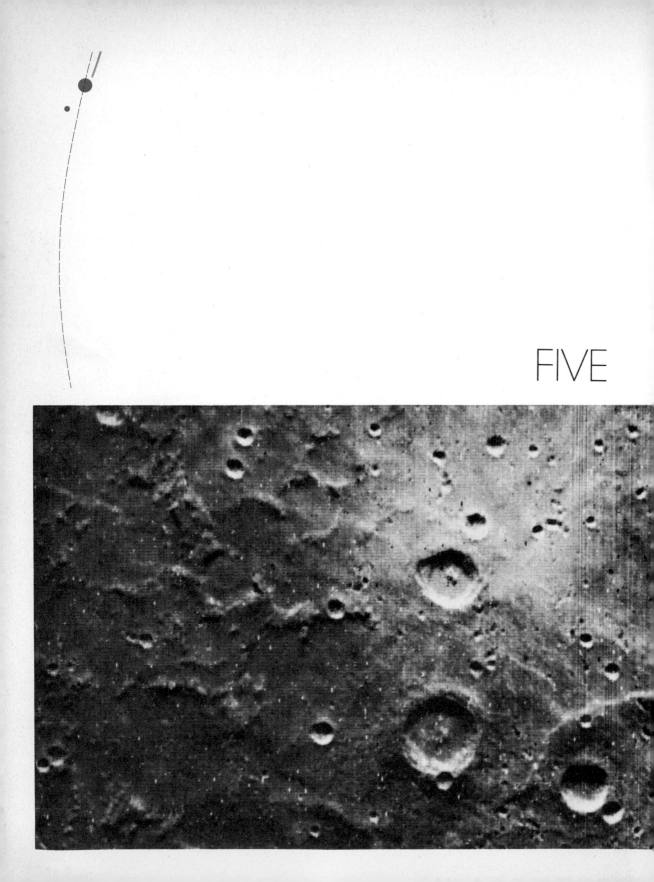

FIVE

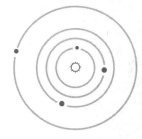

THE PLANETS

While the astronomer is committed to the primary task of describing the physical properties, appearance, and motion of objects in space, he is ever tantalized by the question of their origin and evolution—the conditions governing their formation and the processes whereby they change. What clues will unlock these secrets? As we study the nine planets, 34 moons, and numerous smaller objects that revolve about the sun, let us test several theories proposed for their origin. One theory suggests that the planets were formed much farther from the sun than their present distances indicate and later were captured at random by the sun; this may be referred to as the *random-capture theory.* Another proposes that the sun may once have passed very near another star, and owing to their mutual gravity a long streamer of material was pulled off as they passed. Then the planets formed from the material in the sun's streamer. This could be termed the *encounter theory.* Still another theory suggests that as the sun condensed out of a huge gas cloud, it rotated faster and faster, leaving a flattened disk of material from which the planets formed. This theory of the common origin of the sun and its planets has had numerous variations, one suggesting that rings of material were thrown off to form each planet, another suggesting that smaller condensations (knots) formed in the disk to make each planet. Such planets-in-the-making are called *protoplanets*. Some observers have suggested that the protoplanet might sweep up new material as it moved. This is called *accretion*. Some very general observations concerning planets may help us evaluate these theories.

The Planets in General

All of the planets revolve about the sun in the same direction and along almost the same plane of orbit as the earth. That is to say, their motions could quite accurately be shown on a flat piece of paper much as on the right-hand flip pages starting on page 115. This one observation alone does not seem to suggest a random-capture process, for had the planets been captured at random their motions would likely be oriented in various directions and on various planes (Figure 5.1). Their direction of revolution is counterclockwise as viewed from the "north side" of the solar system, the side toward which the north pole of the earth points. Because they move in virtually the same flattened plane, we on earth see the planets moving, against the background of stars, along almost the same path—that of the ecliptic. The planets appear to move through the 12 constellations known as the signs of the zodiac, and so it is natural to describe a planet as being "in" one of those signs at any given time—for instance, "Saturn is in Gemini."

It is also instructive to describe the position of a planet as it relates to the sun's position. To see what configurations (arrangements) are possible, let us separate the planets into two groups: the *inferior planets*, which have an orbit inside the orbit of the earth (Mercury and Venus), and the *superior planets*, which have an orbit outside the orbit of the earth (Mars, Jupiter, Saturn, Uranus, Neptune, and Pluto).

Figure 5.1 (a) *The orientation of planetary orbits that might be expected if they had been captured at random.* (b) *The actual disklike nature of planetary orbits.*

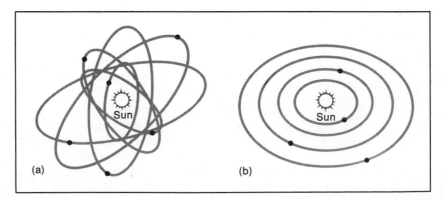

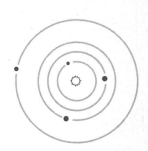

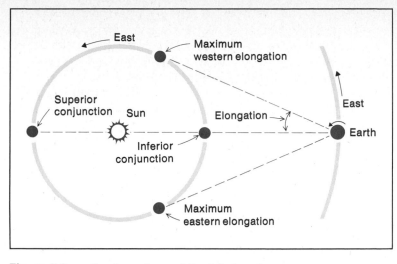

Figure 5.2   *Configurations of the inferior planets.*

Configurations of the Inferior Planets

Whenever a planet is aligned with the sun so that both may be viewed in the same direction from the earth, the planet is said to be in *conjunction* with the sun. In Figure 5.2 we note two possible positions of alignment, one between the earth and the sun, called inferior conjunction, and one on the far side of the sun, called superior conjunction. During such configurations the planet is lost to our view, owing to the glare of the sun. At any time we may describe the position of an inferior planet by the angle it makes with the sun as seen from the earth. This angle is called the *elongation* of the planet, and when this angle increases to its largest value during a given revolution, the planet is said to be at maximum elongation. Since this may occur at either of two different locations, we specify one as a western elongation, if the planet appears to the west of the sun in the sky, and the other as an eastern elongation, if the planet appears to the east of the sun in the sky. The times of maximum elongation are obviously good times to view Mercury and Venus. The right-hand flip pages beginning on page 115 will help you to visualize these various configurations as viewed from the earth.

Configurations of the Superior Planets

When a superior planet is viewed from the earth, there is only one way it may align itself with the sun to produce a conjunction. A superior planet, however, may make any angle up to 180° with the sun as viewed from the earth, and when it is exactly 180° from the sun, a planet is said to be *at opposition*—on the opposite side of the earth from the sun (Figure 5.3). This is a particularly advantageous time to observe the planet, for it is in this configuration that the planet makes its closest approach to the earth, is fully illuminated by the sun, and so appears at its brightest. It is visible all night because it appears to rise at sunset and set at sunrise. Careful observation will also reveal its apparent retrograde motion against the background of stars, for the earth is actually passing that planet at this time.

These same configurations may be applied to the moon. Note that the moon is full when at opposition.

Rotation of the Planets

Seven of the planets rotate (spin on their axes) in a counterclockwise direction. The two planets which differ in their direction of rotation are Venus and Uranus. Their clockwise rotation, because it is contrary to the more common direction, is referred to as *retrograde rotation*.

Spacing of the Planets

At first glance, the distances of the planets from the sun may appear to have no pattern other than that the first four planets are relatively closely spaced and the remaining five are widely spaced; however, in

Figure 5.3 *Configurations of the superior planets.*

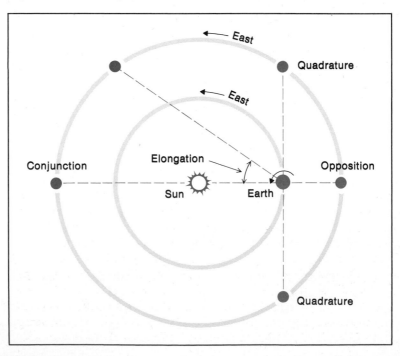

Table 5.1   *The Bode-Titius Relationship*

PLANET	FIRST COLUMN	SECOND COLUMN	BODE NUMBERS	ACTUAL DISTANCES (A.U.)
Mercury	0	4	0.4	0.39
Venus	3	7	0.7	0.72
Earth	6	10	1.0	1.00
Mars	12	16	1.6	1.52
	24	28	2.8	
Jupiter	48	52	5.2	5.20
Saturn	96	100	10.0	9.54
Uranus	192	196	19.6	19.18
Neptune	384	388	38.8	30.06
Pluto	768	772	77.2	39.44

1766 Johann Daniel Titius of Wittenberg, Germany, pointed out a very interesting mathematical relationship. This relationship was later published by Johann Elert Bode (pronounced *Bó-da*) and is often erroneously called Bode's law. In the last column of Table 5.1 you will see the average spacing of the planets measured in astronomical units (A.U.). These numbers may be closely approximated by the process depicted in the first three columns. The first column of figures begins with 0; then comes the numeral 3; and every number thereafter is twice the value of the one preceding it. A second column is formed by adding 4 to the corresponding entries in the first column. A third column is then obtained by dividing each entry in the second column by 10; these numbers (third column) very nearly express the distances to the planets as measured in astronomical units. Although the discoveries of Uranus, Neptune, and Pluto were to come later, the Bode-Titius relationship is here extended to include these planets for our comparison. It is evident that this relationship does not fit the observed distances to Neptune and Pluto; these are listed in the fourth column in the table.

It is also obvious that no single planet corresponds to the 2.8 A.U. figure in the Bode-Titius progression of numbers; as a result, it was thought that an unknown planet might be found at or near this distance. Using Kepler's third law, astronomers quickly computed the period of such a planet and launched a search for it. The search produced no planet, but, within a few years, many small objects had been found having approximately the correct period and orbital distance. These objects, called *asteroids*, will be discussed more fully in Chapter 6. Some observers have felt that because the Bode-Titius numbers so closely approximate the observed distances to the planets (except Neptune and Pluto), a force may have been operative to produce this spacing. Others have felt that the spacing is coincidental. It has also been suggested that Pluto originated as a moon of Neptune, their gravitational interaction having disturbed the original spacing of Neptune.

The Bode-Titius relationship would appear to have more value if it had led scientists to an explanation of why the planets formed at their respective distances from the sun. In fact numerous theories have been set forth which predict entirely different spacings of the planets. Therefore, the Bode-Titius relationship should not be elevated to the level of a physical law but should be treated as merely an interesting way to remember the spacing of the first seven planets and the primary asteroid belt.

Terrestrials or Jovians

The planets fit very naturally into two categories: (1) those which are earthlike, called the *terrestrial* planets, and (2) those which are Jupiterlike, called the *Jovian* planets. The terrestrial planets include, besides the earth, Mercury, Venus, and Mars. The Jovians include Jupiter, Saturn, Uranus, and Neptune. If Pluto were placed in either category it would be in the terrestrial group, but many prefer to consider it separately, more like a moon.

Table 5.2   *Properties of the Terrestrial Planets as Compared to Those of the Jovians[a]*

PLANET	DIAMETER	MASS	DENSITY	ROTATION	TEMPERATURE	ATMOSPHERE
Terrestrials						
Mercury	0.38	0.05	5.2	58^d15^h	110° to 700°K (−163° to 427°C)	None
Venus	0.95	0.82	5.3	-243^d	250° to 770°K (−23° to 497°C)	Significant
Earth	1.00	1.00	5.5	23^h56^m	199° to 311°K (−74° to 38°C)	Thin
Mars	0.53	0.11	3.8	24^h37^m	145° to 293°K (−128° to 20°C)	Very thin
Jovians						
Jupiter	11.23	317.9	1.3	9^h50^m	134°K (−139°C)	Very significant
Saturn	9.41	95.2	0.7	10^h14^m	97°K (−176°C)	Very significant
Uranus	3.98	14.6	1.3	10^h49^m	58°K (−215°C)	Very significant
Neptune	3.88	17.2	1.7	15^h40^m	52°K (−221°C)	Very significant
Pluto	0.23	0.002	1.1	6^d9^h	43°K (−230°C)	Unknown

[a]Diameters and masses are related to those of the earth (as 1). Densities are given in grams per cubic centimeter. Temperatures of the Jovian planets are not meant to be exact but to show a trend.

If the physical nature of each planet is visualized as it compares to the earth, the distinctions between the terrestrials and the Jovians will be quite obvious. Table 5.2 reveals that the terrestrials are small (earth size or smaller). They have low mass but high densities when compared to the Jovians. The higher densities of the terrestrials reflect their higher abundance of heavier elements and their less significant atmospheres. With the exception of Venus, the atmospheres of the terrestrials are not even taken into consideration when computing their diameters and masses (Figure 5.4).

Why should the planets near the sun tend to have little or no atmosphere? Their relatively high temperatures and low masses tell part of the story. As energy was received from the sun, molecules of the atmosphere which may once have surrounded Mercury would have absorbed it, thereby increasing their velocity until they were moving so fast that the gravitational pull of the planet could no longer hold them. They would then have escaped into space. The velocity of escape for each planet is shown in Appendix 6 (page 503), and it is clear that some of the molecules in the atmospheres of the terrestrials could have been heated by the sun to that velocity. The less massive elements such as hydrogen

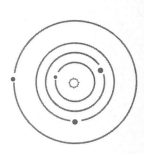

Figure 5.4   *Relative size of the planets.*

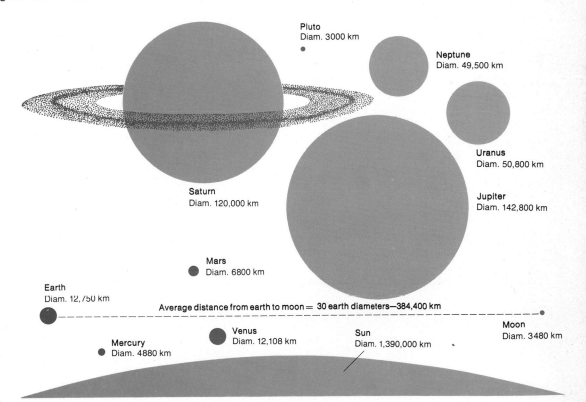

and helium would be most easily accelerated to the escape velocity, so while these elements are very abundant in the universe, today only traces exist in the atmospheres of the terrestrial planets.

While this explanation accounts for the absence of the least massive molecules, it does not account for the low percentage of other gases such as neon and argon. As we come to our detailed study of Venus, we will see that the terrestrials may have lost virtually all atmosphere at one time and that the atmosphere they now possess may have been developed from within the planet, through volcanic action.

The solid (crusty) nature of the terrestrial surfaces contrasts with the uncertain nature of the Jovian surface. The Jovians might be characterized as balls of gas with a liquid and/or solid core beneath a very deep atmosphere. The Jovian atmospheres are believed to be mainly hydrogen with lesser amounts of helium. These planets are able to retain the lighter atoms because of their large surface gravity (high mass) and cool temperatures. Let us now consider some of the distinctive characteristics of each planet.*

Mercury

Mercury is the most elusive of all the planets that can be seen with the naked eye. It appears sometimes as a morning object, rising just before sunrise, and at other times as an evening object, setting just after sunset. This cycle repeats itself just over three times during the year. Ancient observers did not realize that they were seeing the same object alternately in the morning and evening. The Greeks named the planet *Mercury* when it appeared in the evening and *Apollo* when it appeared in the morning. Mercury's elusiveness is explained by the fact that it revolves around the sun at an average distance of only 57 million kilometers, or 0.4 A.U., less than half the orbital radius of the earth. As a consequence, it never appears more than 28° from the sun. Since the earth rotates at the rate of 15°/hr, Mercury appears to rise no more than 2 hr before the sun when it is a morning object and appears to set no more than 2 hr after the sun when it is an evening object. At best, Mercury is seen in the light of dawn or twilight.

The plane in which Mercury travels is inclined 7° to the ecliptic plane, so the planet may appear several degrees above the ecliptic at some times and several degrees below it at others. Mercury's orbit has an eccentricity of 0.206, the second most eccentric orbit in the solar system. This eccentricity is shown in Figure 5.5, which indicates that Mercury's distance from the sun varies between 69 million kilometers at *aphelion*

---

*You will find the physical properties and motions of the planets and their moons summarized in Appendixes 5, 6, and 7. Not all of these properties will be detailed in the following text. Therefore, these appendixes should be read as an integral part of this chapter.

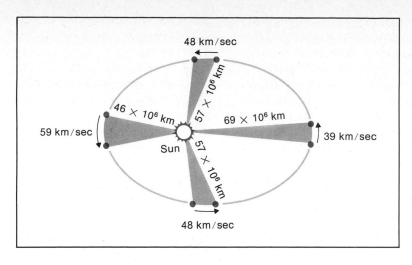
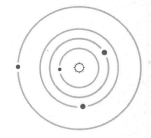

Figure 5.5 *Mercury's orbital speeds and distances.*

(its most distant point) and 46 million kilometers at *perihelion* (its nearest point). Applying Kepler's law of equal areas in equal time to a planet with such an eccentric orbit, we see a wide range of orbital speeds. At aphelion, Mercury travels at 39 km/sec; at perihelion, its speed increases to 59 km/sec. Mercury's average speed in orbit is 48 km/sec. A careful study of Figure 5.5 will help you develop a feeling for the motion of this planet and all planets in general. Each successive planet you study will be farther from the sun and will orbit more slowly.

The *sidereal period* of revolution for Mercury—the time needed for one exact revolution (360°), using the stars as a reference—is 87.96 days. However, we view the planets from a moving platform, the earth. Observationally speaking, a more meaningful period for a planet is its *synodic period*, the time needed for it to move from any given configuration, such as inferior conjunction, back to the same configuration, as seen from a moving earth. Suppose that we begin counting time when Mercury and the earth are in position A in Figure 5.6, an inferior conjunction. Points

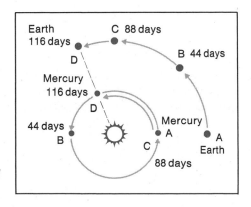

Figure 5.6 *The synodic period of Mercury.*

B, C, and D show the corresponding positions of the planets after 44, 88, and 116 days, respectively. After 116 days, Mercury is again at inferior conjunction as viewed from the earth. The flip pages beginning on page 115 reveal this 116-day synodic period of Mercury.

One of the exciting discoveries of the decade came in early 1965, when R. B. Dyce and G. H. Pettengil, using the 304-m (1000-ft) Arecibo antenna (see Figure 2.51), succeeded in bouncing radar signals off the surface of Mercury. The reflected signals, although very weak, told these researchers that the planet rotates in a period of approximately 59 days. This came as quite a shock, for in 1890 the Italian astronomer Giovanni Schiaparelli had announced an 88-day rotation, and this period had apparently been confirmed by almost every visual observation made from that date until the year 1965. This new discovery was confirmed and refined by subsequent observations, yielding a figure of 58.64 days— almost exactly two-thirds of the revolutionary period of the planet. This means that for every two revolutions of Mercury ($2 \times 87.96 = 175.92$), the planet makes three rotations ($3 \times 58.64 = 175.92$), and the combined effect of these two motions produces a sunrise every 176 earth days. Recalling how tidal action synchronized the rotation and the revolution of the moon with respect to the earth, we see a similar resonance here in that Mercury points first one face and then the opposite face toward the sun on successive times of *perihelion passage* (times when the planet makes its closest approach to the sun).

On March 29, 1974, Mariner 10 flew within 700 km of Mercury and for the first time revealed its densely cratered surface (see the photograph on page 174 and Figure 5.7). At first glance, Mercury resembles the moon in that it has large, relatively smooth areas somewhat like the moon's maria, and it has a variety of crater forms ranging from those which appear to be very old with rounded rims to those which appear very young with sharp rims. We see some deeper craters with central peaks and some large shallow ones with evidence of repeated impacting within their basins.

Based upon measurement using the infrared radiometer on board Mariner 10, the surface temperature has a high of from 600°K to 700°K, depending upon the planet's distance from the sun. A drastic drop to 150°K was noted as the spacecraft began to traverse the shadowed portion. Best estimates of the coldest temperature are as low as 110°K, thus giving a variation of 590°K between the hottest and coldest points. These observations are as expected for a planet close to the sun, with virtually no moderating "blanket" of atmosphere. Furthermore, owing to the combined rotation and revolution rates, a point on Mercury's equator experiences 88 earth days of heating and then 88 earth days of cooling.

The discovery by Mariner 10 which was least expected was the existence of a magnetic field strong enough to deflect the solar wind rather effectively. One does not usually associate a magnetic field with such a slowly turning planet, and while Mercury's magnetic field strength at the equator is only one-fiftieth that of the earth, it provides a definite bow shock and prevents particles from the sun from impinging directly on the surface of the planet. The very presence of a magnetic field, intrinsic to Mercury, raises the question of the source of that field. Based upon the theory that magnetic fields arise from electric currents in a molten core, we must explain how Mercury can have a molten core after 4 billion years of cooling. First we are suggesting that Mercury was once molten to the extent that its more dense components, such as iron, settled toward the core, leaving a silicon-dominant crust. Heat flows only very slowly from core to surface, as is true with the earth; however, Mercury is much smaller than the earth and should have cooled completely, leaving no molten core and hence no magnetic field. Observers are currently endeavoring to explain the apparent contradiction.

In searching for additional clues to the history of the planet, observers are particularly interested in the views of Mercury which were taken by Mariner 10 on its second and third passes (September 21, 1974, and March 16, 1975, respectively; the spacecraft had been placed into a solar orbit with a period of 176 days so that with each two revolutions of

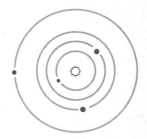

Figure 5.7 *Mercury, as photographed by Mariner 10: (a) A dark, smooth, relatively uncratered area on Mercury which resembles certain areas on the moon—lava flows being suggested. The largest craters shown are approximately 35 km in diameter. (b) Several fresh (sharp-rimmed) craters are seen in older basins. (c) A heavily cratered area of Mercury, showing many low, hill-like structures. The large valley at the bottom is approximately 10 km long. (NASA–JPL)*

(a)   (b)   (c)

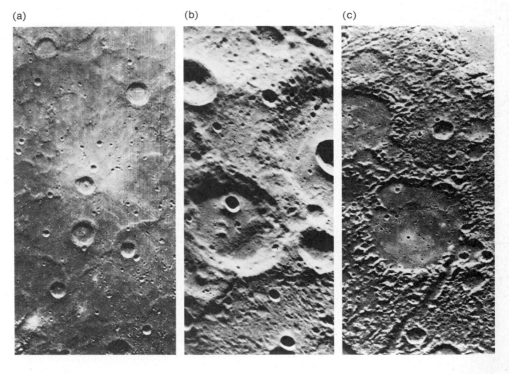

Mercury, the spacecraft encountered the planet again). This required no additional expenditure of fuel, other than for simple pointing procedures. Figure 5.8 (a) is a mosaic of 18 photos showing distinct bright rayed craters in the upper right portion of the photograph. Figure 5.8 (b), the Caloris Basin, its surface cracked and wrinkled, represents a very old major feature on Mercury's surface—evidence of a period of shrinking crust. A surface feature which makes Mercury distinctly different from the moon is its *intercrater plains*, shown in Figure 5.8 (c). If you will compare this illustration with views of the moon in Chapter 4, you will see it is unlike either the moon's heavily cratered highlands or its sparsely cratered maria. The intercrater plains may represent the oldest surface feature on Mercury.

As would be expected, the very low surface gravity and very hot daytime temperatures associated with Mercury have allowed virtually all atmosphere to escape. The light by which we see the planet is reflected directly from its surface. Like the moon, Mercury is not a very efficient

Figure 5.8   Mercury, as photographed by Mariner 10: (a) a photomosaic of Mercury, constructed from 18 photos, taken 6 hr after Mariner 10 flew past the planet on March 29, 1974 (Note the distinct bright rayed craters in the upper-right portion of the photograph); (b) fractured and ridged plains of the Caloris Basin as seen on Mariner's third encounter with Mercury; (c) the cratered terrain shown in this view appears very similar to parts of the moon. The large crater at the top shows conspicuous hills on its floor. (NASA–JPL)

(a)   (b)   (c)

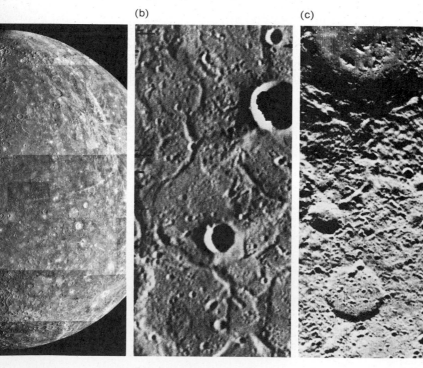

reflector. Its *albedo*, the percentage of reflected light, ranges from about 13% in darker areas to 35% in limited regions where bright material appears to have been ejected during impact [see Figure 5.7 (a)].

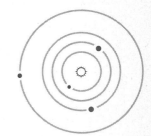

Venus

Even before Galileo turned his telescope toward Venus to see it go through phases like the moon (Figure 5.9), humans had been intrigued by this bright object, which they often saw in the western sky just after sunset and thus called the "evening star." Some humans recognized this planet to be the same object that appeared alternately in the eastern (morning) sky, for as Venus swings around the sun in an orbit not much smaller than the earth's, it appears first on one side of the sun and then on the other (Figure 5.10). Because the earth is also revolving, this synodic cycle requires approximately 19.5 months (584 days).

Early observations revealed Venus to be very nearly a twin of the earth in terms of its diameter and mass and therefore its density. Yet Venus is really vastly different than the earth. Its cloud cover is so dense that no person has ever seen its surface. In fact, it is the only other terrestrial planet with any significant atmosphere. Astronomers who have tried to measure the temperature of the planet by analyzing the light it reflects have gotten only a reading for the top of the clouds (slightly less than 0°C), for it is at that level that the light is reflected. As radio astronomy came into its own in the 1950s, radiation from nearer the surface could be measured in radio wavelengths, and temperatures of the order of 600°K (327°C) were recorded. With this discovery, we began to see Venus as a desert hot enough to melt lead.

In 1962, using the Goldstone antenna (see Figure 2.50), Roland Carpenter and Richard Goldstein succeeded in bouncing radar signals off this planet. To everyone's amazement, they discovered that Venus rotates backward (clockwise) in a period of 243.16 days. Before this finding, speculation had centered around a period of 225 days, synchronized with Venus's revolution. The radar observations were interpreted by noting

Figure 5.9 *The phases of Venus. (Lowell Observatory)*

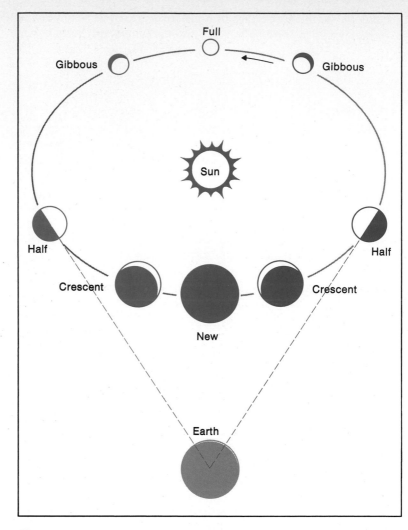

Figure 5.10   *The phases of Venus.*

the Doppler shift due to rotation. A certain wavelength was sent to the planet and a variety of wavelengths returned, for those that were reflected from the approaching limb (edge) were shortened and those reflected from the receding limb were lengthened (see Figure 5.11). Using the amounts by which the wavelengths were changed, it was possible to compute the planet's velocity of rotation, hence its period. If you could live on this planet and you were still dividing time into earth days, Venus's retrograde rotation in 243.16 days and its direct revolution in 225 days would produce a sunrise in the west every 117 earth days. An interesting kind of resonance occurs between Venus and the earth. At each occasion of an inferior conjunction, Venus presents the same face toward the earth, even though it has rotated five times with relation to the sun. While the cause of this resonance is not fully understood, a tidal

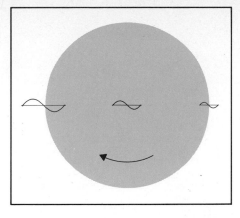

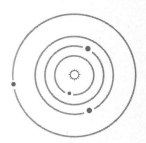

Figure 5.11 *The Doppler effect upon radio signals reflected from Venus.*

effect somewhat like that which synchronized the rotation and revolution of the moon (with respect to the earth) is suspected.

Radar astronomers are continually unfolding a "picture" of the surface features of Venus. Of course, radio waves do not produce a photograph directly, yet the returning signals may be interpreted to produce a topographic map which is as instructive as a photograph. A recent upgrading of the radio telescope at Arecibo, Puerto Rico (see Figure 2.51, page 88) has permitted the use of wavelengths as short as 1 cm and has resulted in a 50-fold increase in resolution of details on Venus. Features as small as 32 km in diameter may now be recognized on the planet's surface. Figure 5.12 shows some of the detail now visualized by means of radar bounce.

Figure 5.12 *Radar view of Venus. The "Alpha" region (bright region in lower right corner) on the surface of Venus is 1350 km across, and it corresponds to rather rough terrain. The darker areas are relatively smooth. Note the bright craterlike structures scattered in the smooth (dark) region of the lower left area. (Courtesy of D. B. Campbell, Arecibo Observatory)*

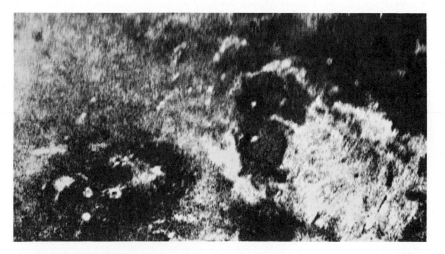

(a)

(b)

Figure 5.13 *These two views represent the only direct photographs ever taken of the surface of Venus. This feat was accomplished by two probes of the USSR which were lowered onto the surface by parachute and survived the very high pressure and temperature of Venus long enough to transmit these television pictures back to earth. Views (a) and (b) were transmitted from Venera 9 and 10 respectively. (TASS, USSR)*

Still another significant phase in the study of Venus began in 1962 with the first successful space probe sent to survey the planet at close range. This probe, called Mariner 2, typified the fly-by approach the United States would take in the study of Venus. A year earlier the USSR had begun a long series of attempts to land probes on the planet's surface. The first Soviet spacecraft are believed to have been crushed by tremendous atmospheric pressures, their communications ceasing before they reached the surface. That problem was overcome, however, and later attempts have produced useful communications for as long as 110 minutes after the craft landed on the surface. Two such landers, called Venera 9 and Venera 10, transmitted back to earth the television pictures you see as Figure 5.13 (a) and (b). Venera 9 landed among numerous rocks of the order of 30 cm across, thought to represent more recent activity—perhaps the debris from a meteorite impact. Venera 10 landed in a smoother, more mature region. Each view was made under existing light scattered by the dense atmosphere, thought to be equivalent to the illumination near twilight's end on earth.

The fly-by and the landing approaches have been equally effective in supplying data on the characteristics of Venus and have complemented each other in confirming various findings. The atmospheric pressure at the surface is approximately 90 times that on earth, a pressure equivalent to that which would be experienced by a diver 823 m beneath the surface of the ocean.*

*This suggests a very interesting phenomenon that would occur in the atmosphere of Venus if it were optically clear. The atmosphere would refract light to such a degree that it would

A United States probe, Mariner 10, which passed within 6000 km of Venus in February 1974, confirmed a very distinct pattern of atmospheric currents when it photographed the planet in ultraviolet light. These upper clouds of sulfuric acid move at a very high rate of speed, rotating in a retrograde motion in a period of 4 earth days. Their equatorial portion seems to lead the way, spiraling off toward each pole (Figure 5.14). The composition of these clouds has been confirmed by infrared spectrograph.

Mariner 10 measured temperatures throughout the Venusian atmosphere and found a range of from 250°K (−23°C) at the cloud tops to 773°K (500°C) on the surface. Why should the surface of Venus be several hundred degrees hotter than its distance from the sun would lead us to expect, and why should Venus have such a dense atmosphere when the other terrestrial planets have almost none? The high surface temperature is attributed to the presence in the planet's atmosphere of CO_2 and H_2O, which together create a "greenhouse effect." You have probably experienced the heating effect that occurs inside an automobile on a sunny day when the windows are closed. The sun's rays enter the car through the windows, are absorbed by the interior, and are then re-emitted at a longer wavelength, primarily in the infrared range. These longer wavelengths are not transmitted out through the windows as readily as the shorter wavelengths are taken in, and a buildup of heat results. This is the same effect that florists use to maintain higher temperatures inside their greenhouses.

Even though the earth's atmosphere contains only trace amounts of CO_2 and H_2O, these gases serve to raise the surface temperature of the earth by at least 45°C. The carbon dioxide (97%) and water vapor (less

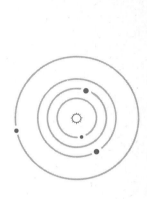

Figure 5.14 *A series of ultraviolet photographs of Venus, taken two days after Mariner 10 flew past the planet on February 5, 1974, during a 14-hr period, showing the rapid rotation rate of the upper cloud deck. The feature indicated by the arrows is about 100 km across.*

(a)　　　　　　　　　　　　(b)　　　　　　　　　　　　(c)

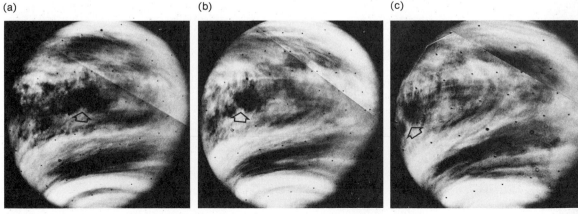

be possible for a beam of light to travel entirely around the planet. One can imagine the distortion of familiar objects that would result. Under ideal conditions, it might even be possible to see the back of your own head while looking straight forward.

Figure 5.15 *The Venus Multiprobe Spacecraft, each component descending into the Venusian atmosphere to perform its own special function. (NASA)*

than 1%) in the Venusian atmosphere absorbs most of the infrared energy radiated from that planet's surface. Although, like any planet, Venus must eventually radiate all of the sun's energy that falls on it, the absorption by these compounds causes the surface and lower atmosphere to be heated to the high temperature indicated before the outflow of energy balances the inflow.

The month of December 1978 represented the most intense period of exploration of Venus. On December 5, the orbiter Pioneer Venus 1 arrived at the planet and began an orbital reconnaissance that lasted for approximately one year. Among its notable discoveries, through means of radar mapping, was a canyon that dwarfs any known in the solar system—1300 km long, 230 km wide, and up to 4.5 km deep. Because of its highly eccentric orbit, this probe actually samples various layers of the planet's upper atmosphere (ionosphere), dipping to within 378 km of the planet's surface on each pass.

On December 9, Pioneer Venus 2 arrived along with its four miniprobes, which had been released earlier to explore both day and night portions of the planet (Figure 5.15). Although not programmed to function on the surface, the "Day" probe actually survived impact and transmitted information from the surface for 67 minutes. The findings that are reported in the following section represent primarily those of the U.S. probes; however they are also confirmed by two additional probes of the Soviet Union, Venera 11 and 12, which plunged through the dense

Venutian atmosphere during the same month. The findings of these craft include continuous lightning below the 32-km level, perhaps producing a steady glow observed by the U.S. night probes.

The composition of the Venusian atmosphere is as follows: 97% CO_2 (carbon dioxide), 2% N (nitrogen), 0.1 to 0.4% H_2O (water vapor), 250 ppm (parts per million) He (helium), 6 to 250 ppm Ne (neon), 20 to 200 ppm Ar (argon), 240 ppm SO_2 (sulfur dioxide), and 60 ppm O (oxygen). While the approximate percentages of CO_2 and N were known from earlier probes, the discovery of primordial argon and neon in abundance levels several hundred times that of the earth's atmosphere was totally unexpected, and may cause a total reevaluation of theories concerning the origin of the solar system. Most theories surmise that the lighter primordial elements, up through argon (and including neon), would have escaped the innermost planets, driven away by the heat of the sun. The most abundant form of argon in the universe is argon-36, so according to the above theory Venus would have a lesser proportion than the earth or than Mars. The findings of Pioneer Venus 1 and 2 indicate a reversed trend—highest abundance on Venus, second highest on earth, and least on Mars. Could Venus have been formed under cooler circumstances than the earth or at a different time? These new findings will keep theorists busy for years. In contrast to our discussion of argon and neon as primordial in origin, we do not consider the major constituents of the atmospheres of Venus and earth as primordial.

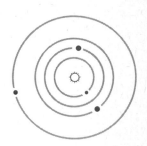

Why does such a very dense CO_2 atmosphere exist on a planet that is virtually the twin of the earth in size and only slightly closer to the sun? If we assume that the hydrogen and helium atmosphere present on the terrestrial planets at the time of their formation would have been driven off by solar radiation, then only as outgassing of other elements occurred (perhaps from volcanoes) would a new atmosphere build up. Carbon dioxide is known to be released by volcanic action. The same processes may have occurred on Venus as on the earth, but the Venusian atmosphere retained a much higher concentration of CO_2. Perhaps the earth once released an equivalent supply of CO_2 and most of it was absorbed by its rocky crust. Within surface rocks, the CO_2 might have combined chemically with calcium (Ca) and oxygen to form calcium carbonate ($CaCO_3$), a basic compound of much of the earth's present surface material. If so, then why did not the CO_2 in Venus's atmosphere combine with its surface material? The answer probably lies in the fact that Venus, being closer to the sun, may have had a slightly higher original temperature than the earth. This higher temperature would have slowed down the absorption process.

Today, life itself plays a vital role in maintaining a rather constant low CO_2 level in the earth's atmosphere. Plants, through the process of photosynthesis, convert CO_2 to oxygen. Marine organisms remove CO_2 from the ocean waters, allowing these waters to absorb more from the air above. Even a slight change in the rate at which CO_2 is used by living organisms might change the earth's ability to support life. A slight increase in CO_2 content in the atmosphere would increase the greenhouse effect and raise the temperature of the earth accordingly. If such a process got out of hand, conditions on earth might one day resemble those on Venus. On the other hand, we might speculate on the possibility of changing the Venusian atmosphere to resemble that of our own planet. It is difficult to visualize any familiar form of life existing on that superheated surface at present. However, the Venusian atmosphere does not lack the essential elements for photosynthesis—light, water vapor, and CO_2. Carl Sagan, one of the foremost experts on Venus, has suggested a possible habitat for life in the clouds of Venus. Were certain microorganisms introduced there, they could adapt and reproduce, and they might change the entire nature of the planet's atmosphere. Through the process of photosynthesis, they would extract H_2O and CO_2 from the clouds and would produce oxygen and organic (carbon-containing) compounds. The removal of CO_2 and water vapor would tend to lessen the greenhouse effect, causing a gradual cooling of the surface. As the temperature declined, water vapor would tend to liquefy on the surface, resulting in a clearer atmosphere rich in oxygen and perhaps more suitable for higher forms of life. While this may all be a bit of fantasy, there are plausible elements in it.

As the Pioneer Venus 2 "bus" and its multiple probes descended through the atmosphere, they found three distinct layers of clouds beginning at the 70 km elevation, each composed of sulfuric acid particles. Only the lower of these three was dense enough to appear as clouds in the usual visible sense (at an elevation of about 48 km). Note that Figure 5.14, page 191, showed the upper cloud formation and rapid rotation only because they were detected in ultraviolet light. These features are not visible in the range of wavelengths to which our eyes are sensitive. Below 48 km, the compounds vaporize and split up to form a "chemical stew," as it is termed by NASA scientists, composed of water vapor, sulfur dioxide, and molecular oxygen. The "sulfurous soup" appears to recirculate back up to the cloud layers where it reforms into sulfuric acid and sulfur, under the influence of incoming ultraviolet radiation.

Over the polar regions of Venus there seems to exist a huge "polar hole," a vortex of descending material which shows heating as it descends. Perhaps there is a general circulation northward and southward from the equator in the upper elevations, only to return through the polar holes to the equator along a more nearly surface route.

Venus has little or no magnetic field, hence the incoming solar wind is not diverted (as by the earth's magnetic field) but "runs" into the planet's ionosphere, at an elevation of approximately 400 km, as compared to the earth's bow shock wave at an elevation of 65,000 km. The turbulence of Venus's bow shock wave produces temperatures in the range of 1 million° C. A strong magnetic field is generated at this point by the interaction of the solar wind with the planet's ionosphere.

Earth

In Chapter 3 we studied our own planet primarily from the point of view of an observer on the earth itself. Before proceeding to the planet Mars and to the question of the existence of life on that planet, let us pose the following question: If an intelligent form of life existed on Mars and possessed telescopes like ours, could that intelligence recognize any life form on the earth? What would there be about the earth that could furnish a clue? Even under ideal conditions, it is unlikely that Martian astronomers would be able to recognize any object or form less than 80 km in its smallest dimension. Even when clear skies prevailed on the earth, only large areas such as the oceans, continents, large lakes, and polar caps and perhaps color variations occurring between the plains and the mountainous regions might be detected. No evidence of freeways, waterways, cities, or farms would be distinguishable. Photos taken of the earth from the proximity of the moon, only about 0.4 million kilometers away, reveal merely the larger continental masses. Mars, at its closest approach to earth, is still 56 million kilometers away. Of course radio communication would be possible, assuming an equivalent technology on the part of the Martians. In fact, one of the most universal "languages" known is that of pictures, and we have been beaming such via television for years.

Mars

More than any other planet, Mars seems to fire people's imagination. Many observers have speculated that some form of life, perhaps as intelligent as our own, might exist on this planet—an idea perhaps due in part to the similarity of Mars to the earth, for Mars does exhibit polar caps and certain seasonal color variations similar to those we find on earth. As early as 1877, Giovanni Schiaparelli, director of the Milan Observatory, asserted that he had seen a network of fine straight lines that seemed to interconnect larger features on the Martian surface. Percival Lowell, an American astronomer, suggested that these lines represented canals, actual waterways built by Martian inhabitants to irrigate their crops and to transport their goods on barges. While this notion of intelligent life on Mars was long ago seized on by science-fiction writers, we must, if we are to give scientific consideration to this matter, look at the evidence offered by recent detailed, close-up observations of this planet.

Before 1965 our only observations of Mars were made from earth-based observatories. Because of the obscuring effect of the earth's atmosphere, details as to the Martian surface features, atmospheric composition, magnetic field, and the like were very sketchy. Figure 5.16 represents a typical result of hundreds of hours spent by astronomers at the eyepieces of telescopes, sketching areas of momentary clearing. Anticipating the close approach of Mars in 1965 (at the time of opposition), scientists prepared a probe that would fly within 9600 km of the planet, photograph its surface, and relay these photos, together with other information, to the earth. The photographs which this probe (Mariner 4) returned to earth were far superior to any produced by earth-based telescopes.

Spurred by this success, scientists refined their equipment and prepared for another opposition of Mars in 1969, when the planet passed within 72 million kilometers of the earth (Figure 5.17). In order to conserve rocket fuel, it was imperative to wait for such opportunities of close approach. Because of the combined motion of the earth and Mars, such oppositions occur only at 26-month intervals—the *synodic period.* Owing to the eccentricity of the two orbits, some approaches are much closer than others, as may be seen in Figure 5.18. Although much was learned during the Mariner 6 and 7 flights in 1969, a more extensive period of observation was planned for the 1971 opposition.

Figure 5.16 *A high-resolution photovisual map of Mars showing the seasonal aspects of late Martian summer, 1969. The excellent quality allowed surface details of less than 30 km to be resolved. This map was produced from measurements of 20 photographs, 13 visual drawings, and 15 telescopic micrometer observations. South is at the top as seen in a telescope. (C. Capen, JPL—Table Mountain Observatory)*

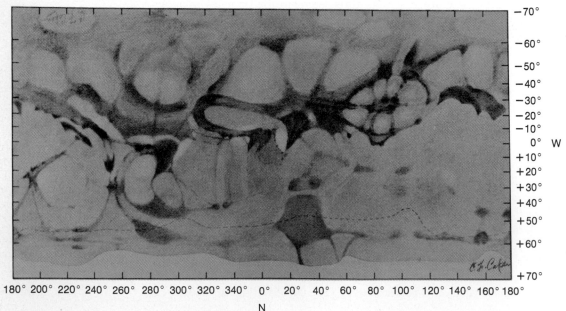

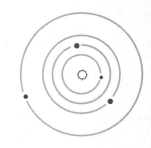

In November 1971, Mariner 9 became the first man-made satellite to orbit another planet. For almost one year, this spacecraft radioed back to earth a continual flow of photos and data about Mars, information which significantly changed man's concept of that planet. Mariner 9's cameras first revealed only a huge dust storm obscuring most of the surface features of the planet, but as the storm subsided, four large volcanic mountains emerged through the dust. As the Martian atmosphere cleared, the full extent of the largest volcano, Olympus Mons, was revealed—600 km across at the base and approximately 23 km high, far larger than any mountain on the earth (Figure 5.19). Three additional volcanoes of major proportions are seen in the same vicinity, indicating a region on the planet which has been active more recently than the cratered portions. If the lava-type material which poured forth to produce these volcanoes was similar to that on earth, large amounts of water would have been released into the Martian atmosphere. Thus, these volcanoes not only may reveal something of the molten nature of at least a portion of the planet but may also have a bearing on the question of life on Mars. They also raise a question of geology. Volcanic action on earth is often associated with the colliding of crustal plates, so it is reasonable to ask if this planet also has crustal plates which move. Figure 5.20 shows a map of Mars, based on Mariner 9 observations.

A second major surface feature is the huge valley which stretches eastward from the volcanic region for a distance of 4000 km, averaging 100 km wide and having a depth of over 6 km in several places. Such a valley, if superimposed upon the earth, would stretch across the entire United States. The discovery of this Mariner Valley, together with the

Figure 5.17 *Mariner 7 photographs of Mars, frames 73 and 74, showing the bright southern cap and the circle of Nix Olympica. The two frames were taken 47 min apart, showing the rotations of the planet. (JPL—NASA)*

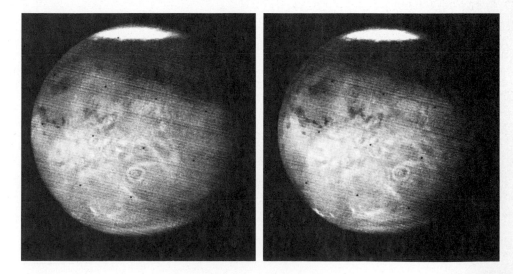

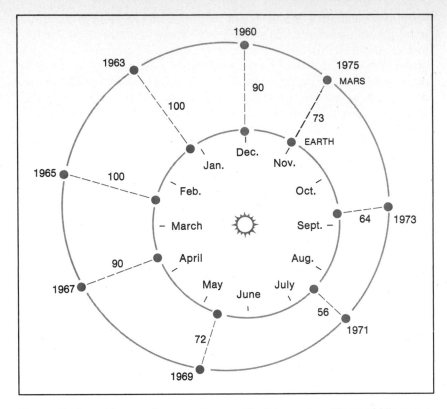

Figure 5.18 *Mars-earth oppositions, with distance in millions of kilometers shown by numbers next to the dashed lines.*

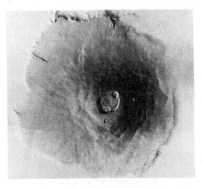

Figure 5.19 *Mariner 9 photo of Nix Olympica, the Martian volcanic mountain: longitude 133°, latitude + 18°. (JPL–NASA)*

details revealed by close-up views (Figures 5.21 and 5.22), has raised many questions about the process of its formation. The general form of the chasm suggests subsidence of the Martian soil due to faulting or plate tectonic motion. Note the fingerlike pit chains in the right-hand portion of Figure 5.22. These probably represent the beginning stages of a larger valley that will develop under continued subsidence. The scalloped edges associated with deeper portions of the valley suggest the enlargement of such pits. On the other hand, the tributarylike complex shown in the lower left-hand portion of Figure 5.21 seems to suggest erosion due to fluid action, and this immediately raises the question of the existence of water on Mars. It is well established that liquid water cannot persist on the surface of the planet under its present lack of atmosphere. The atmospheric pressure on the surface of Mars is less than one-hundredth of that on earth. Under such low pressure, surface water would vaporize in all but the coldest parts of the planet. For liquid water to have persisted long enough to erode a canyon or produce a river bed, a much denser atmosphere must have once been present.

An alternative to liquid water on the surface of the planet is water in a frozen state beneath the surface. The subsidence of soil may have exposed this frozen water, resulting in evaporation and/or seepage, which in turn loosened material to produce the tributarylike structure. Some geologists believe that the side canyons of the Grand Canyon of Arizona were formed as a result of seepage of ground water.

A more convincing example of water erosion on Mars is shown in Figure 5.23, the Amazonis channel. Whereas the bottom of the great chasm does not fall off in one direction, the Amazonis channel does appear to grade downward to allow a flow from south to north. Furthermore the "bars" and braiding of the channel in its northerly portion are very typical of river beds on earth. The entire Amazonis channel is 350 km long and 100 km wide (at its widest), thus requiring a river the size of the Amazon (after which it was named) to produce it.

In the summer of 1975 the United States launched one of its most ambitious and suspenseful space efforts: the search for life on an alien planet. Two identical probes, Viking 1 and Viking 2, went into orbit around Mars after traveling just under one year to reach the planet. The first order of business was to search for suitable landing sites for their respective landers. In that process Viking 1 orbiter photographed large portions of the planet in sharper detail than ever before. Figure 5.24 (a) illustrates this point by showing a river sand bar in the region of Chryse Basin, soon to be the home of the Viking 1 lander. The crater at the head of the teardrop erosion pattern had resisted the onslaught of fluid, thought by most observers to have been water. Also note the flow pattern of material expelled from the crater called Yuty [Figure 5.24 (b)] and the Martian cyclone at latitude 65°N [Figure 5.24 (c)]. This view also reveals a frost-filled crater to the north and patches of the polar cap.

On July 20, 1976, the Viking 1 lander (see Figure 5.25) was released from its Orbiter and descended to Chryse Basin, with the Viking 2 lander following suit on September 3, landing in the Utopia Plain on the opposite side of the planet and farther toward the north polar region, where it would experience the extreme cold of a Martian winter a year later (see Figure 5.20 for these locations). Because the primary purpose of the probes was to search for life and because life, as we know it, is so

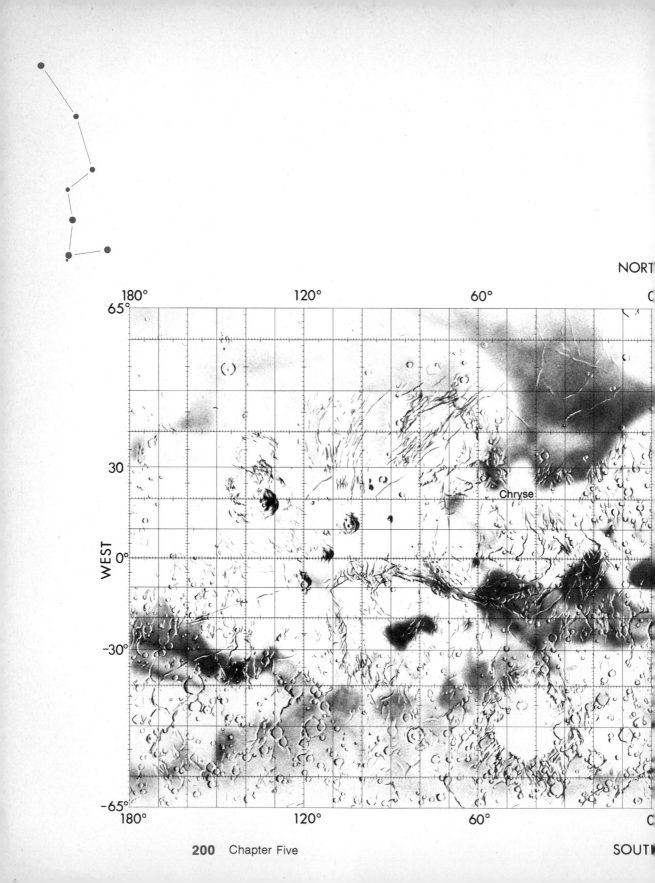

180°   120°   60°   0

65°

30

WEST

0°

-30°

-65°

180°   120°   60°   0

Chryse

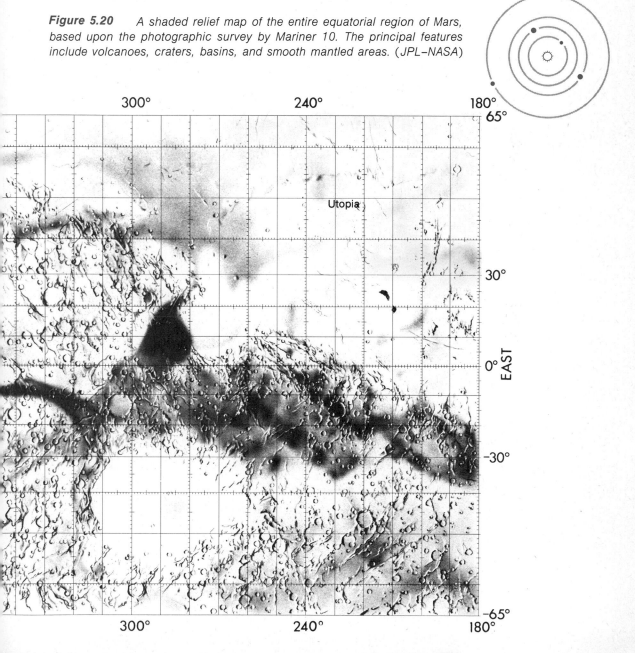

Figure 5.20   *A shaded relief map of the entire equatorial region of Mars, based upon the photographic survey by Mariner 10. The principal features include volcanoes, craters, basins, and smooth mantled areas. (JPL–NASA)*

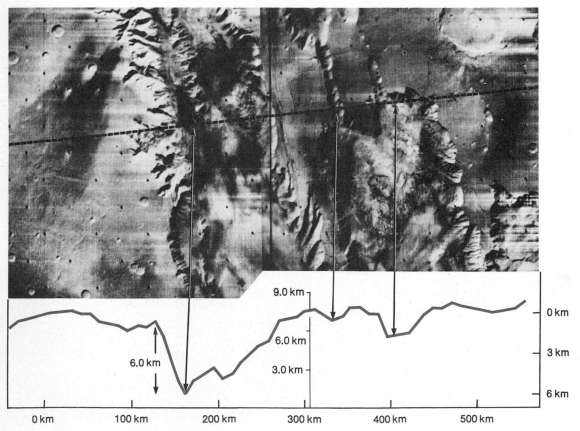

Figure 5.21   The Mariner Valley, as seen by Viking Orbiter 1 from a distance of 4200 km. The principal canyon crosses the bottom half of the picture and shows landslide features probably associated with Mars quakes. (JPL–NASA)

Figure 5.22   A close-up and profile of a portion of the Coprates region centered on longitude 75°, latitude −10°. (JPL–NASA)

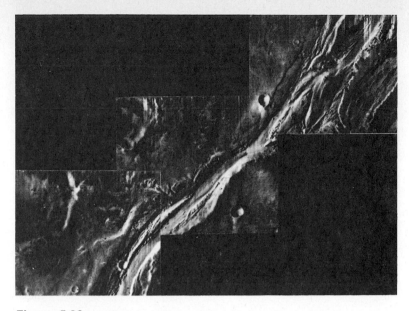

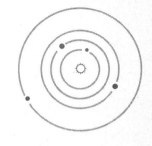

Figure 5.23 *The Amazonis channel, thought to have been formed by running water. The section shown is approximately 100 km long, and the flow direction is thought to be toward the north (upper right). (JPL–NASA)*

Figure 5.24 *(a) Braided channels record the flow of water in the Chryse basin, near the Viking 1 landing site. The shore of the channel is shown to the far right. (b) The crater Yuty, showing a flow pattern in the material that was ejected. (c) A Martian cyclone, as seen by Viking Orbiter 1, near the northern polar region. A frost-filled crater and patches of frost associated with the northern polar region can also be seen. (JPL–NASA)*

(a)　　　　　　　　　　　(b)　　　　　　　　　　　(c)

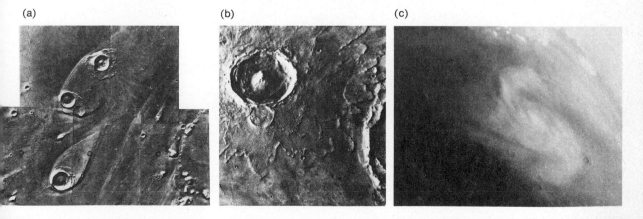

Figure 5.25 *The Viking lander. (NASA)*

closely linked with water, these two sites were natural choices, the first in a basin where sedimentation would likely contain fossil remains of life and the second in a plain where atmospheric water concentrations appeared higher. A secondary, but necessary constraint was a rather large, smooth area in which to land. We will be discussing the life search program; however, let's look first at the many other aspects of Mars these landers saw.

Mars has always appeared reddish in the night sky, whether seen by naked eye or telescope. Now the true color could be recorded at close range: it was found to be on the yellowish-brown side, with the sky a paler shade of this same color (see Color Plate 4). The color is probably due to iron oxides (rust) on the surface and dust suspended in the atmosphere. Mars is known to have high-velocity winds (exceeding 200 km/hr at times) which cause the dust to be held suspended in the atmosphere to an elevation of 40 km. Because the density of the Martian atmosphere is only one-hundredth that of earth's atmosphere, the wind would be rather ineffective in moving anything but dust particles. Wind is only a minor erosion factor on Mars. It very likely took a long time for the drifts shown in Figure 5.26 to be deposited in that form.

The atmospheric composition is also of critical interest, because it contains every gaseous component necessary for life as we know it (see Table 5.3). Although the abundances of surface material cannot yet be accurately determined, X-ray spectrometers on board the Viking landers have identified the following constituents: iron, silicon, calcium, sulfur, aluminum, iron oxides, titanium, rubidium, strontium, and zirconium.

Figure 5.26 *A spectacular view of the Martian landscape, as seen by Viking 1 lander, shows sharp dune crests, which indicate recent wind storms. The small deposits downwind of rocks also indicate wind direction. The large boulder at left measures about 1 m by 3 m. The sun rose two hours before this photograph was taken. (JPL–NASA)*

Table 5.3   *Composition of the Martian Atmosphere*

Carbon dioxide (CO_2)	95%
Molecular nitrogen (N_2)	2 to 3%
Argon (Ar)	1 to 2%
Molecular oxygen (O_2)	0.3 to 0.4%
Xenon (Xe)	Trace
Krypton (Kr)	Trace
Water vapor (near poles)	Trace

Rock types may be identified as ranging from smooth, fine-grain types to very coarse chunks of basalt. Many of the latter are characterized by numerous holes, which may have been created by bubbles of gas associated with their molten origin (see Figure 5.27). An assessment of the fine material on the surface of Mars would be almost 80% iron-rich clay.

Figure 5.27   *The angular blocks in the foreground of this view are only about 4 meters from Viking 1 lander, and they range in size from a few centimeters to a few meters across. (JPL–NASA)*

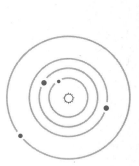

Both probes contained seismometers by which Marsquakes could be detected; however, one detector failed, and so the records of several quakes are a bit sketchy. In November of 1976 a vibration was recorded that had all the characteristics of a quake of magnitude 6 on the Richter scale. Without the additional information that the inoperative seismometer would have provided it was impossible to pinpoint the source of the quake and to determine the interior of the planet from it. However, we may surmise from the many features which show Mars to be active geologically (the volcanic mountains, the Mariner Valley, the Marsquakes, etc.) that the planet must still have a source of internal heat and most likely is differentiated with a denser core.

We mentioned earlier the pattern of changing seasons caused by Mars's tilted axis and its revolution around the sun. Polar caps form in the north when the planet is tipped away from the sun and disappear as spring returns, and for many years scientists have surmised that these caps are simply CO_2 frost (dry ice), which forms when the temperature drops to $-123°C$ ($-190°F$) and sublimates (turns back to vapor) at any warmer temperature. All of this has been confirmed; however, a startling finding has come from Viking 2, situated nearer the north polar region. It is the existence of layers of water ice—permanent polar caps which lie beneath the changing frost caps (see Figure 5.28). The alternate layers of dust show periods when the deposition of ice waned, representing several major periodic changes in climate. The upper dust covering tends to hide the ice caps from earth-based observatories; however, their terraced

Figure 5.28   *Mars's northern polar ice cap as seen by Viking Orbiter 2. The alternate layers of ice and Martian soil are particularly evident along the slopes that are exposed in the upper left portion of this figure. (JPL–NASA)*

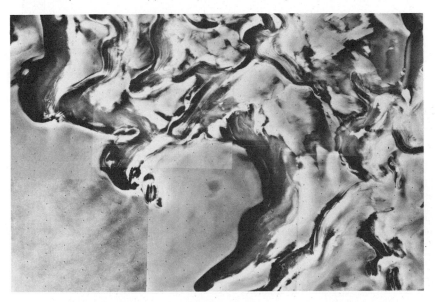

structure is shown quite clearly in maps based on Mariner 9 imaging (see Figure 5.29). Some observers believe that sufficient water is locked in these polar caps to cover the entire planet to a depth of 0.5 to 1 meter, if melted. Water may also exist as permafrost in the soil. Temperatures in the polar regions range down to $-73°C$ ($-100°F$) in the summer season and down to $-125°C$ ($-194°F$) in the winter season.

What is it that motivates humans to search for life elsewhere in the universe, and what basic assumptions do they make about the nature of life? Life as we know it is based on the element carbon. Almost an endless chain of organic molecules can be formed with carbon as their primary bonding agent. Since no other element can play the same role as efficiently as carbon, scientists assume that life elsewhere in the universe would likely be based on carbon. Furthermore, organic (carbon-based) molecules are found in meteorites and in gases between the stars, thus strengthening this view. Out of these considerations another fundamental question arises: is life something that is very likely to occur (in many places in the universe) or is it so unlikely as to occur only on earth? What we have found on Mars should help us answer that question.

Two of the four experiments on Mars (gas-exchange and labeled-release) depended upon the fact that gases are always given off as a by-product of the life process, whether with plants or animals (including

Figure 5.29   *The polar regions of Mars: (a) northern region, October 12, 1972; (b) southern region, February 28, 1972. Note the distinctively different pattern of cratering in the two regions. (JPL–NASA)*

(a)   (b)

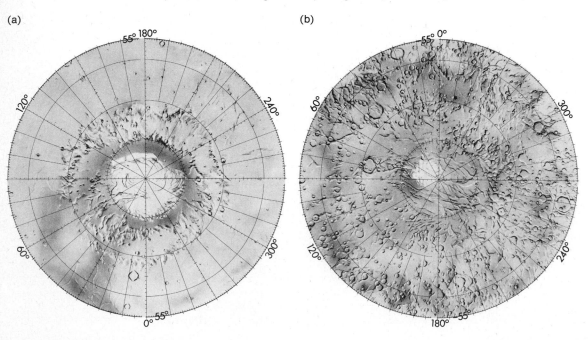

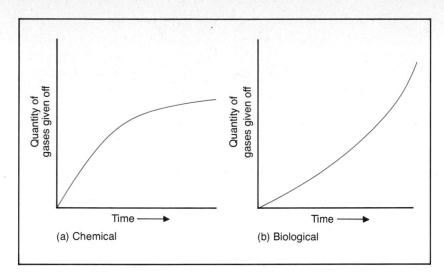

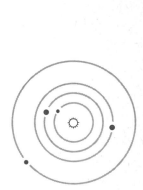

Figure 5.30   *Reaction curves.*

microorganisms). In each case a nutrient solution was added to the Martian soil sample, whereupon a very significant release of gas occurred immediately and then the release rate slowly diminished.While the first reaction was very positive, the conclusion drawn by many observers is that a chemical reaction took place rather than a biological reaction. A typical graph for each type of reaction is presented in Figure 5.30, and the graph associated with the first two Viking experiments resembles that of the chemical reaction. A third experiment (pyrolytic-release) depends upon a living organism's assimilating carbon-14 as a part of its life process. After proper incubation, the soil sample is pyrolyzed (vaporized) and carbon-14 products are sought by a radioactive detector. Seven of the nine samples tested indicated positive results; however, because some of the samples had been heated to as high as 175°C and still produced positive results, it is assumed that living microorganisms would not have survived such heat and therefore the reaction must be chemical.

While the results of these first three experiments appear negative with respect to life on Mars, we should hasten to say that some scientists give a more positive interpretation. They say that the experiments neither proved nor disproved the existence of life on Mars. A heavier blow was delivered, however, by the results of the final experiment—the GCMS (gas chromatograph mass spectrometer) experiment. Organic molecules are generally large and heavy as molecules go. If the soil sample is pyrolyzed and passed through a gas chromatograph, the molecules will separate according to their weight, and they can be identified by the mass spectrometer. No organic molecules were found. One would have thought that organic molecules derived from meteorite impact would have been found. Perhaps they had been destroyed by the ultraviolet radiation of the sun (Figure 5.31).

Figure 5.31   In the search for life, the Viking soil sampler arm was directed to move a rock and sample soil from beneath its original position, with the thought in mind that a living form might have sought refuge there from the sun's ultraviolet radiation. (JPL–NASA)

Our search for life on Mars may not have ended, for the approaches a new exploration might take are being discussed. A rover vehicle might cover a significant amount of surface, or a rocket vehicle might return samples to earth, or perhaps a mission might land astronauts to explore the surface.

We have known since 1877 that Mars has two natural satellites, Phobos and Deimos, but not until Mariner 9 turned its video eye their way did we have any idea of their surface appearance [Figure 5.32 (abc)]. Devoid of an atmosphere, the moons are subject to impact continually, and the Viking orbiter portrait of Phobos [Figure 5.32(b)] reveals this fact very clearly. Note the *striations* (grooves) on the moon's surface, some passing through older craters and others being interrupted by yet younger craters. The history of bombardment, and of shrinkage perhaps, of this moon can be read in its surface. Phobos measures 27 km in its greatest dimension and Deimos only 12 km.

Jupiter

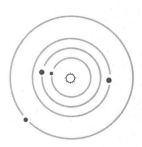

As we begin our study of the Jovian planets, let us recall the vastly different circumstance under which they were formed. Whereas the terrestrials originated close to the sun, in a relatively hot environment, the Jovians, owing to their great distances from the sun, came into being in a very cold environment. As a consequence, the Jovians have retained their original composition, representing essentially the composition of the original nebula from which they were formed. As the Pioneer 10 spacecraft passed Jupiter at a distance of 130,000 km on December 3, 1973 (Figure 5.33), its spectrometers confirmed for the first time the presence of helium (about 15 percent by volume) in the Jovian atmosphere. The main constituent is hydrogen (almost 85 percent by volume), with only traces of methane (CH_4), ammonia (NH_3), and perhaps other elements and compounds. The relatively high percentage of helium represents a fairly close correlation with the estimates of helium produced in the very early stage of the universe itself and with estimates of the helium content of young stars. In fact, some observers have suggested that Jupiter may be an aborted dwarf star, for it radiates approximately twice as much thermal energy as it absorbs from the sun. What is the source of this surplus energy? Could there be a starlike process going on in the core of

Figure 5.32   (a) Phobos, as seen by Viking Orbiter 1 from a distance of 480 km. Note the numerous chains of craters on the surface. (b) The most detailed photograph ever obtained of the moon Phobos. Note the numerous striations on the surface. The origin of these straight-line features is not known. (c) Deimos, as seen by Viking Orbiter 1 from a range of 3300 km. The largest crater measures 1.3 km. (JPL–NASA)

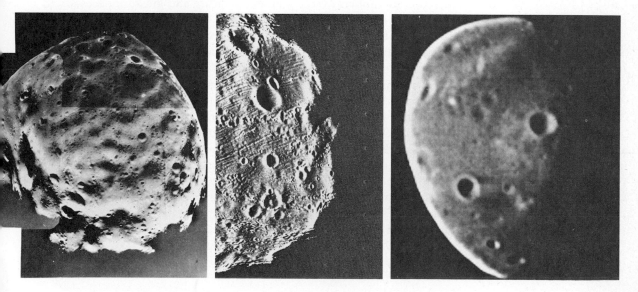

Figure 5.33   *This mosaic of Jupiter was assembled from nine individual photographs taken by Voyager 1. The Great Red Spot is approximately 35,000 km long, whereas the smallest features visible in this view are approximately 140 km across. (JPL–NASA)*

Jupiter? No, for even with its great mass, 318 times that of the earth, Jupiter could not create the high temperature and pressure necessary to initiate the thermonuclear process characteristic of a star. However, it may generate thermal energy (heat) through continuing contraction. Whenever a body contracts, some of its gravitational energy is converted to thermal energy. In the case of Jupiter, a contraction of only 1 cm per year is sufficient to release enough energy to explain this added thermal radiation.

While the Pioneer 10 spacecraft was still 12 million kilometers from Jupiter, its sensors detected that the magnetic field of the planet was strong enough to deflect the solar wind (the outflow of charged particles from the sun). As the craft made its pass through the belts of high-energy electrons and protons entrapped in the magnetic field, its instruments were saturated almost to their limit with this radiation—more than 100 times what would be lethal to humans. The integrated picture of Jupiter's magnetic field that emerges shows a total magnetic energy 400 million times that of the earth. The field is tilted 10° with respect to the planet's axis of rotation and is highly flattened and dimpled in toward the poles (Figure 5.34).

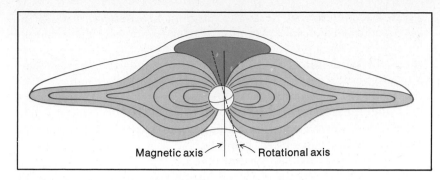

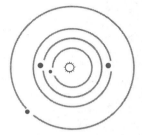

Figure 5.34   *The magnetic field of Jupiter as determined by the Pioneer 10 fly-by.*

We know by direct observation that Jupiter is enveloped by a thick layer of dense clouds, which perpetually obscure any hint of the planet's interior. Scientists, however, can derive a model of the interior by simply considering factors such as composition and the influence of gravity in creating pressure and temperature variations with depth. The electronic computer assists the astronomer in developing such a model (Figure 5.35). The portion we could call gaseous (the atmosphere) occupies a relatively thin layer only 1000 km deep. In the next major layer, hydrogen has been converted to a liquid state, primarily by very high pressures. In still a deeper major layer, owing to ever-increasing pressures, hydrogen is converted to a liquid metallic state; the protons of the atoms are stripped of their electrons, creating a fluid which is a good conductor of electricity. Within this layer may occur electric currents capable of creating Jupiter's strong magnetic field. Finally, we speculate that further increases of pressure (to the order of 100 million times the earth's atmospheric pressure) create a relatively small central core, perhaps twice the diameter of the earth. (Jupiter overall is eleven times the earth in diameter.) Core temperature is estimated to be of the order of 50,000°F.

In March and July 1979, Voyager 1 and 2 respectively flew by Jupiter to produce the most revealing picture of the planet ever made and to study four of its moons in close detail. Atmospheric turbulence exceeded the scientists expectation, as did the clarity with which motions in that atmosphere were revealed. But let's start with the most obvious division of the atmosphere, that of the banding effect of the bright *zones*, thought to be higher elevations, and the alternate darker *belts*, thought to be regions wherein our view reaches to a lower level (Figure 5.36). The bright zones are likely composed of a cold strata of ammonia crystals, whereas the darker belts of reddish-brown hews may consist of ammonium hydrosulfide and various polymers. This basic structure is thought to be a result of Jupiter's very short rotation period (approximately 9 hr 50 min). A point in the equatorial region must move at about 33,000 km/hr due to rotation alone. Recalling the Coriolis effect, which causes wind and ocean currents on earth to circulate in a specific direction (counterclockwise in the northern hemisphere), we can see that effect to a high degree on Jupiter. Many astronomers believe that the Coriolis effect has stretched out cloud motions that would normally be north and south into eastwest bands. The clouds also have a very turbulent vertical motion, owing to the heat the planet radiates. The combined effect is that of a cyclone which surrounds the entire planet. The higher zonal regions can be thought of as continually dumping their material back into the belts. The belts and zones have different velocities, such that certain portions lag behind.

    Superimposed upon this pattern are a multitude of "swirls" and jet streams which have speeds of several hundred km/hr plus or minus the rotational velocity of the planet (Figure 5.37). A major variation within the banded structure is the Great Red Spot (see Color Plate 5). The Voyager time lapse photographs of the largest "hurricane" in the solar system (13,000 × 19,000 km) leave no question as to the counterclockwise rotation of this feature. Bright white regions were seen to be

Figure 5.35   *A model of Jupiter's interior. The planet is mainly liquid hydrogen. (NASA)*

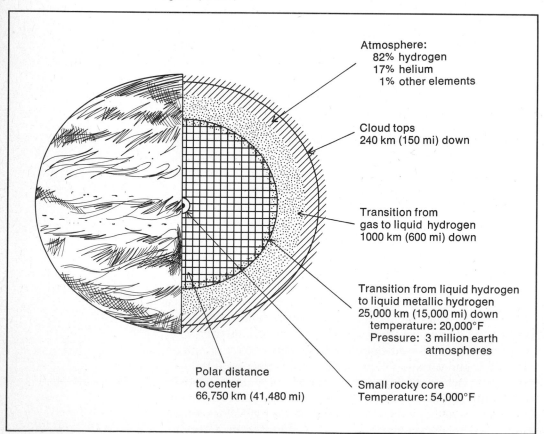

Atmosphere:
82% hydrogen
17% helium
1% other elements

Cloud tops
240 km (150 mi) down

Transition from
gas to liquid hydrogen
1000 km (600 mi) down

Transition from liquid hydrogen
to liquid metallic hydrogen
25,000 km (15,000 mi) down
temperature: 20,000°F
Pressure: 3 million earth
atmospheres

Polar distance
to center
66,750 km (41,480 mi)

Small rocky core
Temperature: 54,000°F

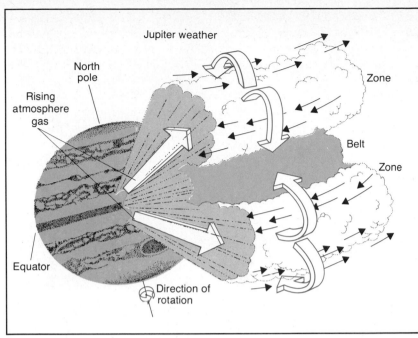

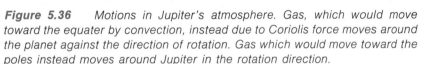

Figure 5.36 *Motions in Jupiter's atmosphere. Gas, which would move toward the equater by convection, instead due to Coriolis force moves around the planet against the direction of rotation. Gas which would move toward the poles instead moves around Jupiter in the rotation direction.*

drawn into the huge vortex, swirled around, stretched and deformed, only to be ejected and reformed into another white region. The temperature of the Great Red Spot is measurably cooler than its surroundings and therefore is presumed to be of a slightly higher elevation.

The 13 natural satellites of Jupiter fall into three very obvious categories. If you look at the listing of these moons in Appendix 7, you will notice that the first five are all within 2 million kilometers of the planet, and they orbit in direct motion very near the equatorial plane of the planet. The next four orbit in direct motion at a distance of approximately 12 million kilometers and have inclinations of 25° to 30°. The last four moons orbit in a retrograde direction at distances of 21 to 24 million kilometers and have highly inclined orbits. This suggests that the first five moons were formed from the same cloud of gas and dust as was Jupiter, their direction and plane of revolution reflecting the rotation of the original nebula. The remaining moons were probably captured, and their motions reflect the conditions under which this capture occurred. The four larger moons, called the *Galilean satellites* (after Galileo, who first observed them with the aid of his early telescope), can be easily viewed by the amateur with only a very modest telescope.

Pioneer 10 sensed an ionosphere around Io and traces of sodium on its surface. This discovery was followed by a ground-based observation of

Figure 5.37   The Great Red Spot of Jupiter, together with one of the white ovals that can be seen in earth-based telescopes. The details of turbulance to the left of these two features had never been seen prior to these Voyager 1 views. (JPL–NASA)

Figure 5.38   The sodium cloud surrounding Io and preceding it in orbit around Jupiter. (JPL–NASA)

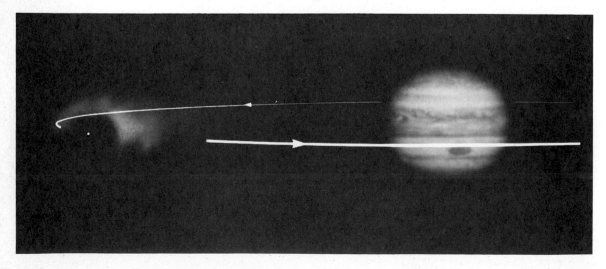

a sodium cloud which surrounds Io (see Figure 5.38). In fact the sodium cloud may fill the entire orbit of Io. This orbit exists within the radiation zone of Jupiter; hence the surface of Io is continually bombarded by high-energy particles. Because the surface of Io shows deposits of sodium and potassium (and perhaps other salts), this bombardment will continually release these elements to the region of Io's orbit. Without that continual release, the atoms would soon be lost to space. This speculation has been dramatically confirmed by Voyager 1, for Io reveals by its color the presence of numerous salts. Its splotchy appearance, including areas of white, yellow, orange, and red color, resembles a glorious pizza, baking and bubbling in a hot oven. See Color Plate 6(a) and 7(b). Io is literally bathed in the radiation of Jupiter, causing salts on its surface to sputter off into space. A flux tube connects Jupiter to Io, allowing an electrical current of a million amps to flow between the two objects. More remarkable still is the discovery of Io's active "geology," including scarps, canyons, irregular depressions, and even active volcanoes spewing out gas and dust at velocities of 1 km/sec (Figure 5.39). Measurements of Io's density (3.5 g/cm^3) confirm its rocky consistency, reflecting about 60 percent of the light that falls on it.

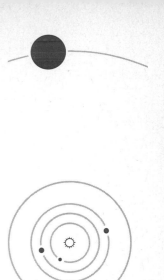

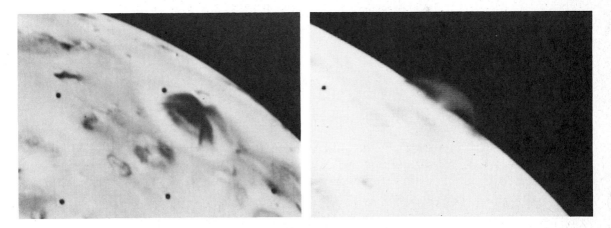

Figure 5.39   *Two active volcanoes on the moon Io as seen by Voyager 1. The plumelike structures rise more than 100 km above the surface. (JPL–NASA)*

Europa, second of the four large Galilean satellites, also appears to be of rocky composition with a density of 3.0 g/cm^3. Europa differs from Io in that its surface appears covered by a layer of ice up to 100 km thick. A faint network of crisscrossing lines, up to 130 km wide and thousands of kilometers long, have been seen. The albedo of this moon is also about 60 percent. See Color Plate 6(b).

Next in order is Ganymede, a dimmer brownish sphere with markings that remind one of maria on the moon. Voyager showed clear evidence of craters on this moon, some of which are brightly rayed. See Color Plate 6(c). The density of this moon is only

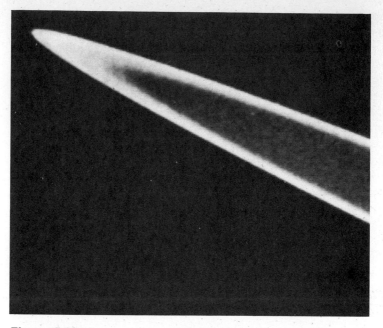

Figure 5.40 *This striking view of Jupiter's ring was recorded by Voyager 2 on July 10, 1979. (JPL–NASA)*

1.9 g/cm³, indicating quite a different composition from that of Io and Europa. While it may have a rocky core, it is probably covered by water and/or water ice. Its darker surface reflects approximately 50 percent of the light that falls on it.

Last of the four Galilean satellites is Callisto, least dense of all, 1.8 g/cm³. It must be composed largely of water ice with little or no rocky substance. In Color Plate 6(d), one may detect a huge 2600-km basin surrounded by concentric rings. This feature may have been created by a huge impact on the icy surface. Its dark surface is not fully understood, yet it has an albedo of 60 percent.

In addition to its close approach to the four Galilean satellites, Voyager 1 photographed the tiny innermost moon, Amalthea. It appears reddish in color and highly elongated. Most unexpected was the discovery of a very thin ring around Jupiter at a distance of 57,000 km above its visible surface (Figure 5.40).

As Pioneer 11 passed Jupiter, the gravitational field of this planet produced a slingshot effect on the spacecraft, changing its trajectory by almost 90° and accelerating it on a new route that brought it near the planet Saturn in 1979.

Saturn

The most beautiful of all the planets, Saturn is second in size and mass and has a distinctive system of rings (Figure 5.41). In December 1972 the first radar echo was received from Saturn by means of the 64-m Goldstone

Figure 5.41 *Saturn, as photographed with the 2.6-m Mount Wilson telescope. (Hale Observatories)*

antenna. Richard Goldstein and George Morris pointed a 400-kilowatt beam toward the planet and after $2\frac{1}{4}$ hr the echo returned, revealing a much more highly reflective surface than had been expected. The signal was not reflected effectively by the planet proper, confirming its gaseous state; the signal was reflected with about 60 percent efficiency from the rings, as compared to the 1.5 percent efficiency of Venus. Whereas the rings were previously thought to be ice crystals, they are now interpreted as being ragged icy chunks at least 1 m in diameter or icy spheres at least a few centimeters across. Ice is a particularly good reflector. When we view the rings of Saturn in a telescope, they appear as solid disks, but different parts of the rings have vastly different Doppler shifts, thus confirming that they are composed of many separate bodies, each orbiting the planet in the manner predicted by Kepler's laws. The inner particles orbit the planet in less than 4 hr; the outer ones require 14 hr. It is known that the rings lie entirely within a certain distance from the planet called the *Roche limit*. Typically, within this radius gravitational tidal forces would prevent the formation of a moon. On the other hand, if a moon wandered into this region, it would be torn apart by these same gravitational tidal forces.

Ground-based observations suggest four subdivisions of the rings: the *subcrepe ring* (closest to the planet), the *crepe ring*, the *bright ring*, and the *gray ring* (Figure 5.42). Between the bright and gray rings a gap exists, called Cassini's division, within which few particles move. The rings may appear bright or dim, depending upon such factors as the density to which the ring is populated with particles, the size, composition, and surface of these particles, and the frostlike covering.

On September 1, 1979, Pioneer 11 passed Saturn just beyond its gray ring and returned to waiting observers more information about this planet than had been gathered in more than 350 years since Galileo first recognized its beautiful ringed form. The spacecraft actually passed through one of three newly discovered but sparcely populated outer rings which lie beyond those shown in Figure 5.42. The following picture of Saturn emerges from the data collected. Saturn has a magnetic field of approximately the same strength as that of earth—weaker than had been expected. Furthermore the magnetic field is not tilted with respect to the axis of rotation, as is true of most planets. The north magnetic pole of Saturn is in what we would term the southern hemisphere—a case of reversed polarity. A model of the planet's interior is really quite similar to that of Jupiter (see Figure 5.35). Its rocky inner core may be the size of the earth; however its overall density is only one half that of Jupiter. Saturn is flattened, due to rotation, so that its polar diameter is about 12,000 km less than its equatorial diameter. Saturn is composed primarily of hydrogen and helium, with traces of methane and ammonia found in

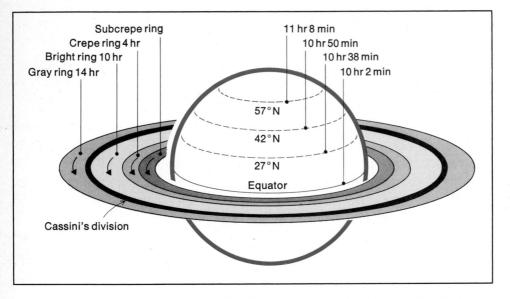

Figure 5.42 *The rotation periods of the clouds of Saturn at various latitudes and of selected particles in the rings.*

its atmosphere. Like Jupiter, Saturn radiates two and a half times as much thermal energy as it receives from the sun.

Careful scrutiny of the body of Saturn will reveal a banded appearance, somewhat like that of Jupiter yet without the high level of contrast characteristic of that planet. A high altitude ammonia haze may tend to wash out details.

Saturn has 10 moons in all, the latest having been discovered in 1966 when the rings were seen edge-on. This moon, called Janus, orbits very near the outer edge of the ring structure and is usually lost in its glare. Twice during the 29.5-year revolution period of Saturn the rings are presented in an edge-on view to earth observers; because the rings are so thin, they seem to disappear except in the larger telescopes. When this occurred in 1966, astronomers identified Janus. They had previously suspected its existence after noting perturbations (disturbances) that created a small gap in the rings. Several of Saturn's moons create gaps, although only Cassini's division is readily visible in a small telescope. (The ability of an observer to identify Cassini's division using a 7.5-cm refractor is a good test of its resolving power and therefore of its quality.)

Largest moon in the solar system, Saturn's Titan created a great deal of interest because it is known to have an atmosphere whose density is somewhere between that of Mars and earth—an atmosphere that includes methane. Scientists speculate that amino acids, precursors of life, may have been formed in Titan's atmosphere. While the observations of Pioneer 11 were not definitive in regard to this possibility, it certainly opens the way for further exploration of both Saturn and Titan by Voyagers 1 and 2 in 1980 and 1981 respectively. Voyager 2 also has the potential for a pass near Uranus in 1986. Such an ideal alignment of these planets will not occur for another 45 years. The Voyager 2 cameras will provide pictures of better resolution than those of any previous probe. This probe is designed to perform tasks automatically rather than depending upon ground-based commands. One can appreciate the necessity for this approach by considering that the time delay in communicating with a probe near Uranus is almost 6 hr—3 hr to get there and 3 hr for a responding signal to return.

The Voyagers carry a very interesting cargo in addition to all their scientific instruments—a record upon which photographs have been electronically recorded depicting our person, our science, and our place in the universe. It includes greetings in 60 languages and presents the music of many cultures. The cartridge and needle by which it may be played are also enclosed. As the Voyager craft leaves the solar system, this message may reach beings of another star system, perhaps on their way to finding us.

Uranus

Uranus is the first planet of our study that was not known to the ancients. It is not usually considered to be visible to the naked eye; however, under ideal conditions it may be seen quite easily using only binoculars. Uranus was discovered almost by accident. While it had been seen and its position charted as early as 1690, it was only in 1781 that Sir William Herschel recognized it as a planet; he also discovered two of its larger moons in 1787. Herschel, who was appointed court astronomer to King George III in 1782, had one of the best telescopes of his time, which made his accomplishment possible (Figure 5.43). Other observers had termed the object a star, but he was able to recognize its disklike appearance (Figure 5.44).

Uranus orbits the sun at an average distance just under 20 A.U. and in a period of 84 earth years. If a human could spend an entire lifetime

Figure 5.43   *Sir William Herschel's 1.25-m reflecting telescope, which he built himself. (Yerkes Observatory)*

Figure 5.44   *Uranus, with three moons. (Lick Observatory)*

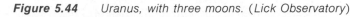

on Uranus, he or she would die at the age of only one Uranus year. One of the most distinctive characteristics of this planet is the tilt of its axis, which is 98°. If we imagine a planet whose axis is not tilted at all, we can say that its equator is in the same plane as its orbit. Now imagine that this same planet rotates in a counterclockwise direction, a point on its equator traveling in an eastward direction. Tilt the planet's axis of rotation by 90°; the axis now lies in the plane of its orbit. Continue to tilt the axis of the planet 8° more, making a total of 98°. In what direction will it now be rotating? To answer this question, use your finger to represent the planet rotating in a counterclockwise direction with no tilt at all, then gradually tilt your finger until you have reached the 98° position. You will see that your finger is now rotating in a clockwise direction; likewise the rotation of Uranus is clockwise (retrograde).

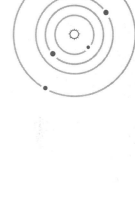

Uranus has five moons, all of which orbit in the same plane as the planet's equator and in the same direction as the planet's rotation. Because of the planet's extreme tilt, the plane in which its moons are seen circling it is sometimes perpendicular to and at other times aligned with our line of sight.

On March 10, 1977, a most unexpected observation occurred. A group of observers flying in NASA's Kuiper Airborne Observatory were preparing to witness Uranus as it passed in front of a faint star. This is called an occultation observation, and its value lies in the fact that as a planet passes in front of a star, details concerning the planet's atmosphere, surface profile, and size may be determined. This is especially true if simultaneous observations are made from different locations and the results combined.

As the group was observing Uranus, and before the occultation began, there were several dips in the intensity of the light coming from the star. The first was thought to be a momentary malfunction of the equipment, but by the time five dips had been recorded they were no longer attributable to malfunctions. The planet then occulted the star for 25 minutes and, as Uranus passed on by, five distinct dips again occurred. The comparative time spacing of the dips on either side of the planet correlated very well. The interpretation that emerged is that Uranus has at least five rings. Because each dip in light output lasted only for one to four seconds, the rings are interpreted as being very thin (see Figure 5.45). Additional observations have indicated that eight or nine thin rings may exist. It is questionable that these rings will ever be seen by ground-based observations; however, the space telescope, described in Chapter 2, and/or the Voyager 2 probe, programmed to pass Uranus in 1986, should provide confirmation. Certainly this new discovery makes the Voyager fly-by even more enticing.

Neptune

The discovery of Uranus stimulated observation of that planet and its orbital motion. If Uranus were the only planet moving around the sun,

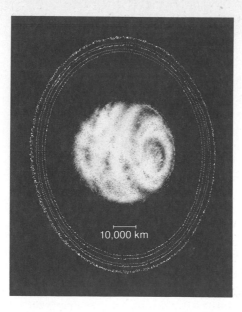

Figure 5.45
The rings of Uranus.

its orbit would consist of an ellipse along which it moved so as to sweep out equal areas in equal times, just as Kepler's laws predict. However, Uranus is not alone in the solar system, and its motion is perturbed by every other planet. Observationally, the algebraic sum of these perturbations shows up as deviations in Uranus' position. Not only is this what one would expect, but the total (net) deviation can be predicted by computing the individual deviations using Newton's law of gravity.

When the perturbations of all known planets had been computed and other necessary corrections made, Uranus still deviated from the predicted position by an amount large enough to raise the suspicion that another perturbing object must be present. This was the conclusion of two astronomers, John Adams of England and Joseph Leverrier of France. Without knowledge that the other scientist was even working on the problem, each man came to the same conclusion: a planet existed beyond Uranus which was responsible for its perturbations. Assuming the theory of gravity to hold for all objects, each astronomer had asked, "Where would an object have to be located and what would be its mass, in order to create the unexplained perturbation seen in Uranus?" Based upon the answer to this question, each man had tried to persuade an observational astronomer to look for a new planet. Adams was put off at Cambridge, but Leverrier's request was met by Galle of the Berlin Observatory, and he found the new planet in his first evening of searching, very near the location that both Adams and Leverrier had predicted. The date was September 23, 1846. This discovery represents one of the most significant triumphs in astronomy to that date, further verifying Newton's description of universal gravitation and demonstrating its predictive value.

There is always a temptation to name a planet (or other astronomical objects) for its discoverer and there was a move to honor Leverrier; however, he chose to carry out the theme already begun and named the

planet Neptune, after the mythological god of the ocean—"king of the deep." In less than one month's time, Neptune's larger moon (diameter 4000 km) had been discovered, and it was called Triton after Neptune's mythological son (Figure 5.46). Neptune has two moons, but the unusual feature of Triton is its retrograde motion. Triton is the only moon of significant size that has a retrograde motion in relation to the rotation of the parent planet. All the moons of Uranus orbit in retrograde, but that would be expected, because the planet itself rotates in retrograde.

Using Appendix 6, compare the size, mass, density, rotation, and albedo of Uranus and Neptune. Note their almost twinlike nature.

Pluto

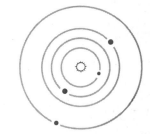

The orbital motions of both Uranus and Neptune were followed with much interest by numerous observers. In the early 1900s Percival Lowell, founder of Lowell Observatory in Flagstaff, Arizona, concluded that Neptune could not account for all the perturbations in the orbit of Uranus. By mathematical calculations he predicted that still another planet might be found in one of two possible locations. In 1930 a planet was found by Clyde Tombaugh, a colleague at the observatory, who had photographed these regions and later analyzed his plates on a device called the *blink microscope*. Pluto was found by this technique within 6° of one of the positions predicted by Lowell (Figure 5.47). Some observers would discredit his calculations, saying that he had too little information with which to work and that therefore the almost perfect coincidence of prediction and discovery was only accidental. In keeping with the

Figure 5.46 *Neptune and its largest moon, Triton. (Lick Observatory)*

tradition, Pluto was named for the mythological god of the under-world—but note the "coincidence" that the first two letters of the name are also the initials of Percival Lowell and that the symbol for Pluto, ♇, is constructed from these initials (see Appendix 5).

Pluto has the most unusual orbit of all the planets. Besides being the most eccentric, it is inclined 17° to the plane of the earth's orbit (the ecliptic). Pluto orbits the sun at an average distance of 40 A.U., or 5.5 billion kilometers, but it travels within less than 30 A. U. at perihelion and out to almost 50 A.U. at aphelion. The planet appeared at aphelion in 1865 and will again appear at this point in 2113. It will appear at perihelion in 1989. The orbit of Pluto is so eccentric as to bring it within the orbit of Neptune during the interval from 1979 to 1998 (Figure 5.48). During this period Neptune will be the farthest known planet from the sun. A collision between Neptune and Pluto is impossible, for while their paths appear to intersect, the planes of orbit do not coincide. Pluto requires almost 248 earth years for one revolution; hence it has not moved very far against the background of stars since its discovery.

Because Pluto appears only as a tiny speck in even the largest telescopes, astronomers have had to guess at its physical properties: diameter, mass, density, and surface composition. Until very recently, many observers have passed Pluto off as being somewhat like the earth's moon, having a hard crust, no atmosphere, low albedo, diameter about half that of earth, and a density of 4 to 5 g/cm³. Recent photometric measurements of Pluto's spectrum in the infrared portion of the spectrum reveal a surface covering of frozen methane at a temperature of 43°K (−230°C). When this condition is duplicated in the laboratory, an

Figure 5.47   *Discovery of Pluto. These plates, taken 6 days apart, show the motion of a new planet among the stars (see arrows): (a) January 23, 1930; (b) January 29, 1930. (Lowell Observatory)*

(a)   (b)

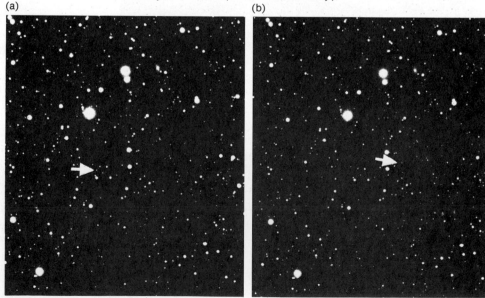

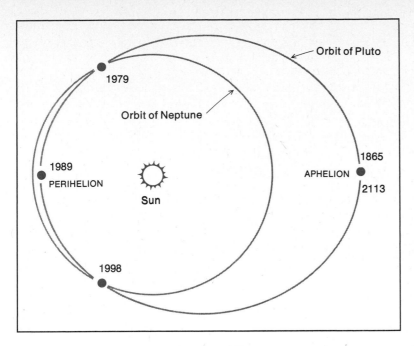

Figure 5.48   *The orbits of Pluto and Neptune.*

albedo of 40 to 60 percent is found. Based upon this finding, astronomers would conclude that Pluto is significantly smaller than previously thought—slightly smaller than our moon. The reasoning here is that, if Pluto is a better reflector than previously thought, it need not be so large in order to be as bright as it appears. If not so large, then the planet is not as massive as thought. Further, if it is characterized by frozen gases, then it is not so dense as thought—perhaps only slightly more dense than water, with a mass of about $\frac{1}{500}$ that of the earth.

You can see from this discussion how many factors are interrelated, one factor being inferred from another. Because we cannot see surface details on Pluto, we cannot determine its rotation directly; however, a study of its periodic changes in brightness suggest one rotation in a period of 6 days 9 hr 17 min.

One of the most surprising discoveries about Pluto is that it has a moon. When first photographed the moon appeared as just a blur and the observer could not be sure of what it was (Figure 5.49). The confirmation came by inspecting plates taken earlier. These revealed the same blur on different sides of the planet, with a revolution period of approximately six days. The moon is called Charon after the mythological ferryman of souls across the river Styx into Hades.

Planet X

Several different objects have been referred to, at least temporarily, as Planet X—the unknown or new planet. In the early nineteenth century

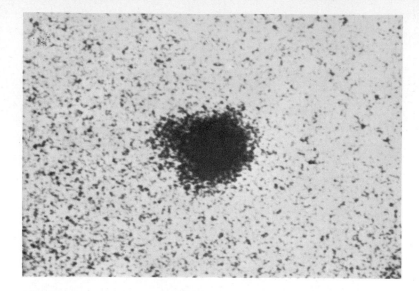

Figure 5.49   *The discovery plate of Charon, moon of Pluto, as indicated by the "bulge" on the upper left side of the planet's image. The discovery of this moon has led to a complete reassessment of the planet's diameter, mass, and density. (U.S. Naval Observatory)*

Joseph Leverrier had predicted the existence of a planet between the sun and Mercury (based upon the observed perturbations of Mercury). He called the planet Vulcan, keeper of the fires of Hades. Vulcan was never found, and the perturbation of Mercury was explained another way.

In 1971 Joseph L. Brady of the University of California did an extensive computer study on the perturbations of Halley's comet and concluded that a planet more massive than Saturn must orbit the sun beyond the orbit of Pluto. He predicted planet X's period of revolution to be 464 years, and he predicted its location in the constellation Cassiopeia. Extensive photographic searches have revealed no such planet, however. It is now thought likely that the irregularities in the flight of Halley's comet are due to nongravitational effects, which will be explained in Chapter 6.

In November of 1977 you may recall reading in the newspaper about still another Planet X. Although its discoverer, Charles Kowal, had not intended his newly found object to be called the tenth planet, the press proclaimed its discovery far and wide. Contrast the speed with which even the casual thought of an astronomer is disseminated today with the almost total lack of communication a century or two earlier. As a group, scientists communicate with each other primarily through the scientific journals. An astronomer reports his research as soon as he has any assurance of its validity, and within a very short time other astronomers are reacting and checking the original researcher's work. This line of communication is vital to the advancement of science. In the case of

Kowal's Object, however, other lines and modes of communication came into play. The newspapers turned his object into "Planet X." Most of them failed to identify the highly speculative nature of his discovery.

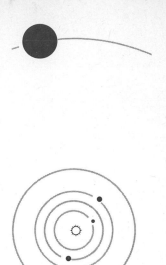

What do we actually know about the Kowal Object, now called Chiron (after the mythological Centaur, son of Saturn)? Once it had been discovered by means of the blink microscope (the instrument used for Pluto's discovery), Charles Kowal and other astronomers began a search on photographic plates which were taken before its discovery. They succeeded in identifying it on numerous plates, like that shown in Figure 5.50. Its flight has been traced backward in time approximately 80 years and in space $1\frac{1}{2}$ revolutions around the sun, its period being 50.68 years. Chiron orbits at an average distance of 13.7 A.U. but has a more eccentric orbit than Pluto (eccentricity = 0.3786). At aphelion, Chiron moves almost as far from the sun as Uranus, but at perihelion it moves within Saturn's orbit.

Nothing we have related thus far would keep Kowal's Object from being a planet; however, the crucial factor is its size. Judging from its faintness on the photograph plates, its size would be that of an asteroid or a comet, 50 to 300 km in diameter. The latest thinking favors the asteroid description. As you will see in Chapter 6, most asteroids orbit between Mars and Jupiter or in highly eccentric orbits nearer the sun. Therefore, an interesting speculation is that Chiron may be a first of another zone of asteroids, or it may simply represent an asteroid which was perturbed by Jupiter into this more distant orbit. Some of the outer moons of Jupiter

Figure 5.50   *The motion of Chiron, as recorded during a 75-min. exposure on the Palomar 48-in. Schmidt telescope.* (*Hale Observatories*)

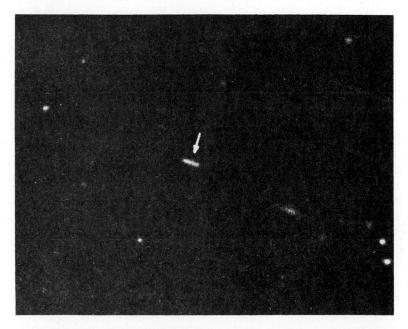

and Saturn, by their motion, suggest capture by these giant planets, and they may have been asteroids as well.

Origin and Evolution of Planets

We opened this chapter with a question about how the solar system came into being. Do the generalizations we have made or the distinguishing characteristics we have considered point to an answer? We have already suggested that the random-capture theory does not appear to fit the orderly revolution of all nine planets in almost the same plane. We are also inclined to eliminate the encounter theory, not because it is inconsistent with planetary motion, but rather because of the low probability that the sun passed close enough to another star to have pulled off material from it. Owing to the great distances between stars, it is estimated that no more than ten such encounters have occurred in the life of the Milky Way galaxy (10 to 15 billion years), and astronomers believe that many solar systems exist in our Galaxy.

By 1950 several variations of the *protoplanet* theory had been suggested. This theory may not be the "last word" on the subject, but in many respects it is more plausible than its forerunners. It suggests that the planets were formed about the same time as the sun and out of the same nebular material. The process is thought to have begun when a huge cloud of gas and dust, many times larger than the diameter of the entire solar system, became unstable. Owing to gravitational attraction, a condensation (knot) occurred toward one point within the gas cloud. This contraction we would term the *protosun*—the sun in the making. Any such cloud is thought to have some component of rotation and, much like a spinning ice skater pulling in her arms, the cloud rotated faster and faster as it contracted. As a result, it left some of its gas and dust in a flattened disk surrounding the protosun, and within this disk other condensations formed several protoplanets, which continued to move in the same direction and in the same plane as the original disk.

At the same time the planets assumed a rotation and tilt of axis that reflected the combined motion of the molecules and dust particles from which they were formed. The rotation and tilt of the planets show great variation, yet there is a general tendency favoring small angles of tilt and counterclockwise rotation, just what we might expect within a disk revolving in a counterclockwise direction. The same general process of condensation on a smaller scale surrounding each protoplanet may have acted to form the moons. While there is no absolute correlation between the mass of a planet and its number of moons, a tendency does exist for the more massive planets to have more moons.

As the material that composed the disk became more concentrated in the protoplanets, the space between the sun and these objects cleared, allowing the sunlight to reach them. This heat was naturally most intense on the inner planets, driving off the lighter elements and perhaps most of the atmosphere they had, leaving only the more dense core of heavier elements. At the distance of Jupiter or beyond, the effect of heating

would be greatly reduced, allowing the lighter elements such as hydrogen and helium to remain as part of their atmospheres. This model seems to fit what we have seen concerning the atmospheres. However, we may also surmise that the terrestrial planets went through a period of numerous impacts with objects in space and, further, that they had active volcanic periods, at which time they developed a secondary atmosphere composed of gases expelled from the planet itself.

We do not know the answer to one of the most fundamental questions: Why does the sun rotate so slowly? If the sun and the planets were formed from the same rotating cloud of gas and dust, we would expect the sun to rotate faster and faster as it condensed (like the spinning figure skater); this relationship is called *conservation of angular momentum.* If the sun had followed this pattern, it would rotate in a period of only a few hours, not its present 24.6 days. Perhaps it rotated much faster when it was first formed, but by some force, such as magnetism, it was slowed to its present rate. Alternatively, the solar system may have lost a large portion of its original mass. The sun contains over 99 percent of the mass of the solar system, yet it has less than 3 percent of its angular momentum.

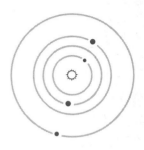

Questions

1. What basic properties distinguish the terrestrial planets from the Jovian group?

2. The rotation of Mercury was once thought to be 88 days, a period equal to that of its revolution. It is now known to be _____ days, as recently measured by radar.

3. What properties of Mercury cause it to be one of the hottest and at the same time one of the coldest planets?

4. Why is the albedo of Venus so much greater than that of Mercury or of the earth?

5. How many hours after sunset is Venus visible when at maximum eastern elongation?

6. During which phase does Venus appear brightest? Why?

7. The maximum surface temperature of Venus is probably greater than that of Mercury. How is this possible when Venus is farther from the sun?

8. Which configuration of Mars brings it closest to the earth: conjunction, quadrature, or opposition?

9. Where and when is Jupiter seen when at opposition?

10. Which method of observation yields the most detailed information about Mars?

11. Are the straight-line figures referred to as "canals" actually visible at times on Mars?

12. True or false: The volume of Jupiter is greater than the combined volume of all the remaining known planets.

13. List the Jovian planets.

14. How many moons has Jupiter? How many are larger than the earth's moon?

15. What property of Saturn is demonstrated by the fact that it would float in water?

16. What problems would be encountered in maintaining human life on any of the Jovian planets?

17. What principal characteristic distinguishes Uranus from other planets?

18. What kind of observations led to the discovery of Neptune?

19. For a period of 19 years, beginning in 1979, Pluto orbits nearer the sun than does Neptune, yet it is not possible for these planets to collide. Why?

20. What ideas about the origin of the planets are suggested by the fact that they orbit the sun in almost the same plane as does the earth?

21. Show by a drawing how Galileo's discovery that Venus goes through all possible phases from new to full and back to new again disproves the Ptolemaic model of the planetary system.

22. Suppose you could stand on the visible surface of Saturn at a point near its equator. Describe the apparent motions of particles in each ring from your point of view.

23. Find the semimajor axis of a planet which has an aphelion distance of 160 million kilometers and a perihelion distance of 100 million kilometers.

24. Find the period of a hypothetical planet that orbits the sun at an average distance of 8 A.U. (You may wish to refer to Chapter 1.)

25. Speculate on how life on the earth might have been affected if Jupiter had been a dim star which together with the sun formed a binary system of stars.

26. What are several fundamental differences between stars and planets?

27. Why do observers believe that Venus lost its original atmosphere and then developed a secondary atmosphere? How could it develop a secondary atmosphere?

28. What characteristics of Venus were revealed or confirmed by probes which either landed or flew by the planet?

29. What are the possible dangers of releasing ever-increasing amounts of carbon compounds into the earth's atmosphere by the burning of fossil fuels?

30. Group the planets into categories according to the predominant elements that are in their atmospheres: (a) hydrogen, (b) nitrogen, (c) carbon dioxide, (d) other elements.

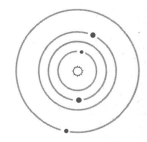

Suggested Readings

ARVIDSON, R. E., BINDER, A. B., and JONES, K. L., The surface of mars. *Scientific American* 238 (3), 76 (1978).

BEER, ARTHUR (ed.), *Vistas in astronomy*, vol. 10, pp. 175–206. Elmsford, N.Y.: Pergamon Press, 1968.

HOROWITZ, NORMAN H., The search for life on mars. *Scientific American* 237 (5), 52 (1977).

KUIPER, GERARD P., and MIDDLEHURST, BARBARA M. (eds.), *Planets and satellites*. Chicago: University of Chicago Press, 1961.

LEOVY, CONWAY B., The atmosphere of mars. *Scientific American* 237 (1), 34 (1977).

LEWIS, JOHN S., The chemistry of the solar system. *Scientific American* 230 (3), 50–65 (1974).

MARINER MARS 1971 PROJECT: FINAL REPORT, vol. VI (Science results). Pasadena, Calif.: Jet Propulsion Laboratory—NASA, July 15, 1973.

MARINER 10 VENUS ENCOUNTER. *Science* 183, 1289–1321 (1974). (Many authors; a variety of reports.)

MELBOURNE, WILLIAM G., Navigation between the planets. *Scientific American* 234 (6), 58 (1976).

MURRAY, BRUCE C., Mars from Mariner 9. *Scientific American* 228 (1), 48–69 (1973).

NEWBURN, R. L., JR., and GULKIS, S., A survey of the outer planets— Jupiter, Saturn, Uranus, Neptune, Pluto, and their satellites. *Space Science Reviews* 3, 179–271 (1973).

PIONEER 10 MISSION TO JUPITER. *Science* 183 (4122), 292–324 (1974). (Many authors; a variety of reports.)

SCHRAMM, D. N. and CLAYTON, R. N., Did a supernova trigger the formation of the solar system? *Scientific American* 239 (4), 124 (1978).

Scientific American, *The solar system* (readings). San Francisco: W. H. Freeman & Company Publishers, 1975.

SHKLOVSKII, I. S., and SAGAN, CARL, *Intelligent life in the universe*. San Francisco: Holden-Day, 1966.

VEVERKA, JOSEPH, Phobos and Deimos. *Scientific American* 236 (2), 30 (1977).

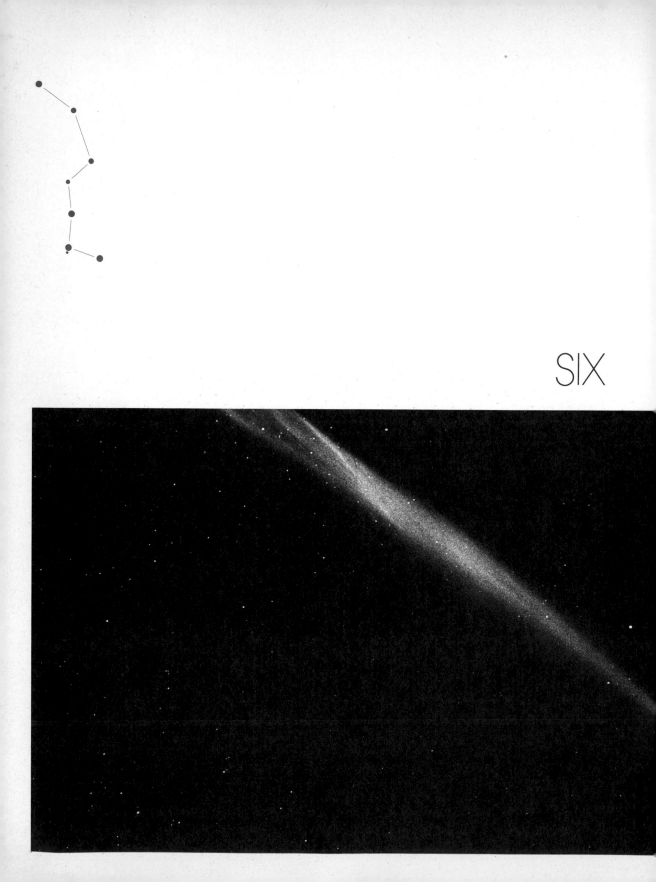

SIX

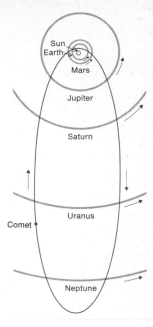

A comet in motion (Halley's)

ASTEROIDS, COMETS, AND METEORS

The asteroids, comets, and meteors constitute the lesser members of the solar system. The orbits of these objects reveal their kinship to the planets, for they all show the effects of the sun's gravitational field.

When the Bode-Titius relationship was first published in 1772, no object was known to orbit the sun at 2.8 A.U.; however, the very fact that the other numbers in this relationship so nearly agreed with the actual distances to the known planets suggested the existence of an object at that distance. Searches were initiated but remained fruitless until 1801, when quite by accident Giuseppe Piazzi, director of the observatory at Palermo, noticed a small starlike object that was not recorded on his charts. Watching it for several nights, he noticed a slight shift among the stars. At first he thought that the object was a comet; however, after continued observation and computation, it was identified as a small planet orbiting the sun at an average distance of 2.76 A.U. Thinking he had found the lost planet, Piazzi called this small object Ceres. Within a year, Heinrich Olber, a German astronomer, found another object in a very similar orbit and named it Pallas. This second discovery gave the clue that there might be many more such orbiting objects and in fact marked the beginning of a long series of discoveries of asteroids (see Table 6.1).

Table 6.1   *Discovery of Asteroids*

YEAR	NAME	DIAMETER
1801	Ceres	1000 km
1802	Pallas	600 km
1804	Juno	250 km
1807	Vesta	540 km
1845	6 known	
1890	300 known	

Asteroids

By the late nineteenth century important new photographic techniques had been developed. In order to take a time exposure of a given star field, the camera must be turned slowly to compensate for the rotation of the earth and thus produce sharp star images. Any object that is moving in relation to those stars will produce a streak on the film. By means of this photographic technique several thousand asteroids have been definitely identified. As new discoveries are made, they are numbered in order: (1) Ceres, (2) Pallas, . . . , (1566) Icarus, and so on. While most asteroids orbit the sun between Mars and Jupiter, a number have rather eccentric orbits. Of particular interest are Hermes, Apollo, Adonis, Icarus, Geographos, and Eros, because their orbits cross that of the earth, thus bringing them relatively close to the earth at times. In 1932 Apollo passed within 3.2 million kilometers of the earth; in 1936 Adonis passed at a distance of less than 1.6 million kilometers; in 1937 Hermes came within 0.8 million kilometers. Icarus and Geographos passed within 6.4 million kilometers in 1968 and 1969, respectively.

Asteroid Icarus is of special interest because it passes closer to the sun than any other known asteroid—within 27 million kilometers—and then moves out beyond the orbit of Mars (Figure 6.1). Of all the known asteroids, it has the shortest period (409 days) and the highest eccentricity of orbit (0.83), and it orbits in a plane that is inclined 23° to that of the earth (Figure 6.2). Icarus is thought to be only about 0.5 kilometer in its largest dimension, as measured by radar bounce.

Icarus was discovered by Walter Baade when he photographed a star field using the 1.22-m Schmidt telescope at Mt. Palomar. Because of its sun-grazing orbit, it was called Icarus after the mythological young man who flew too near the sun. With his father, Daedalus, he attempted to escape an island prison by flying away on artificial wings. Daedalus, a master craftsman, had fashioned two sets of wings and attached them to their bodies with wax. He warned his son not to fly too high, for the wax would be melted by the sun's heat. The boy was so thrilled at being free and able to fly, however, that he soared higher and higher. The wax melted, and Icarus fell to his death in the sea. When the motion both of

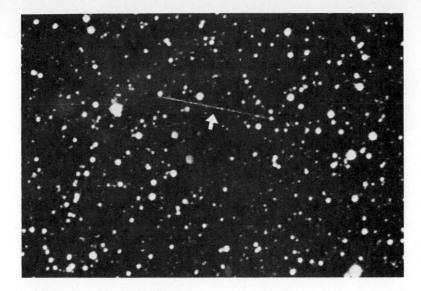

Figure 6.1 *The asteroid Icarus, showing motion among a field of stars. (Hale Observatories)*

asteroid Icarus and of the earth is considered, it is only every 19 years that they approach each other within a few million kilometers. We may summarize the motions of asteroids, as illustrated in Figure 6.2, into three categories: (1) those which orbit in nearly circular orbits in a belt centered on about 2.8 A.U., (2) those which have very eccentric orbits, passing near the earth, and (3) two groups of asteroids called the Trojan groups, which seem to be "trapped" near the orbit of Jupiter about 60° ahead and behind the planet.

Even within the asteroid belt itself, one should not picture these objects as "filling" the space. In fact, they are so spread out as to seldom come within a million kilometers of each other. Pioneers 10 and 11 passed through the belt without being struck by an asteroid.

Since 1970, astronomers have been applying new techniques to determine the physical properties of asteroids. These techniques include polarization, infrared radiation, and spectral studies. The combination of these observations provides measures of albedo, diameter, and surface composition. Typically, the new assessment gives diameters larger than previously thought (see Table 6.1). Furthermore, observers have begun to classify asteroids according to surface composition. Some are basalts, like the lava flows of the lunar maria; some contain significant amounts of carbon; some are light in color with higher reflectivities (10 to 30 percent, with a few even higher), others dark with reflectivities closer to 5 percent (a few as low as 1 to 2 percent). There appears to be a pattern of distribution within the asteroid belt, the inner particles tending toward the lighter, stony type and the outer particles tending toward the darker, carbonaceous type.

Origin of Asteroids

Some observers have suggested that the asteroids were produced when a single planet exploded; however, if that were the case, we would expect a more random arrangement of types. Perhaps it is more reasonable to expect that numerous smaller bodies formed throughout the solar nebula. While some of these bodies were gravitationally joined to others to form the planets, some remained separate (these would be the larger asteroids, thought to be spherical in shape). Others of these bodies may have experienced a disruptive force and were torn apart (forming the smaller, irregularly shaped asteroids). Some of these early bodies later collided with the planets to produce the cratering we still see on their surfaces.

Figure 6.2 *The orbits of selected asteroids.*

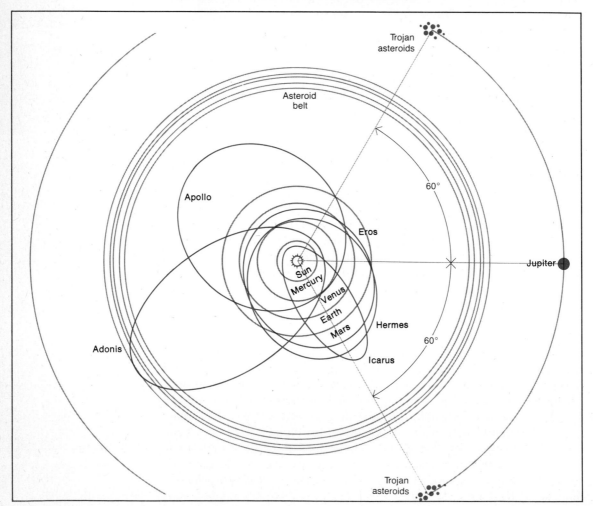

A fundamental question follows from the identification of asteroid types. Will the meteorite types, which we will study near the end of this chapter, correspond to the asteroid types? Are the asteroids the source of meteorites?

Comets

A comet presents one of the most spectacular apparitions in the sky. Often unheralded, it appears first as a small source of light, somewhat fuzzy and often resembling a nebula or distant galaxy. As the comet moves closer to the sun, however, a vaporous tail is driven backward from its head by the sun's radiation; this tail always extends away from the sun. The length of the tail continues to increase as the comet nears the sun. Such comets presented a fearsome sight to ancient observers. The tail was thought to contain poisonous gases and to be within the earth's atmosphere.

If a comet moved within the earth's atmosphere, then it would streak across the sky like a meteor. It is, however, millions of miles from the earth even at its closest approach, and so, it has only the slightest apparent motion in a period of a few hours. Thus, whenever a bright comet appears it will likely be seen for several weeks. Most comets are not bright enough to be seen with the naked eye and so they go unnoticed by the casual observer.

Comets, like the planets and asteroids, are orbiting the sun and so are members of the solar system. Their very eccentric orbits distinguish them from the other members, as does the fact that the planes of their orbits are often inclined to the ecliptic. A comet may appear at any point in the sky and move in direct or retrograde motion. Comets seem to be randomly oriented in the system. Since they move in orbits primarily dictated by the gravitational influence of the sun, their paths represent a conic section, usually that of an ellipse. The comparison between the earth's orbital shape and that of the comet Kohoutek may be seen in Figure 6.3. Some comets have very long periods, of the order of 10,000 to 50,000 years (or longer), and they have very eccentric orbits that take them to distances of several thousand astronomical units from the sun. Other comets have short periods, ranging from 3 to 100 years, and their orbits are less eccentric and are generally confined to distances less than that of Pluto's orbit. If a long-period comet approaching the sun is perturbed by the gravitational field of a planet that it happens to encounter, the shape of its orbit may be altered so that it becomes a short-period comet (Figure 6.4). Jupiter has perturbed the orbits of a number of comets (45 to 50—the *Jovian group*), which now orbit the sun in a period of from 5 to 10 years.

It was not at first realized that the same comet was being observed over and over again; however, in 1705 the British astronomer Edmond Halley published the orbits of 24 comets and noted the similar periods for

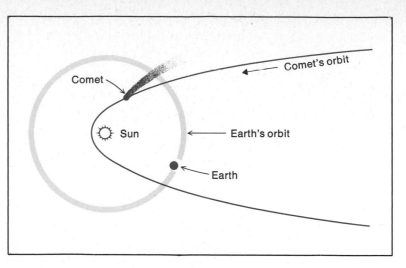

Figure 6.3 *The highly eccentric orbit of the Comet Kohoutek (1973f) compared to the almost circular orbit of the earth. The shape of the comet's orbit is almost like that of a parabola, typical of long-period comets.*

the comets of 1531, 1607, and 1682, each date separated by 75 or 76 years. He predicted the return of this comet in 1758. The comet did in fact reappear on Christmas night of that year and has since been called Halley's Comet (Figure 6.5). The period of the comet is known to vary between 74 and 79 years, owing to perturbations by the Jovian planets and by expulsion of gases from its nucleus. It is interesting to note that Halley's Comet appeared in the year of Mark Twain's birth (1835) and reappeared in the year of his death (1910), just as he had prophesied. The comet's next passage near the sun is expected in 1986. (The flip pages beginning on page 235 depict the flight of Halley's Comet.)

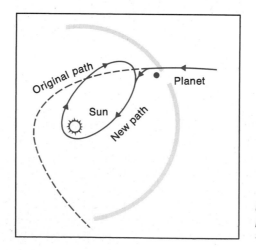

Figure 6.4 *A comet may be perturbed by a planet so that the shape of its orbit is changed.*

Origin of Comets

The comets that become visible are not likely to have been inter-stellar wanderers that only by chance entered the gravitational influence of the sun; rather, they are members of a "comet cloud" surrounding the solar system at some distance beyond 50,000 A.U., perhaps extending a third of the way to the nearest star, α (Alpha) Centauri, which is located at a distance of 260,000 A.U. Examples of periodic comets that orbit to these great distances include the Pons-Brooks comet, the Giggs-Mellish comet, and others with aphelion distances of 30,000 to 60,000 A.U. Members of the great comet cloud are thought to be perturbed by major planets or passing stars, causing them to alter their orbit, bringing them close to the sun. Further perturbations may then entrap them in a smaller orbit, which makes their return more predictable. This theory of a comet cloud was set forth in 1950 by J. H. Oort of Leiden University Observatory. In 1973 we viewed a comet which tends to confirm Oort's theory, for calculations based upon the almost parabolic orbit of comet Kohoutek

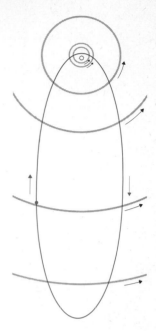

Figure 6.5   *Halley's Comet of 1910. The two photos were taken a day apart; by examining the position of the comet relative to the background of stars, you can detect the motion of the comet. (Lick Observatory)*

(a)

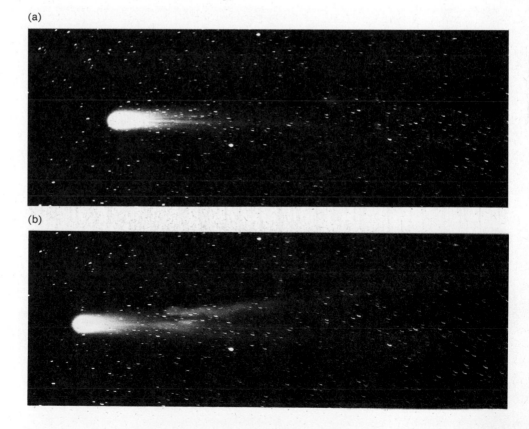

(b)

show that it came near the sun from a great distance (several thousand A.U.). On December 28, 1973, it passed within 21 million kilometers of the sun (Figure 6.6).

Many astronomers believe that comets were formed as some of the earliest members of the solar system, before the sun and planets, while the solar nebula was still a very large cloud of gases and dust. Our study of comets therefore may give us clues to the origin of the solar system. What are comets like, and of what are they made?

Physical Nature of Comets

When a comet is far from the sun, it consists of only a nucleus that we might characterize as a "dirty snowball," or a swarm of dirty snow-balls. Such a description implies a mixture of rocky and metallic particles bound together by frozen carbon dioxide, methane, ammonia, and water molecules. As the nucleus approaches the sun, the solar radiation causes the frozen material to vaporize (*sublimate*), thus producing a large gaseous region surrounding the nucleus called the *coma*. The thinness of

Figure 6.6   *The Comet Kohoutek (1973f), as observed on January 12,1974. (Hale Observatories)*

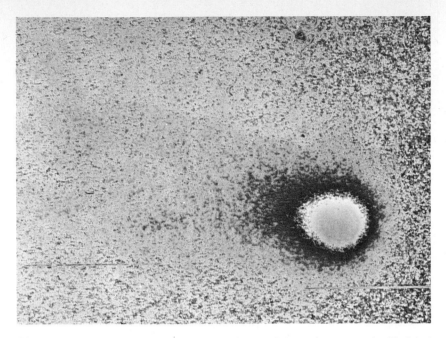

Figure 6.7 *The hydrogen halo of the Comet Kohoutek, as seen by Skylab 4 in the far ultraviolet wavelengths. (NASA)*

the coma is revealed by the stars seen shining through it. Although the nucleus is very small—often only a few kilometers in diameter—the coma may grow larger than the earth. When a comet passes quite near the sun, its temperature may rise to several thousand degrees. At this time, not only can the absorption spectrum of reflected sunlight be seen, but also a bright-line spectrum is produced by the comet itself. A comet's emission spectrum results primarily from the ultraviolet radiation of the sun, exciting the gases to fluorescence, revealing the presence of carbon, nitrogen, oxygen, and hydrogen, in the following radical forms: CH, NH, CN, NH_2, OH, C_2, C_3. Observers suspect that these derive from the breakdown of ammonia (NH_3), cyanogen (C_2N_2), methane (CH_4), and water (H_2O) ices which compose the nucleus of the comet, along with rocky and metallic particles, including iron, nickel, sodium, calcium, and carbon. The Skylab crews had a particularly good opportunity to photograph comet Kohoutek and to analyze its composition spectroscopically because of their freedom from atmospheric interference. They found a hydrogen "halo" engulfing the entire comet as shown in Figure 6.7.

Most comets produce a tail only after approaching the sun within about 2 A.U. The gas and dust liberated by the sun's heat experience a force away from the nucleus in this sublimation process; however, these same molecules and particles also experience two other fundamental forces that ultimately determine their direction of motion. One force is due to the sun's radiation, the other to the solar wind, an outflow of

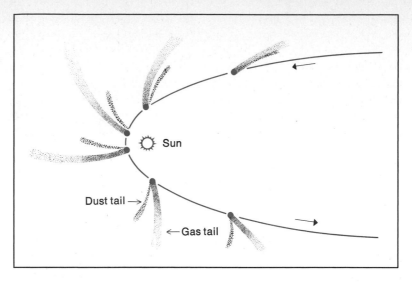

Figure 6.8 *The tail of a comet, showing the separations of the gas and dust tails.*

particles from the sun. The gas molecules or atoms or ions present in the tail appear to be most directly influenced by the solar wind. The reason may be in part that the solar wind carries with it a magnetic field, which in turn would carry off the ions (charged particles) from the coma, forming a tail. On the other hand, dust particles are larger and appear to

Figure 6.9 *Comet Arend-Roland of 1957, showing an antitail, which certain comets develop. (Hale Observatories)*

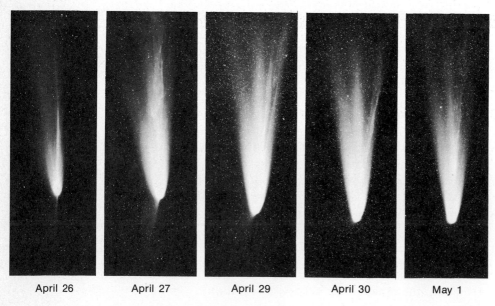

April 26   April 27   April 29   April 30   May 1

be influenced primarily by the sun's radiation. Thus we might expect comets to develop two tails: a gas tail that tends to point more directly away from the sun, and a dust tail that tends to curve and lag behind the gas tail. The separation of these two tails will become more apparent as the comet recedes from the sun, for at that time that the gas tail precedes the head of the comet and the dust tail tends to some degree to follow the head (see the lower portion of Figure 6.8). In the case of the comet Arend-Roland (Figure 6.9), the dust tail appears to point toward the sun; however, this illusion is created by our angle of view. The comet is receding from the sun, its gas tail leading the way, and the dust tail merely curves backward enough to appear as an antitail. Note also the change in structure of the gas tail in these views.

The tails of some comets have reached the tremendous length of more than 160 million kilometers, a distance greater than that from the sun to the earth. The nebulosity of the tail is revealed by the fact that stars may easily be seen through it (Figure 6.10). Only when we see the tail of a comet at right angles to its length do we see it full length. If we

Figure 6.10   *Halley's Comet of 1910, showing the growth of the tail as the comet approached the sun in April and May and then the decline of the tail as the comet receded from the sun in June. (Hale Observatories)*

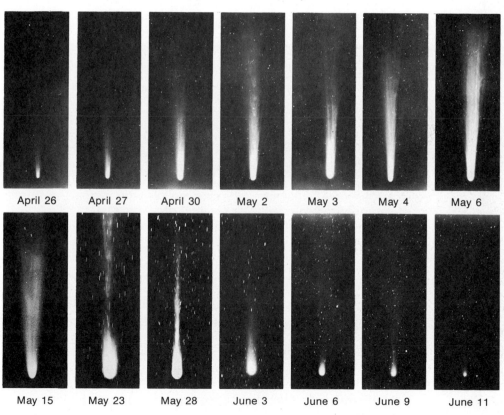

April 26   April 27   April 30   May 2   May 3   May 4   May 6

May 15   May 23   May 28   June 3   June 6   June 9   June 11

view the comet from a direction in line with its tail, then no tail is apparent at all.

Lifetime of a Comet

We know from direct observation that Halley's comet has been returning to its near passages of the sun about every 76 years from the year 239 B.C., when it was first recorded. Yet with every such passage a comet loses thousands to millions of tons of material. How long can it last? In a few cases we have failed to find a comet when it was due to return, and in its place we have experienced meteor showers, produced by the disintegrated particles which once composed its nucleus. This may indicate the death form of a comet. We believe that it continually injects particles into its orbit, producing the predictable meteor showers (a full discussion follows on page 251).

As a comet passes its closest approach to the sun (perihelion), it experiences such tremendous tidal forces that astronomers are hard pressed to explain how the nucleus "hangs" together at all. Indeed, a number of comets have broken into two, three, or four parts. A recent example of this phenomenon was recorded (Figure 6.11). As Comet West passed the sun, it broke into four parts. This will certainly shorten its lifetime (Figure 6.12).

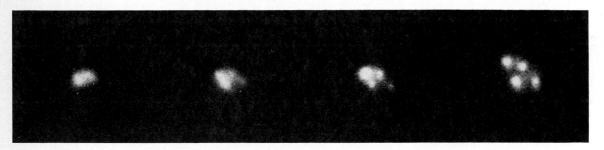

Figure 6.11 *Comet West broke into four parts, over a 10-day period, as it passed near the sun. (New Mexico State University)*

How Are Comets Discovered and Recorded?

It is evident from the limited list given in Table 6.2 that the names of many persons have been immortalized by their discovery of a comet. Comets are listed as they are discovered in any given year by assigning the letters a, b, c, d, . . . in the order of their discovery: a comet listed as 1967d is therefore the fourth to be discovered in 1967. In addition, the designation p/Tempel II, for example, indicates that this comet is a periodic (p) comet, merely returning to the sun, and that it was originally discovered by Tempel as his second comet (II). Those comets lacking the

Figure 6.12 *This spectacular photograph of Comet West shows a great deal of structure in both gas and dust tails. (Lick Observatory)*

"p/" designation represent comets newly discovered in 1967. Are the discoverers' names those of professional astronomers alone? No! There is no such criterion for the naming of comets. The only requirement is that the name be that of the person who saw the comet first. Lubos Kohoutek, a professional astronomer, found comet 1973f while searching photographic plates for asteroids; most amateurs do not have access to such plates, however. Still, comets are sometimes found by visual observation,

Table 6.2   *Comets Sighted in 1967*

DATE		NAME
Jan. 3	1967a	p/Tuttle
Feb. 4	1967b	Seki
Feb. 11	1967c	Wild
Feb. 12	1967d	p/Tempel II
June 5	1967e	p/Reinmuth II
June 29	1967f	Mitchell-Jones-Gerber
Aug. 7	1967g	p/Finlay
Aug. 3	1967h	p/Encke
Aug. 8	1967i	p/Schwassmann-Wachmann II
Oct. 5	1967j	p/Wolf
Oct. 5	1967k	p/Wirtanen
Oct. 5	1967l	p/Arend
Oct. 5	1967m	p/Borrelly
Dec. 28	1967n	Ikeya-Seki

especially in dark regions of the sky where no photographs are being taken. Remember that comets may appear at any point in the sky and do not always follow the ecliptic as did comet Kohoutek.

The tools with which a comet search may be conducted include a telescope of short focal length—say, one with a focal ratio of $f/4$ and a low-power eyepiece. This combination will provide a large, bright field of view, allowing the observer to recognize the rather faint, diffuse image that a comet presents at early stages of visibility. Since the image of a comet often resembles that of a nebula or distant galaxy, an adequate sky map is essential. Known diffuse objects, such as nebulae, clusters, and galaxies, will be shown on such a map and a newly apparent comet will not. It was this need to be able to recognize the different types of objects that led Charles Messier, a French observer whose primary concern was comet searching, to compile a catalogue of known nebulae, clusters, and galaxies. Such objects kept getting in his way as he sought comets. The Great Galaxy in Andromeda, as an example, still bears his catalogue number M31; in fact, Messier is better known for his catalogue of nebulae than for his comet discoveries. Together with the more recent *New General Catalogue* (NGC) and *Index Catalogue* (IC), this list provides an essential tool for cometary searchers today. (The *Messier Catalogue* is given in Appendix 9.)

If a certain diffuse (fuzzy) object is seen that does not appear on the sky map, check the object for any apparent motion against the background of stars. A comet will show such motion, but this motion may not be evident unless the observer stays with the object for hours or even days. Make drawings and notations of its position among nearby stars. Note its right ascension and declination from a sky chart. If in fact the object does display movement and is diffuse in appearance (asteroids also display movement but are not diffuse), you have probably discovered a comet. What does one do next? To record your discovery, send a telegram to the Central Telegram Bureau, Smithsonian Observatory, Cambridge, Massachusetts 02138, stating your name, the date and time of observation (Universal Time), the nature of the object (diffuse), its right ascension and declination, the direction of motion, and the estimated magnitude. This will set into motion a series of events by which your observation will be verified and the orbit calculated. If this is a periodic comet that you have merely rediscovered, it will of course not bear your name; but if this is indeed a new comet and your telegram stating the facts of your observation arrives first, your name will be given to that comet.

While comets are often seen after they have moved close to the sun, thus developing a tail, it is not easy to spot the dim image of a comet that is just becoming visible at a distance of more than 500 million kilometers from the sun. For your observations, find a location as far away from city lights as possible and direct your attention to the darker portions of the sky. Good hunting!

Meteors

Unlike the comet, which seems almost fixed among the stars throughout an evening's viewing session, the *meteor* is a phenomenon that lasts only a few seconds. The word *phenomenon* is used here to stress that what you see shooting across the sky is not the object itself but rather a streak of light caused by the passage of a small solid body into the earth's atmosphere. The object itself, called a *meteoroid* while still in flight, experiences the heating effect of friction between itself and atoms that make up the atmosphere. As a result of this heat the air is ionized, the atoms losing electrons. As the ionized atoms "find" free electrons with which to recombine, downward transitions of these electrons occur, and light is produced.

Occasionally, the object that produced the meteor survives its flight and falls to earth. Such an object is called a *meteorite*. Since most meteoroids are only the size of a grain of sand or a pebble at most, they are destroyed in flight. The rare meteoroid of larger size often produces a spectacular display called a *fireball*, or *bolide meteor*. The streak of light persists, and often explosions and other loud noises accompany the event. Objects that exhibit this rapid motion and that sometimes deposit material on the earth are obviously very close by; in fact, they are usually within 160 km of the observer during their visible flight.

The meteorite itself is of great interest to the astronomer, for until recently it had represented the only sample of extraterrestrial material with which scientists could work directly. These objects may carry clues to the origin of the solar system and to life elsewhere in it. Careful analysis of certain meteorites has shown that they contain as many as 18 different amino acids and also hydrocarbons. Six of the 18 amino acids are of the type normally found in living cells, specifically in proteins. Figure 6.13 shows a photomicrograph of a section of meteorite in which investigators at the University of Chicago have found inclusions (white spots) containing a form of oxygen (isotope ^{16}O) so pure as to suggest its origin in interstellar material before the birth of the solar system. On earth, other isotopes of oxygen (^{17}O and ^{18}O) are usually found together with ^{16}O.

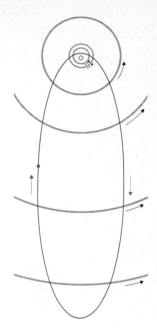

Observing Meteors

Are meteors predictable? Where should one look to see a meteor or meteor shower? It is usually possible to see a few meteors on almost any night by simply assuming a reclining position and staring at the sky. Your eye will pick out streaks of light quite easily (Figure 6.14). However, there are some guidelines for viewing meteors. In general, the best time is

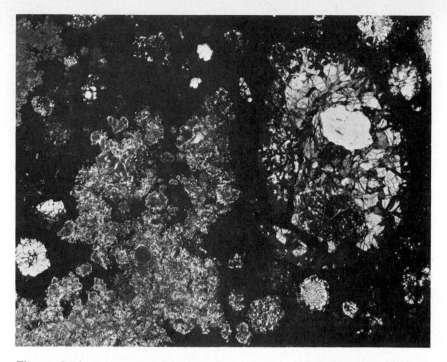

Figure 6.13   *A photomicrograph of a section of the Allende meteorite showing inclusions (white) of pure oxygen 16. (Courtesy of Richard J. Kjarval, Graphic Arts Facility, Physical Science Division, University of Chicago)*

Figure 6.14   *Meteor trail near Pleiades. (Yerkes Observatory)*

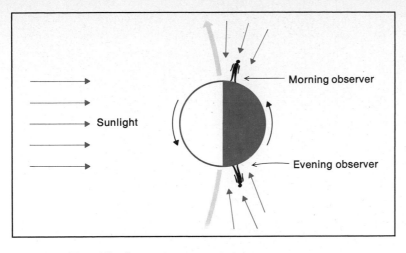

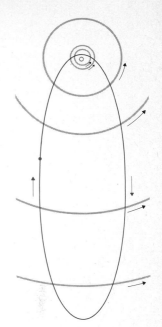

Figure 6.15   *Viewing meteors.*

during the early morning hours before dawn. The brilliance of a meteor depends largely on the speed with which it enters the earth's atmosphere. In the evening hours, the meteors you see would be overtaking the earth and hence would be entering the atmosphere quite slowly (15 km/sec). In the early morning hours, the meteors you see are meeting the earth head-on, in effect adding the earth's speed to their speed, which may result in relative speeds as high as 70 km/sec (Figure 6.15).

Meteor Showers

On certain nights more than the usual number of meteors will be evident. Streaks of light may appear in many parts of the sky, yet if we extend imaginary lines backward from the direction in which the meteors move, these lines will appear to intersect in a definite point called the *radiant*. We are seeing a *meteor shower*, many particles entering the earth's atmosphere within a short period. Although the meteors appear to radiate from a given point (Figure 6.16), this is merely an optical illusion; the individual particles that produce the shower are actually traveling in parallel paths. Just as parallel railroad tracks appear to converge at a point in the distance, so these parallel streaks of light seem to emerge from a single point. An analysis of the orbits of these particles reveals that they are associated not only with each other but also with a comet that passed along the same orbital path at an earlier time. The particles are very likely debris left from the comet itself. Since the earth crosses the orbit of a given comet at about the same time each year, these meteor showers are predictable, and the list provided in Table 6.3 can be used for any year. Each shower is named for the constellation in which its radiant is located.

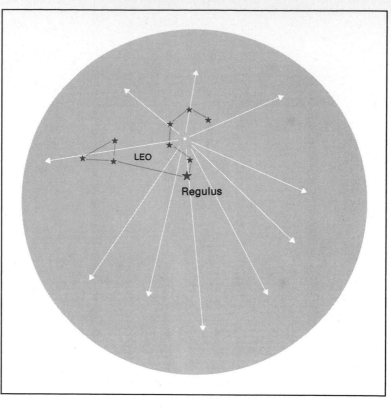

Figure 6.16   *The radiant point of the Leonids.*

Table 6.3   *Meteor Showers*

| APPROXIMATE DATE | NAME OF SHOWER | ASSOCIATED COMET | RADIANT | | DEC. |
| | | | R.A. | | |
			HR	MIN	DEG
Jan. 1–4	Quadrantids	—	15	20	+52
April. 19–23	Lyrids	1861 I	18	4	+33
May 1–6	May Aquarids	Halley	22	16	−2
July 26–31	Delta Aquarids	—	22	36	−11
Aug. 10–14	Perseids	1862 II	3	8	+58
Oct. 9–11	Draconids	Giacobini-Zinner	17	40	+55
Oct. 18–23	Orionids	Halley	6	8	+15
Nov. 1–15	Taurids	Encke	3	40	+17
Nov. 14–18	Leonids	1866 I	10	00	+22
Dec. 10–16	Geminids	—	7	32	+32
Dec. 21–23	Ursids	Tuttle	14	28	+75

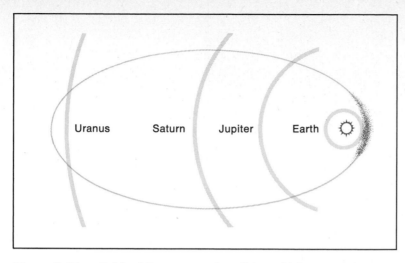

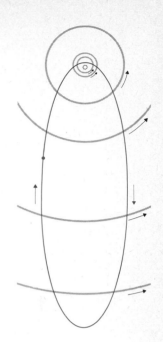

Uranus   Saturn   Jupiter   Earth

Figure 6.17 *Orbit of the swarm of particles which produces the Leonids.*

A given shower may be better one year than the next because the particles that produce it tend to travel in swarms rather than being evenly distributed along the orbit of the comet (Figure 6.17). When the earth passes through a swarm, a spectacular display results. On November 13, 1833, the earth passed through a swarm that was later associated with the comet of 1866 I, and it is estimated that more than 30,000 meteors per hour were visible. An outstanding display of this shower, the *Leonids*, occurs in a cycle of 33 years, the same as that of the comet that is its ultimate cause. Further verification of the 33-year cycle came when spectacular showers were observed in 1932 and 1965.

More prevalent than the shower meteors are the sporadic meteors that enter the earth's atmosphere almost continuously. They seem to bear no specific relationship to the comets and may appear in any part of the sky at any time.

Physical Properties of Meteorites

A typical meteoroid is a small, porous, easily fragmented body that disintegrates in flight. The average meteoroid recorded by the camera has a computed density of about 0.26 g/cm³, which is somewhat like that of pumice stone. The exceptional meteoroid that reaches the earth is usually much more dense—approximately 2 to 5 g/cm³. Such meteorites fall into one of three main categories, which the specialist subdivides into several additional types. The stones (aerolites or chondrites) constitute about 92 percent of all meteorites; they are very much like ordinary stones, containing iron, silicon, carbon, magnesium, aluminum, and other metals (see Figure 6.18). The stony-irons (siderolites) represent 1 to

Figure 6.18 The world's largest stony meteorite fell in the Jiling province of northeastern China in 1976. The large fragment shown weighs 1,770 kg (almost two tons). The fusion crust shows the effects of heating due to friction as it passed through the earth's atmosphere. (*Thomas Tsung*)

Figure 6.19 A polished slice of a stony-iron meteorite from Brenham County, Kansas. The dark material is olivine. (*Ronald Oriti, Griffith Observatory*)

Figure 6.20 *The polished section of an iron meteorite, showing Widmannstätten lines. (Ronald Oriti, Griffith Observatory)*

2 percent of all meteorites, and a polished slice of this type shows a matrix of stone and iron (see Figure 6.19).

The *irons* (siderites), representing 5 to 6 percent of all meteorites, are composed largely of iron with varying amounts of nickel (1 to 20 percent). Why, one may wonder, do museums and planetariums usually display a predominance of irons and stony-irons? The answer is that these two types are much easier to recognize when one is looking for meteorites in the countryside. The "stones" are so similar to ordinary rocks that they often go unnoticed. The irons are quite easily picked out as "suspect" because of their high density; however, this is not a proof by itself. The irons, upon cutting, polishing, and etching with a mild acid, reveal a very interesting pattern of lines and geometric designs (Figure 6.20). These are called *Widmannstätten lines,* and no natural rocklike material found on earth possesses such a pattern. The irons contain from 1 to 20 percent nickel as well. Nickel does not appear in this proportion along with iron in terrestrial materials.

One particular subdivision of the stones, called carbonaceous chondrites, represents one of the oldest types of meteorites—approximately 4.6 billion years old. The polished section in Figure 6.13, on page 250, shows the round chondrules for which the chondrite meteors were named.

Orbit and Origin of Meteorites

The facts just mentioned point to an origin of meteorites outside the earth. The temperatures that the irons have experienced to produce the evident separation of materials of different density must have been higher than anything they experienced on the crust of the earth or during their flight to earth. We have seen from the discussion of meteor showers the close relationship between meteors and comets. It has been reported that of the more than 400 meteors for which the Harvard Observatory has computed the original orbit, the large majority possessed a cometlike orbit, and a cometary origin is quite likely. Several had orbits suggesting asteroidal origin, and radioactive dating of meteorites indicates an age of 4 to 5 billion years. Some observers have suggested that before or during the creation of the planets, other objects of lesser size formed. The centers of these objects experienced very high temperature, perhaps from short-lived radioactivity, creating a molten core. Materials of higher density gravitated to the center and, upon gradual cooling, remained in that separated state. The object later experienced a collision and broke into many smaller chunks, the core producing the iron meteorites, the outer portions producing the stones, and the intermediate layers the stony irons.

Meteorite Craters

If one stands on the rim of the Barringer Crater, near Winslow, Arizona, one cannot help but wonder what cataclysmic event could have caused this great natural depression in the earth's surface. What evidence do we have of its origin? Will an answer to this question shed any light on the origin of similar features on the moon? If we assume that the Barringer Crater is the result of the high-speed impact of a massive object from outside the earth, then we should expect that upon impact the object itself suffered some damage. It might have partly fragmented, leaving a portion buried deep within the ground; it might have fragmented entirely, distributing its mass over the area surrounding the crater; or it might have vaporized completely, leaving no visible trace. The fact that more than 30 metric tons of meteoric material have been found around the Barringer Crater strongly suggests its impact origin; however, no central object has yet been found (Figure 6.21).

We should also expect to see evidence of the high temperature and high pressure that would naturally have accompanied such an impact. Surrounding a number of craters there exists a very dense form of silica called *coesite*. Silica of this density is not generally found, and it is believed to result from the tremendous shock wave that accompanies the impact of a large object on the earth. Under this sudden increase of pressure, rocks do not always shatter along their normal fracture lines, and the shatter structure of the crustal materials that surround the impact area very often shows a special form. High temperature may exhibit itself in the formation of glasslike objects from constituents of the soil. Tektites, to be discussed in the next section, may be a byproduct of such an impact.

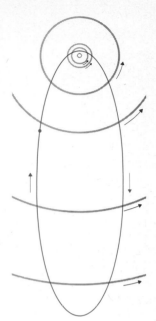

Figure 6.21   *The Barringer Crater, near Winslow, Arizona. (Meteor Crater Museum)*

With these ideas in mind the scientist sets out to test specific features of the earth's surface. Where does he look? Unlike the moon, the earth has only a very limited number of sites that resemble a crater structure. The most obvious sites have been tested, and many reveal their impact origin by meeting at least two of the criteria mentioned above. Less obvious sites are sometimes revealed by satellite photography or by a very careful study of topographic maps that show the contours of a given region. While a certain site may look almost flat today, at an earlier time its crater form may have been very obvious. The change would have taken place gradually as a result of erosion. Atmospheric factors such as wind, rain, and temperature variations have effectively obliterated the once obvious surface features. Lacking an atmosphere, the moon craters are subject to almost no erosion and therefore remain almost unchanged for thousands of years.

The age of a crater can be roughly inferred from its appearance. The Barringer Crater, the result of a relatively recent event, is probably no more than 30,000 years old. On the other hand, the Vredefort Crater of South Africa, almost indiscernible now, is probably 250 million years old. The impact force that produced the Vredefort is estimated to be 500,000 times that which produced the Barringer. The Vredefort Crater has a diameter of almost 10 km, whereas the diameter of the Barringer is about 1.3 km.

In this century there have been two large falls, both in Siberia—one in 1908, the other in 1947. Even at the rate of one large fall for every 1000 years, the earth should have experienced 1000 such events in the past million years, yet no such frequency is indicated by the number of known craters. Perhaps many eroded craters lie undiscovered.

Large-impact craters are caused primarily by iron or stony-iron meteorites or asteroids, for only these falling objects have sufficient mass and density to survive their fiery flight to earth. The meteorite believed to have produced the Barringer Crater is estimated to have had a weight of over 100,000 metric tons. To blast a hole in the ground that large would have required the discharge of more than 10 million tons of TNT, buried several hundred feet underground.

The Siberian Event

On the morning of June 30, 1908, an event took place in the Tunguska Valley of Siberia that still has scientists speculating. Witnesses several hundred kilometers away saw a fireball that seemed to "fill much of the sky" and gave off a heat which they clearly felt. It was accompanied by a terrifying sound and then exploded with such force as to shake the ground and break windows at that distance. When scientists visited the site, they found no crater; instead, the blast had simply leveled everything for several tens of kilometers in every direction, with the exception of a small stand of trees in the center of the devastated area. This would indicate an aerial detonation with the power several hundred times that of the first atomic bomb.

This destruction does not appear to be the result of a huge iron meteorite hitting the earth, for if that were the case the evidence would most likely be a huge crater much larger than Barringer. In fact, no meteorite fragments have been found to this day in the vicinity of this event. An alternate suggestion characterizes the object as being very massive, yet far less dense than the iron meteorite and able to vaporize readily without leaving a distinct remnant. A body composed of loosely packed snow (water ice) could meet all these requirements, and if the Tunguska object was of that nature, it may be likened to the nucleus of a small comet—too small to have been recognized in the usual visual sense. Comets generally release dust as they pass near the sun, and it is true that several unusually luminous nights had been observed in Europe just prior to the event, thought to have been caused by dust in the atmosphere. An atmospheric pressure (shock) wave was detected at numerous meteorological stations around the earth with some indication that the wave traveled around the earth more than once. As such a body approached the earth, it would set up a shock wave which travels faster than the speed of sound, creating very high temperatures in the atmosphere ahead of itself. This high temperature could account for the charring of the trees.

Others have suggested a nuclear blast or the collision of an antimatter body with the matter body (earth). When antimatter and matter come into contact, both are annihilated with a great release of energy; however, events of either of these types would have also produced radioactive telltale signs, which were not found. Still others have suggested that a superdense black hole passed through the earth at that time. Can you add your own possible explanation?

Tektites

No subject has aroused more diverse opinions than the origin of tektites. Before we present two of these divergent views, let us consider the factors which seem to be fairly well agreed upon. Tektites are small, black, glassy objects which have been found only in restricted land and sea areas (Figure 6.22), the largest being the Southwest Pacific area, including Australia, Indochina, Indonesia, and the Phillipines. Lesser "falls" have been discovered along the Ivory Coast of Africa, in Germany, Czechoslovakia, the United States, and the USSR. All tektites show evidence of having been heated to a molten state, perhaps at their time of origin and/or as they passed through the earth's atmosphere. When sliced in a thin section, a tektite appears a translucent green or brown. Typically, tektites are homogeneous (crystal-free) structures which appear to have cooled very rapidly. They are lacking in water and have unusually high silicon content. Note the typical teardrop and/or dumbbell shapes in many of the samples from Thailand (Figure 6.23). These are sometimes called "splash" forms, suggesting that they cooled while in flight, having been thrown from a meteorite impact or volcano. The surface markings on these samples are evidence of the frictional heating that resulted from their passage through the earth's atmosphere. Part of the surface material was undoubtedly lost in flight—a process called *ablation*. All surfaces of the *indochrinites* from Thailand show this effect. Contrast the shape of the *australites* (Australian tektites, Figure 6.24). These "buttons," as they are often called, show a molten origin as all tektites do, but they show a second period of heating during a controlled, tumble-free, flight through the earth's atmosphere. The ablation effects show up on one side only. In fact, the models shown in Figure 6.24 were

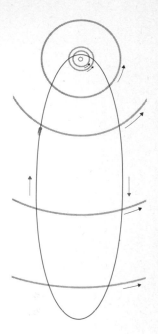

Figure 6.22   *Tektites are found only in a limited number of locations on the earth's surface. Each dot represents a field over which tektites were strewn.*

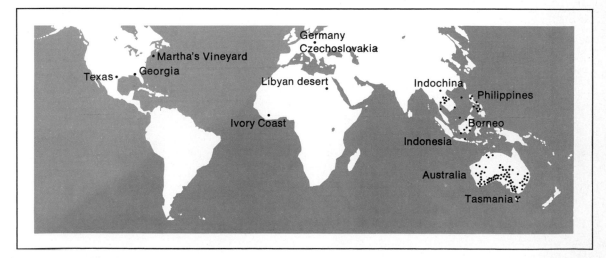

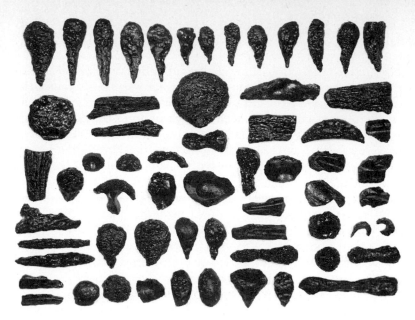

Figure 6.23 *Some tektites from Thailand. (Ronald Oriti, Griffith Observatory)*

artificially manufactured (by NASA scientists) by placing a sphere of glass in a wind tunnel and applying heat to one side from a blow torch. The models so closely resemble the true australites that it is not easy to distinguish them.

Radioactive dating of the various tektite types also seems quite well established; the general group called Australasian tektites (from Australia, Indochina, Indonesia, the Philippines, and nearby) are approximately 700,000 years old; the Ivory Coast tektites are approximately 1.2 million years old; the Moldavites (from Czechoslovakia) are approximately 15 million years old; and the North American tektites are approximately 35 million years old. Thus we are not looking for a single event but for a number of events which correspond to these dates.

Figure 6.24 *Australites—tektites from Australia. (NASA)*

Opinion diverges when we try to determine the place of origin of the tektites and the mechanism whereby they were placed into flight. Researchers have suggested that they may have come from as far away as the moon, but probably no farther, for had they come to earth from the distance of Mars, for instance, their dispersal on the surface of the earth would have been far greater. Let us consider four possibilities; then perhaps you will think of still more. Tektites may have been thrown from the moon as a result either of meteorite impact or of volcanic activity. The same two possibilities occur on the earth as well, yielding four distinctly different hypothetical origins.

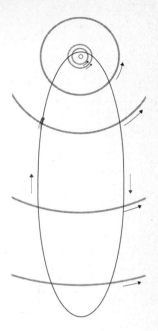

One group of researchers finds the most difficult question posed by theories of terrestrial origin, whether by impact or by volcanic activity, to be the mechanism that would impart sufficient velocity to these blobs of glass to sustain them in flight long enough to produce all the observed effects. Owing to the lesser surface gravity of the moon, this question is less problematic in theories of lunar origin; the accelerated blobs could more easily escape the moon's gravitational field. If they originated by impact of a meteorite on earth, furthermore, we must ask: where are the craters which should be left by the impact? Even for the youngest tektites from Australia there are no signs of remnant craters. On the other hand, craters have been found near the tektite falls in Germany (the Ries Basin) and the Ivory Coast (Bosumtwi Crater in Ghana). The dating of these craters shows fairly close agreement with the ages of their respective tektites. This seems to strengthen the possibility of terrestrial origin, but further testing is called for.

When we compare the composition of tektites with that of earth rocks or earth soil, we find significant differences in silicon content and in water levels. Typically terrestrial rock is lower in silicon and higher in water content than tektites. The water could easily have been lost in the heating process, however. When lunar rock samples were returned by the Apollo crews, none seemed to correlate with tektite composition. On the other hand, the glassy portions of the lunar soil can be shown to be very similar. One set of observers see this as strong evidence for lunar origin, whereas another group interpret the craters found in Germany and the Ivory Coast as equally strong evidence for terrestrial origin. Certainly the question of origin cannot be easily settled.

Micrometeorites

In addition to meteorite and tektite falls, approximately 3000 metric tons of other material fall on the earth each day from extraterrestrial sources. Most of this material may be termed *micrometeorites:* individual particles so small that they are not heated as they pass through the earth's atmosphere and therefore produce no visible streaks of light. This form of deposit may actually serve to enrich the soil in certain lean areas of the earth's surface, yet the percentage of material which is added to the earth's mass, even in a million years, is negligible.

Questions

1. The first asteroids discovered had orbits that lie between the orbits of which two planets?

2. What is the general range of size of the asteroids?

3. Why are astronomers particularly interested in the asteroids which have very eccentric orbits that carry them close to the sun and across the orbits of several planets?

4. New asteroids are discovered most easily on a photographic plate that has been exposed for an hour or more. During the exposure, the telescope is moved to follow the stars. How is an asteroid recognized on such a plate?

5. If all the material of the known asteroids could be combined into one "planet," how would that planet compare in size and mass to the earth?

6. Does a comet have a tail during its entire orbit around the sun? Explain.

7. The planets all orbit the sun in the same direction and along almost the same plane. Is this also true of comets? Explain.

8. Explain what is meant by the "Jovian group" of comets.

9. What is thought to be the primary nature of the head of a comet?

10. Why does the tail of a comet always extend away from the sun?

11. May only professional astronomers have their names associated with a new comet? Explain.

12. Describe the usual appearance of a comet when it is first discovered.

13. When is the best time to observe meteors? Why?

14. What is the range of size of most meteoroids as they begin their fiery flight into the earth's atmosphere?

15. Meteor showers are thought to be associated with what other astronomical object?

16. Why is the occurrence of a meteor shower predictable?

17. List the different types of meteorites.

18. Most meteorites that have been dated by radioactive elements have been found to be which of the following: (a) much

younger than the earth, (b) about the same age as the earth, (c) much older than the earth?

19. Why have we not found a meteor crater for each large meteorite that has hit the earth in the last 1 billion years?

20. True or false: The composition of tektites proves their origin on the moon. Explain.

21. What evidence does the astronomer have for thinking of asteroids as objects of irregular shape?

22. Why is it true that comets often develop two tails? How would you recognize the composition of each tail?

23. In addition to the theory that proposes a comet cloud surrounding the solar system, can you think of any other theory which might explain the origin of comets?

24. In what ways do Kepler's laws apply to comets? How is the period of a comet related to its average distance from the sun?

25. Why is it impossible to predict the approximate number of meteors that will be seen during a given meteor shower, even if that same shower has been observed many times before?

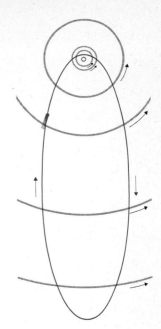

Suggested Readings

CAMERON, I. R., Meteorites and cosmic radiation. *Scientific American* 229 (1), 64–73 (1973).

GINGERICH, OWEN, Tycho Brahe and the Great Comet of 1577. *Sky and Telescope* 54 (6), 452–458 (1977).

HARTMANN, WILLIAM K., Cratering in the solar system. *Scientific American* 236 (1), 84 (1977).

JACCHIA, LUIGI, The brightness of comets. *Sky and Telescope* 47 (4), 216–220 (1974).

JACCHIA, LUIGI, A scientist's comet. *Sky and Telescope* 47 (3), 153–158 (1974).

LAWLESS, J. G., FOLSOME, C. E., and KVENVOLDEN, K. A., Organic matter in meteorites. *Scientific American* 226 (6), 38–46 (1972).

MIDDLEHURST, BARBARA M., and KUIPER, GERARD P. (eds.), *The moon, meteorites and comets*. Chicago: University of Chicago Press, 1963.

O'KEEFE, JOHN A., The tektite problem. *Scientific American* 239 (2), 116 (1978).

SCIENTIFIC AMERICAN, *The solar system* (readings). San Francisco: W. H. Freeman & Company Publishers, 1975.

WHIPPLE, FRED L., The nature of comets. *Scientific American* 230 (2), 48–57 (1974).

WOOD, JOHN A., *Meteorites and the origin of the planets*. New York: McGraw-Hill, 1968.

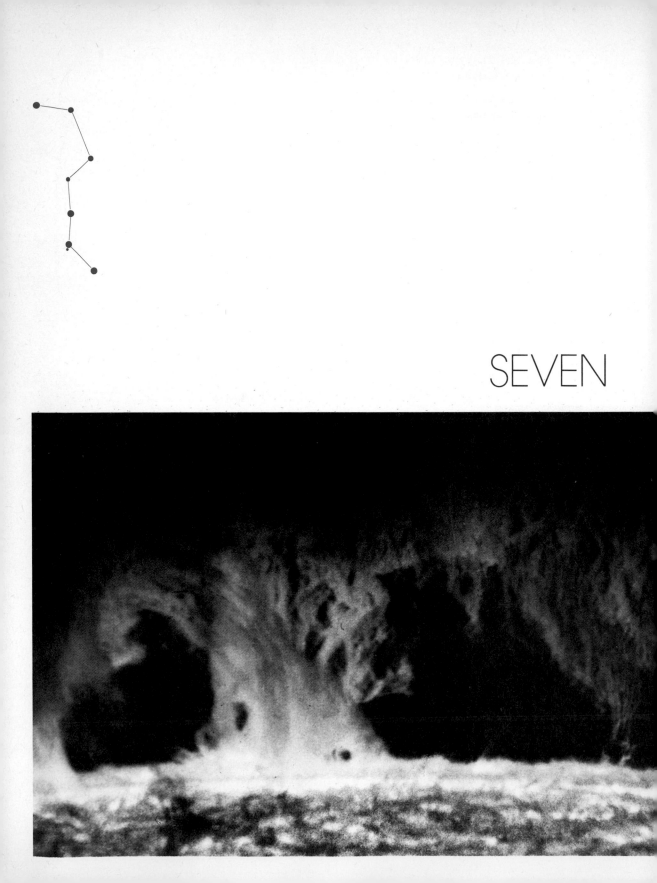

SEVEN

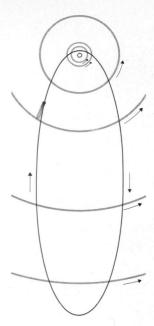

THE SUN

To speak of the sun as a star is to recognize its true nature, a nature which is drastically different from that of the planets, the moon, the meteors, and the comets. The sun is a tremendous seething inferno, generating its own energy from within. Most of the sources of energy on the earth may be traced back to the sun. Fossil fuels, such as coal and oil, are derived from once-living organisms that were dependent on the sun for their existence. Life as we know it today depends upon energy from the sun for its continuance.

In considering our dependence on solar energy, we must realize that the earth is situated at a very strategic distance from the sun. The average distance of the earth from the sun is called the *astronomical unit.* Modern methods, using radar, have greatly simplified the task of finding this distance; however, if a radar signal were directed toward the sun, its echo would be very difficult to detect against the background of other radio signals that the sun itself emits. Therefore we use an indirect approach. A radar signal is directed toward a given planet or asteroid, and the time required for the echo of that signal to be heard is noted. Since the observer knows the speed with which the radar signal travels, he may then compute the distance to the object from the time required for the

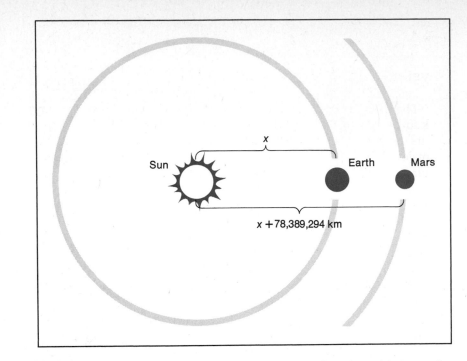

Figure 7.1 *Average distance between the sun and earth, and between the sun and Mars. The period of Mars (P_1) is 1.88 years; the period of the earth (P_2) is 1 year.*

two-way trip. Knowing both the distance and period of an object that orbits the sun, he can then calculate the astronomical unit using Kepler's third law, which says that the square of a planet's period is proportional to the cube of its average distance from the sun. For instance, suppose that the average distance between the earth and Mars at opposition is found by radar observation to be 78,389,294 km; then the computation of the astronomical unit would take the following form (see Figure 7.1):

$$\frac{P_1{}^2}{P_2{}^2} = \frac{r_1{}^3}{r_2{}^3}$$

$$\frac{(1.88)^2}{(1)^2} = \frac{(x + 78,389,294)^3}{x^3}$$

$$x = 149,597,890 \text{ km} \qquad \text{(details of the solution have been omitted)}$$

Here P_1 is the period of Mars and P_2 is that of the earth; r_1 is the average distance of Mars from the sun and r_2 (equals x) is the average distance of the earth from the sun. Thus, 145,597,890 km is the internationally adopted value of the astronomical unit (A.U.).

Because the sun is approximately 300,000 times closer than the next nearest star, we may determine its diameter rather easily, as we shall now show.

Physical Characteristics of the Sun

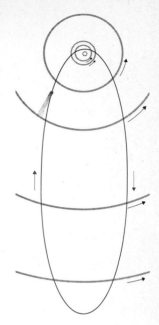

The determination of the diameter of the sun follows directly from our knowledge of its distance from the earth as well as the apparent angle that its diameter makes with our eye. Let us suppose that the sun makes an angle of 0.532° on a day when its distance is 149,598,000 km. The calculations that follow exemplify a very useful technique which requires only a simple proportion and arithmetic to solve for the diameter of a distant object.

Visualize the earth at the center of a huge circle of radius 149,598,000 km (see Figure 7.2). Think of the sun's diameter as a portion of the circumference of that circle.

Note that the angle of 0.532° compares to the entire angle at the center of the circle (360°) in the same way that the diameter of the sun (d) compares to the entire circumference of the circle ($2\pi r$):

$$\frac{0.532°}{360°} = \frac{d}{2\pi(149{,}598{,}000 \text{ km})}$$

Figure 7.2   *Determining the diameter of the sun.*

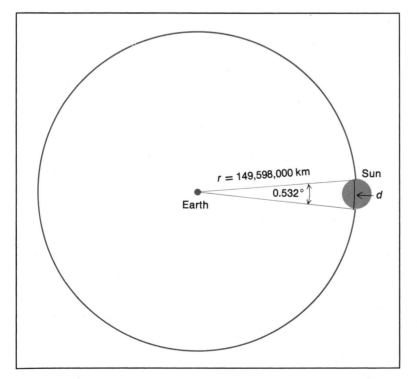

Solving for d:

$$d = \frac{0.532(2\pi)\,(149,598,000)}{360} = 1,390,000 \text{ km}$$

This gives a diameter of approximately 1,390,000 km, almost 110 times that of the earth. The immense size of the sun may be visualized by thinking of 110 earths placed side by side across the sun's diameter; however, we may find it more productive to visualize its volume.

The volumes of two spheres compare as the cubes of their radii (or diameters), so the volume of the sun must exceed that of the earth by a factor of $(110)^3$ or 1,331,000: over a million earths would fit into the space occupied by the sun. The mass of the sun is determined by the gravitational effect that it has on the earth, causing the earth to "fall" toward the sun at all times, thus maintaining the earth in its orbit. Using Newton's laws of motion, we find the mass of the sun to be approximately 2×10^{33} g, more than 300,000 times the mass of the earth. When mass and volume are compared, we see that the average density of the sun is only one-fourth that of the earth, or 1.4 g/cm³. This density is similar to that of certain rocks on the crust of the earth. Does this suggest that the sun is of a crusty nature, rigid in shape and form? No. The sun must be characterized as a ball of gas that under certain conditions of temperature and pressure can have an increased density without being a solid.

The Sun: A Hydrogen Fusion Reactor

We often use the expression, "The sun is burning." Were the sun merely burning, however, it could not begin to produce the vast amount of energy that radiates from its surface. The astronomer now realizes that the true source of the sun's energy is a nuclear reaction similar to that of the hydrogen bomb but taking place under controlling conditions. At the temperature (15 million degrees Kelvin and up) and pressure (1 billion atmospheres) found at the core of the sun, atoms tend to fuse together and in the process release energy. The hydrogen atom supplies the "fuel" for this reaction. Hydrogen is the simplest atom in the universe, consisting of one *proton* (positively charged particle) in the *nucleus* and one *electron* (negatively charged particle) moving about the *nucleus*. For our present purpose we may ignore the electron, for we are concerned primarily with nuclear changes. One would think that two positively charged protons would repel each other, and under ordinary conditions they do. However, under the conditions of high temperature and pressure in the core of the sun, two protons may be forced so close together that a nuclear binding force becomes more effective than their electrostatic repulsion. When two or more nuclear particles "stick" together, a new isotope or element is formed. The first step in this chain reaction occurs when two hydrogen nuclei fuse to form a special type of hydrogen called *deuterium* ($_1^2$H). This may be pictured as follows, where $\oplus$ represents a proton and $\textcircled{N}$ represents a neutron:

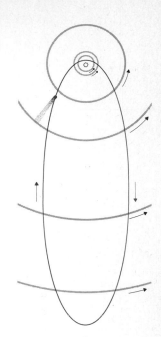

$$\oplus \; + \; \oplus \longrightarrow \; \oplus \; \textcircled{N} \; + \; \text{positron} \; + \; \text{neutrino}$$

or

$$_1^1\text{H} + {}_1^1\text{H} \longrightarrow {}_1^2\text{H} + {}_1^0e + \text{neutrino}$$

A positron ($_1^0e$) is viewed as a particle similar in mass to an electron but with a positive charge. The neutrino is a particle with no mass or charge. Because of their character, neutrinos are very hard to detect. The second step occurs when the deuterium atom fuses with still another hydrogen atom to form helium-3 ($_2^3$He) plus a gamma ray:

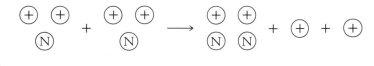

or

$$_1^2\text{H} + {}_1^1\text{H} \longrightarrow {}_2^3\text{He} + \text{gamma ray}$$

The third step occurs when two such helium-3 ($_2^3$He) atoms fuse to form an atom of helium-4 ($_2^4$He) and two hydrogen atoms:

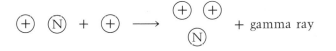

or

$$_2^3\text{He} + {}_2^3\text{He} \longrightarrow {}_2^4\text{He} + {}_1^1\text{H} + {}_1^1\text{H}$$

Since two hydrogen atoms are given back in the final reaction, only four hydrogen atoms were used in forming the one helium-4 atom; however, the mass of the helium-4 atom is less than the total mass of the four hydrogen atoms. It is this difference in mass, which has been converted into energy, that constitutes the source of energy for the sun and for stars in general.

In order to specify the mass of a given nucleus, we adopt the following convention: let one-twelfth the mass of a carbon-12 nucleus be called 1 *amu* (atomic mass unit). On this basis each hydrogen nucleus has a mass of 1.008 amu, making a total of 4.032 amu for four such nuclei. The helium-4 nucleus has a mass of 4.003 amu, hence 0.029 units have been converted into energy. How much energy? The relationship between mass and energy was expressed by Albert Einstein in the equation $E = mc^2$, where E is the amount of energy (in ergs), m is the mass that is converted (in grams), and c is the velocity of light (in centimeters per second). The velocity of light has been found to be approximately 3×10^{10} cm/sec. This number squared would equal 9×10^{20}. Therefore Einstein's equation, stated in words, says that the total amount of energy produced is equivalent to 9×10^{20} (that is, 900,000,000,000,000,000,000) times the mass that is converted, using the units mentioned above. The sun converts almost 4.5 million metric tons of mass per second, producing energy at a rate of almost 4×10^{33} erg/sec. Considering only the hydrogen in the core of the sun, the sun could continue this reaction for more than 10 billion years. This nuclear reaction, called the *proton-proton cycle*, is initiated and sustained by a central core temperature of about 15 million degrees Kelvin. At higher temperatures a reaction called the *CNO cycle* may become more important, but the net result is almost the same. Four hydrogen nuclei react to form one helium-4 nucleus, the difference in mass having been converted into energy. Carbon, nitrogen, and oxygen merely serve as catalysts for the reaction. (The CNO cycle is considered in detail in Chapter 12.)

We have distinguished thermonuclear fusion from the superficial phenomena of burning; however, astronomers typically speak of this fusion process as *hydrogen burning*. We will adopt this convention for the balance of the book; *helium burning*, then, will mean helium fusion to form still heavier elements.

Neutrinos from the Sun

In the very first step of the proton-proton cycle we saw the production of a neutrino, a particle which possesses neither mass nor charge. Virtually all neutrinos, because of their massless nature, pass directly from the core of the sun to the earth without interacting with the other material of the sun. Thus, if these particles could be detected, astronomers would in effect be "looking" into the core of the sun. However, because of their character, neutrinos are difficult to capture. A very interesting attempt at such captures has been made over the past ten years. In a rock cavern 4,850 feet below the surface of the Homestead Mine in Lead, South Dakota, a 100,000 gallon tank of tetrachloroethylene (C_2Cl_4) was installed under the auspices of the Brookhaven National Laboratory (see Figure 7.3). It was theorized by the researchers that neutrinos could be expected to react with the chlorine-37 atoms to produce a radioactive form of argon that is easily detectable. Because similar reactions might also be initiated by other particles (i.e., free protons or neutrons) that penetrated the tank, the underground shielding was employed. It was predicted that 100,000,000,000 neutrinos would penetrate every square centimeter of the earth's surface per second, passing through the earth almost unimpeded. The aim was to detect some interaction in the tank of C_2Cl_4. To date, the capture rate has been less than one-third that predicted. These experimental results pose a serious

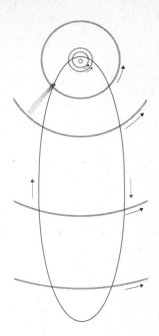

Figure 7.3 *The Brookhaven solar neutrino experiment, located 4850 feet underground in the Homestake Gold Mine in Lead, South Dakota. The tank contains 100,000 gallons of tetrachloroethylene (C_2Cl_4), with which neutrinos react to produce radioactive argon-37. Using this device, astronomers seek evidence of the thermonuclear process going on in the core of the sun. (Brookhaven National Laboratory, Upton, New York)*

question for the solar astronomer. Could the well-established proton-proton theory be wrong? Could the sun's core be cooler than predicted by solar models? Or do we simply fail to understand the neutrino? Some observers have suggested that the thermonuclear reaction in the core of the sun is not as constant as we had thought and may simply be at a low in a long cycle. Certainly continued experimentation in this area is required.

Layers of the Sun

The fusion of hydrogen into helium is continually building a helium core within the sun, with the immediate layer surrounding that core being the layer of thermonuclear reactions (hydrogen "burning"). Tremendous quantities of energy are being produced and must be transported to other layers and ultimately to space. Astronomers believe that the major transport mechanism within the body of the sun is radiation. Energy generated in the core of the sun travels through the *radiation zone* by means of a multiple absorption and reemission process. A photon which leaves the core does not radiate in a straight line to the surface of the sun but rather experiences multiple collisions, scatterings, absorptions, and reemissions. If you could follow a single photon through such a myriad of random reactions, you would find that up to a million years is required for that single reaction to reach the next layer and become visible to the outside world. Thus, as we view the sun in visible light, we are not seeing the core but rather its upper layers from which the photons are finally emitted. Contrast this with the fact that neutrinos

Figure 7.4   *Layers of the sun, shown with approximate temperature distribution.*

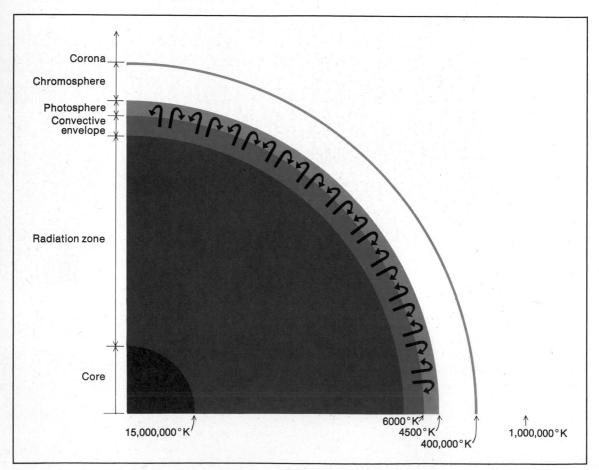

come directly to earth from the core of the sun, and you will appreciate why astronomers are so interested in detecting them.

The sun's next layer, called the *convective envelope*, transports energy only by means of the motion of the material within it, hot gases rising to deliver energy to the visible surface of the sun and then the cooled gases falling to be reheated (see Figure 7.4). You are probably familiar with the phenomenon of convection in the heating of a room. If the heater is located on one side of the room, the air which it warms will tend to rise, flow across the ceiling, and fall on the far side of the room as it cools. Thus a natural circulation pattern is set up. We can imagine the violent motion which must take place in order to transport the vast quantities of energy being emitted from the radiation zone through the convective envelope—but what evidence do we have of this turbulent motion? When the sun is photographed in white light, its "surface" takes on a spotty appearance called the *granulation of the photosphere* (Figure 7.5). The bright spots are interpreted as rising columns of hot gases, the darker regions as cooled gases returning to lower levels to be heated once again. This interpretation is borne out when we inspect the spectrogram

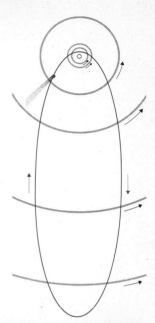

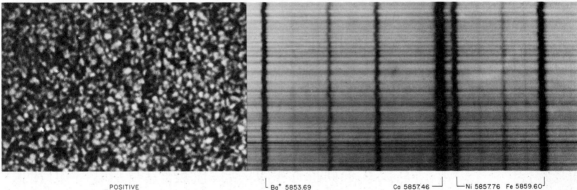

POSITIVE   Ba⁺ 5853.69   Ca 5857.46   Ni 5857.76 Fe 5859.60

Figure 7.5   *The granulation of the photosphere. The rising columns of gas cause a Doppler shift in the spectrum in one direction, whereas the falling gases cause a shift in the opposite direction, producing the waviness of the spectral lines. (Aerospace Corporation; Sacramento Peak Observatory)*

accompanying Figure 7.5. The waviness of the spectral lines is due to the Doppler shifts caused by the motion of the gases; blue shifts are associated with the rising spots that are approaching the observer; red shifts are associated with the cooler gases receding from the observer.

The layer immediately above the convective envelope is called the *photosphere*. This layer represents the visible surface of the sun in a white-light photograph (Figure 7.6); some variation in depth, however, is shown by the darkening near the sun's limb (edge). When we look at the central portion of the solar image, we can see through the photosphere to its lower levels, where temperatures are of the order of 6000°K. When we

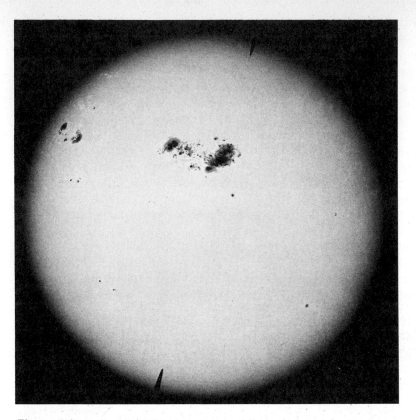

Figure 7.6 *A white light photograph of the sun showing limb darking due to the cooling trend in the upper levels of the photosphere. (Hale Observatories)*

look near the limb of the sun, we are seeing the higher levels of the photosphere, where temperatures are 1000°K to 1500°K cooler; the limb therefore appears darker to us. This temperature decrease not only explains limb darkening but is an essential factor in the formation of the solar spectrum. The spectrum of the sun is an absorption spectrum (Figure 7.7), and it is within this photospheric layer that absorption takes place, according to the elements that are present. In other words, if we could look at the sun's spectrum at the lower boundary of the photosphere, it would likely be of a continuous nature, but as we view it from the upper boundary, it is an absorption spectrum. As a result it reveals the composition of the photosphere and lower chromosphere, not of the core or other interior layers.

The next layer of the solar atmosphere, the *chromosphere* (colored sphere), extends approximately 10,000 km above the photosphere. The chromosphere is normally invisible, owing to the brilliance of the photosphere; however, when the central disk of the sun is just covered by the moon during an eclipse, the reddish light of the chromosphere may be seen. When the light from this layer is allowed to enter the spectrograph, its bright-line spectrum is revealed (Figure 7.8). The chromosphere is composed largely of hydrogen gas, which produces a bright red line in its spectrum; hence it takes on its characteristic reddish hue. In addition to hydrogen, the bright lines reveal the presence of sodium, calcium, magnesium, and ionized helium. No longer must the astronomer wait for the occurrence of a solar eclipse to study the chromosphere, for he may artificially cover the solar disk near the focal plane of his telescope and thereby photograph only the chromosphere itself. Such a device is called a *coronagraph*. The spectrum of the chromosphere reveals a temperature gradation from 4500°K near the photosphere to 400,000°K in its upper levels.

The third principal layer of the solar atmosphere is the *corona* (Figure 7.9). Like the chromosphere, the corona is normally invisible, owing to the brilliance of the photosphere; however, if it is viewed during a total solar eclipse or photographed using the coronagraph, one can see its extended luminous region, largely as a result of the scattering of light by particles in the corona. The visible portion is only a very small part of

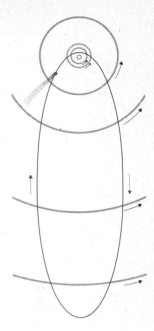

Figure 7.7 *The solar spectrum, with certain lines identified with known elements. (Hale Observatories)*

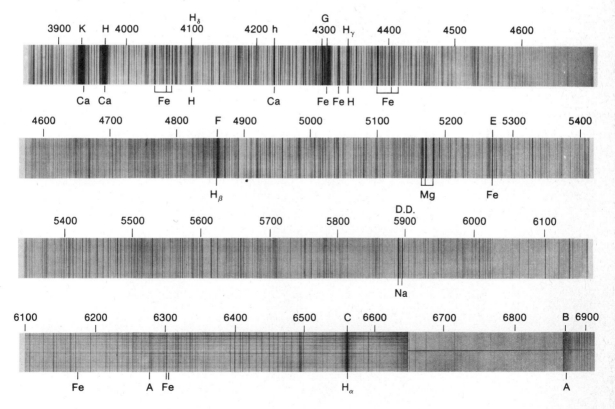

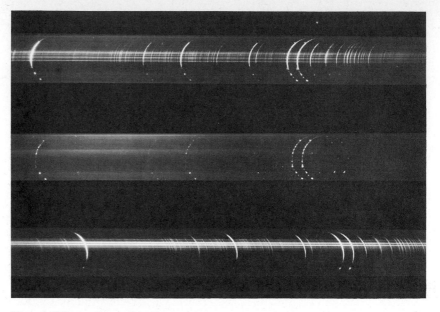

Figure 7.8 *The bright line spectrum of the chromosphere, photographed during a total solar eclipse. (Hale Observatories)*

the corona, for the latter is now known to extend beyond the earth itself. In fact, most of the planets are continually bathed in its outflow of material. We have noted that the sun sends out in every direction many forms of electromagnetic disturbances, such as light, radio, and X rays; however, this is not the full extent of its emission. There also appears to

Figure 7.9 *The corona: (a) during a sunspot minimum; (b) during a sunspot maximum. (High Altitude Observatory; Yerkes Observatory)*

(a)

(b)

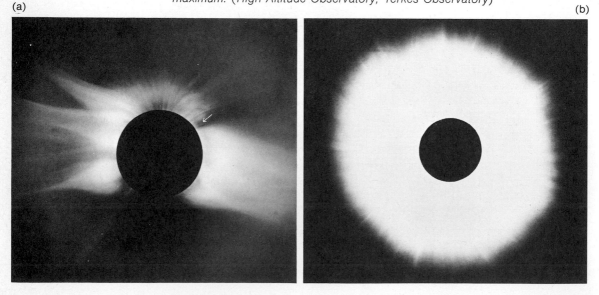

be a continual outflow of particles, including negatively charged electrons and positively charged protons, the components of the hydrogen gas that is the basic constituent of the solar atmosphere. While the corona is more static near the sun, farther out near the planets it constitutes an expanding envelope and therefore produces a rapidly moving stream of charged particles called the *solar wind*. We have seen its effect on the tail of a comet, always driving it away from the sun. At the earth's distance from the sun, the solar wind has a velocity of about 1,450,000 km/hr. The amount of material that is removed from the sun, in the form of this solar wind, is approximately 1×10^9 kg/sec, which in terms of the sun's total mass represents a negligible amount of material. As these charged particles approach the earth's magnetic field, they are deflected into a spiral path around the earth and eventually spiral toward the north or south polar regions, where they react with the atoms of the earth's atmosphere to produce the northern or southern lights.

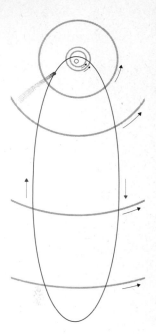

The temperature of the corona can be determined by the degree to which elements in this layer are ionized. Iron, for instance, has 26 electrons in each atom under normal conditions of pressure and temperature. In the corona of the sun, the spectrum reveals that as many as 13 of these electrons have been lost, indicating a temperature of about 2 million degrees Kelvin.

This very high temperature demands a special explanation, for none of the usual mechanisms of heat transfer will suffice. One theory suggests that acoustic (sound) waves are generated in the convective envelope of the sun and as they travel outward they produce a shock wave whereby energy is transferred to the corona. While this theory is very plausible, no direct evidence of such waves has been found, hence alternate ideas are also being pursued. For instance, it is possible for magnetic fields, by the electric currents they develop, to heat the corona, even as we would use an electric current flowing in a resistor to heat our homes. This latter theory has definite merit in light of the discovery of coronal "holes."

It is only recently that astronomers have recognized the significance of certain "gaps" in the corona of the sun. Such gaps are especially evident when the sunspot cycle is near a minimum. In Figure 7.9 (a) you will note that virtually no coronal plumes extend from the polar regions; this is quite normal. However, note the dark region marked by the small white arrow. This is called a "hole" in the corona. It is marked by relatively cool temperature, very low density, and open magnetic-field lines. Magnetic regions of the sun normally are connected to regions of opposite polarity nearby. In the case of the coronal "hole" no region of opposite polarity exists nearby and its field lines remain open. The interesting result appears to be an unusual outpouring of the solar wind from that region, characterized by very high velocities. When such a stream of charged particles reaches the earth, the earth's magnetic field may be greatly disturbed. Such disturbances are called geomagnetic storms. This is only one of many ways in which the earth may be affected by solar activities.

The Solar Spectrum

The electromagnetic radiation of the sun exhibits itself in many forms: the light that we see, the infrared and ultraviolet light on either side of the visible spectrum, radio, X rays, and so on. Figure 7.10 shows a plot of the average output of the sun in these various wavelengths. The total area under the curve denotes the amount of solar energy received by the earth at the outer edge of its atmosphere. This amount, the so-called *solar constant*, is equivalent to approximately $2 \, cal/cm^2/min$. One calorie (cal) of energy is sufficient to raise the temperature of $1 \, cm^3$ of water $1°C$. While the amount of energy the earth receives from the sun is tremendous, we must remember that the earth intercepts less than one-billionth of the total energy that is radiated from the sun. The remainder is received by other planets or dissipated in space.

As early as 1814, the German scientist Joseph Fraunhofer recognized that the rainbowlike spectrum of the sun was crossed by several dark bands, which he designated by capital letters. In Figure 7.7 (page 275) you will see several identified as Fraunhofer lines: the H and K lines of Ca, the F line of H_B, the D lines of Na, and the C line of H_α.

Because of its high intensity, the visible spectrum of the sun is the easiest natural spectrum with which we may work; it may be spread out over a distance of 13 m or more, thus revealing more than 30,000 absorption lines. It is presumed that most of these lines were formed in the photosphere of the sun. It should be pointed out, however, that the earth's atmosphere may also create some of the lines, in which case they are called *telluric lines* and are usually associated with such elements as nitrogen, oxygen, and water vapor, which are found in the earth's atmosphere. Telluric lines may be distinguished from the spectral lines of the sun in that they show no Doppler shift, whereas the spectral lines of the receding and approaching limbs of the sun show a Doppler shift due to rotation.

How may the astronomer determine which elements are indicated by the some 30,000 lines in the solar spectrum? The spectrum of a known element must be placed alongside the solar spectrum to see if all the lines of the given element match their corresponding lines in the solar spectrum. In Figure 7.11 you will see that the lines of iron do in fact match a given set of spectral lines in the solar spectrum, from which we conclude that iron is present—in a vaporized form—in the photosphere of the sun. Following this procedure, approximately 70 of the 92 natural elements that occur on the earth have been found to exist within the solar atmosphere. One should not be misled into thinking that the abundance of the heavier elements is very great, however, for hydrogen and helium still constitute the major portion of the solar atmosphere. The total of all other elements present represents less than 1 percent of that atmosphere, as indicated by the relative weakness of their spectral lines (See Figure 7.7). Table 7.1 lists the more abundant elements. We will see, as we study

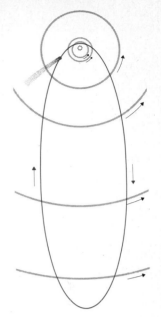

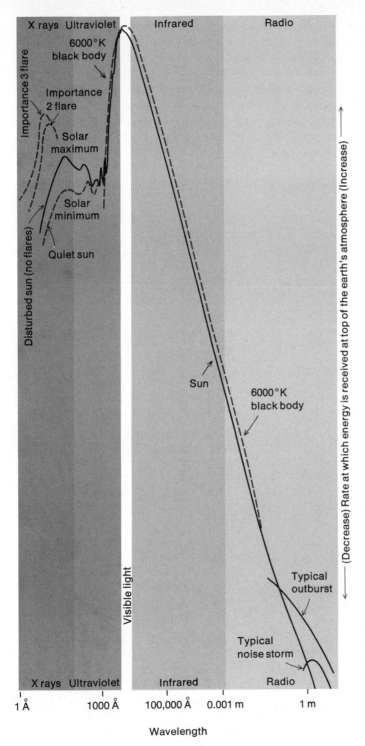

Figure 7.10 *The solar spectrum. The total solar energy received at the top of the earth's atmosphere is represented by the total area under the curve. This is called the solar constant and amounts to approximately 2 calories per square centimeter per minute. (Compiled by H. H. Malitson, NASA–Goddard Space Flight Center)*

other stars, that the abundances of the heavier elements run even less in some cases.

The spectrum of the sun also gives an indication of temperature. The photosphere radiates maximum energy at a wavelength of approximately 4700 Å (angstroms), indicating a surface temperature of 6000°K. This follows directly from a consideration of the energy radiated by a blackbody at various temperatures. A *blackbody* is any object that absorbs all the energy that falls on it and then radiates all the energy that it has received. As the temperature of the blackbody increases, not only does the total energy radiated increase, but the wavelength at which maximum energy occurs also changes (Figure 7.12). The sun's energy distribution best fits the blackbody curve corresponding to a temperature of 6000°K. These two facts may be specified rather accurately by the Stefan-Boltzmann law, which states that the total energy flux (*F*), which is represented by the area under the output curve, is proportional to the temperature (Kelvin) raised to the fourth power.

$$F = kT^4$$

This relationship indicates that a star with twice the surface temperature of the sun would radiate 16 times as much energy per unit of

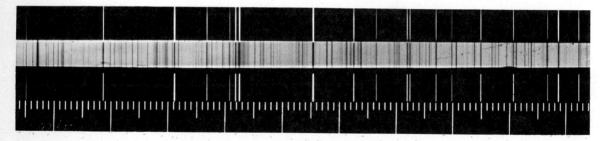

Figure 7.11   *A small portion of the solar spectrum. The central bands with dark lines represent the spectrum of the sun. The bright lines above and below the solar spectrum represent the spectrum of iron (photographed for comparison). (Hale Observatories)*

Table 7.1   *Abundances of Selected Elements in the Solar Atmosphere*

ELEMENT	PERCENTAGE
Carbon	0.035
Nitrogen	0.010
Oxygen	0.067
Magnesium	0.003
Silicon	0.004
Sulfur	0.002
Iron	0.003

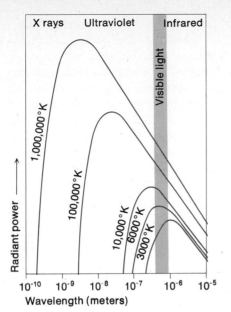

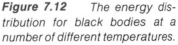

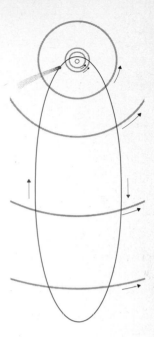

Figure 7.12 *The energy distribution for black bodies at a number of different temperatures.*

surface area. [Note: $(2)^4 = 16$.] It is also apparent from Figure 7.12 that hotter stars "peak out" in their radiation at shorter and shorter wavelengths. The details of this relationship, called Wien's law, are presented in Chapter 8, page 322.

Spectroheliograph

Solar activities cannot be easily seen in photographs taken in the ordinary light of the sun. A *spectroheliogram* is a photograph of the sun taken in the light of only one narrow portion of the spectrum. The word itself reveals this meaning, for if we take it apart in reverse order, "gram" means picture, "helio" means sun, and "spectro" means that we use the spectrum to obtain the picture. The *spectroheliograph* is the instrument by which such a picture is taken. This may be accomplished either by use of a filter that allows only the desired wavelength to pass or by use of a more complicated arrangement of spectrograph and a moving slit. If we choose to use the light in the first red line of the hydrogen spectrum—the hydrogen-alpha (H_α) line—we obtain a picture of the sun that represents the upper photosphere [see Figure 7.13 (b)]. If we want to view the lower chromosphere, we may use the light of the K line of ionized calcium (Ca II). The temperature within the bright regions, as viewed in the light of calcium, may reach 20,000°K, even though these regions are usually directly over a sunspot, a relatively cool region of the photosphere [see Figure 7.13 (c)].

With the increasing use of the Orbiting Astronomical Observatories, astronomers can now study the sun in ultraviolet wavelengths. In order to interpret these results, as seen in Figure 7.13 (d, e), we must first recognize how they were obtained. We have just referred to Ca II, which stands for calcium atoms that have been ionized once (that is, which have lost one of their electrons); Ca I stands for neutral calcium, atoms that have the same number of electrons as protons. The representation of the sun shown in Figure 7.13 (d) was derived from information received from an ionized form of oxygen, namely, O VI. This is oxygen which has been ionized five times. In such an atom, certain electron transitions are possible that are not possible in neutral oxygen. One such transition produces an ultraviolet wavelength of 1032 Å, and it was at this wavelength that the information was received. Furthermore, a temperature in the range of 350,000°K is necessary to ionize oxygen to this degree, and such a temperature exists in the upper chromosphere; hence we have a "picture" of the sun at that level. It is not a photograph in the usual sense, but rather a plot showing the hot spots (lighter areas) and cooler regions (dark areas).

Suppose we want to look at yet a higher level of the sun. We choose a wavelength produced by an ionized atom at still higher temperature,

Figure 7.13   *The sun: (a) in ordinary light; (b) in the light of H_α; (c) in the light of Ca II; (d) in ultraviolet radiation of O VI; (e) in ultraviolet radiation of Si XII; (f) X-ray photograph. [a-c, Hale Observatories; f, from L. P. VanSpeybroeck, A. S. Krieger, and G. S. Vaiana, Nature 227, 818–822 (1970)]*

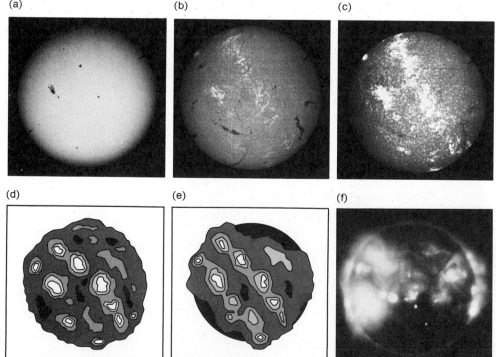

(a) (b) (c)

(d) (e) (f)

say, 2,000,000°K. Silicon XII (Si XII) is such an atom, emitting an ultraviolet line at 499 Å. We can now "see" the sun at a level well into the corona [Figure 7.13 (e)].

Rockets fired above the earth's atmosphere have given us still another view of the sun. Revealed in X-ray wavelengths [Figure 7.13 (f)] are perhaps the most energetic regions of the sun, areas of the corona where very high temperatures prevail.

Thus, by expanding their use of the entire spectrum, astronomers have gained a multilevel view of the sun.

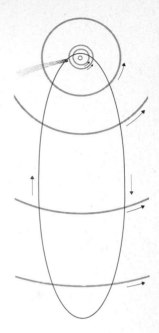

Magnetic Fields of the Sun

It will be quite apparent, in the remainder of this chapter, that solar activities are controlled by magnetic fields evident on and near the sun's visible surface. A strong suggestion of such a magnetic field, surrounding a pair of sunspots, is seen in the spectroheliogram of Figure 7.14. The dark filaments resemble the pattern formed by iron filings near a magnet. More direct evidence can be obtained, however, by use of the *Zeeman effect*. When light is created in the presence of a magnetic field, the lines of its spectrum are broadened or split. The degree to which the lines are split indicates the strength of the magnetic field, and the direction of optical polarization of the lines indicates the polarity of the field at any point (see Figure 7.15).

Figure 7.14   *A spectroheliogram showing dark filaments in the vicinity of a sunspot group. Strong magnetic fields are suggested in this Big Bear Solar Observatory photograph. (Hale Observatories)*

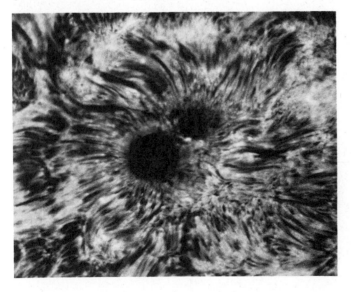

Figure 7.15   *The Zeeman effect. The splitting of spectral lines results from the presence of a magnetic field on the region of radiation. (Yerkes Observatory)*

If the sun is scanned on any given day, a plot of its magnetic properties may appear as in Figure 7.16. The bright areas represent the stronger fields, and the slant of the region indicates its polarity. The strength of the magnetic fields near sunspots measures about a thousand times that of the general field of the sun. In fact, it is very likely that the strong magnetic field creates the sunspot.

Figure 7.16   *The magnetic fields of the sun. The brighter areas represent the stronger fields, and the slant of a region indicates the polarity of that region. (Hale Observatories)*

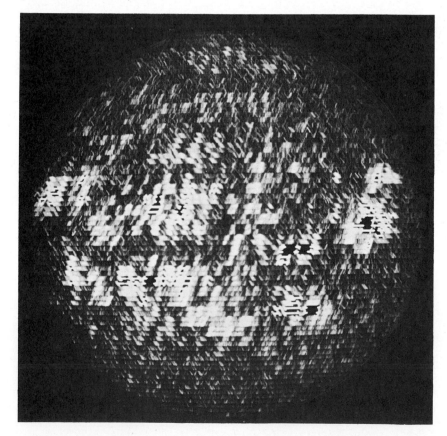

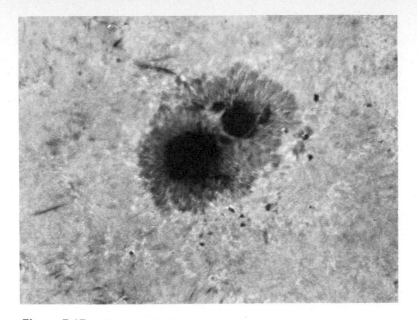

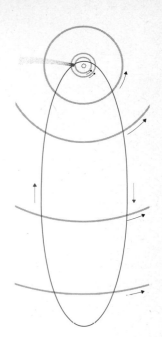

Figure 7.17 *Details of a sunspot group photographed in white light. The darkest portion is called the umbra and the gray portion is called the penumbra. Sunspots represent cooler regions on the visible surface of the sun. This same sunspot group, photographed in the red light of H_α is seen in Figure 7.14. (Hale Observatories)*

Sunspots

The *sunspot* is a region on the visible surface of the sun that consists of gases cooler than those surrounding it. The very dark central portion called the *umbra* has a temperature of approximately 4300°K. The gray *penumbral* region surrounding the umbra has a temperature of approximately 5000°K, compared to the general surface temperature of 6000°K. If one could isolate the sunspot from its surroundings, it would appear bright. This is to say that a sunspot is dark only in contrast to its surroundings (see Figure 7.17).

In 1842 an amateur astronomer, Heinrich Schwabe, suggested that sunspots increase and decrease in number in a somewhat regular cycle. Professional astronomers soon confirmed that a cycle does exist. Can you determine the period of that cycle as you inspect the record shown in Figure 7.18? Rather than looking at any one cycle, note that seven cycles are shown in a period of approximately 78 years (1876–1954). This yields an average period of 11.2 years (78 ÷ 7). As obvious as this cycle appears to us, we should not conclude that the sunspot phenomenon is quite as regular as it appears between 1874 and the present date.

Maunder Minimum

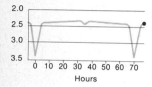

Between the years 1645 and 1715, almost no spots were observed on the sun. That in fact none had appeared seemed so unlikely that alternate explanations were sought: for example, perhaps no one was recording sunspots during that period. In 1893 an extensive search of the records was made by E. Walter Maunder of the Royal Greenwich Observatory. Maunder's conclusion that the spots were simply missing during that time has more recently been substantiated by the work and publications of John A. Eddy of the High Altitude Observatory. You are probably aware that the growth rings of older trees allow the scientist to "read" the history of local climatic changes over many years. A Dutch scientist, Hessel DeVries, investigated the content of carbon-14, a radioactive form of carbon, within the annual tree rings. Carbon-14 is created in the upper atmosphere of the earth, and its formation correlates with sunspot activity. His studies show a sharp rise in carbon-14 content during the period we now identify as the Maunder Minimum. The high carbon-14

Figure 7.18   *The butterfly pattern of sunspots from 1874 to 1953. The top view shows the plotting of the sunspots according to their latitude on the sun. The middle graph shows the number of sunspots. The lower graph shows the variations in the magnetic fields on the sun's surface. Note here the very close correlation between the sunspot cycle and the magnetic cycle. All three graphs reveal the 11-year cycle of these phenomena. (Royal Greenwich Observatory)*

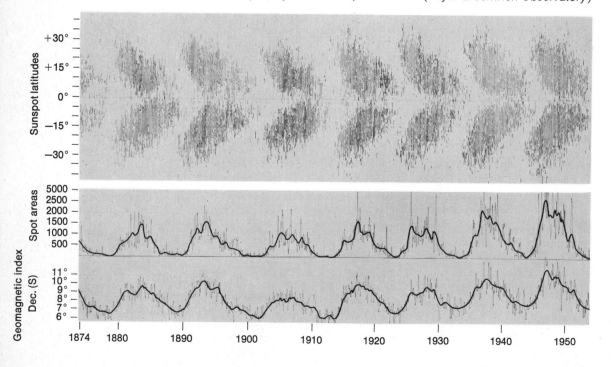

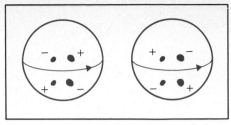

Figure 7.19 *Successive 11-year cycles of magnetic polarity of sunspots.*

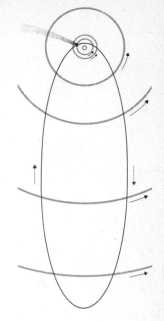

content correlates with low activity on the sun, because it is then that the earth experiences its least protection from cosmic rays. At the same time, the total energy received from the sun appears to be at a low, for glaciers appear to have grown during this time. To illustrate this point, the Maunder Minimum seems to correlate with what many observers would term the Little Ice Age, a time of extreme cold, particularly in Europe and Greenland.

As many as 12 periods like that of the Maunder Minimum are indicated in the tree-ring story. Today there is a tremendous interest in the relationship between solar activities and terrestrial weather; however, we must be forewarned that the variations we have been studying are very long-term ones. They have little direct relationship to the 11-year sunspot cycle or to variations evident in our lifetime. We are in danger of developing myopic vision.

Let us trace the appearance of sunspots over a period of 11 years. First, consider the sun when it is devoid of spots—a "quiet" sun—as it appeared in 1944–1945. Next we see a few sunspots appearing in the northern and southern hemispheres at about 30° latitude. As the number of spots continues to increase, we observe that they also appear closer to the solar equator. Completing the 11-year period, the spots diminish in number as they move near the equator. A plot of sunspots, according to their latitude, as time goes by reveals the butterfly pattern shown in the upper band of Figure 7.18. Note also that the correlation between sunspot activity and magnetic activities on the sun is confirmed in the lower two bands of Figure 7.18.

Sunspots often appear in pairs, of which one possesses a positive and the other a negative polarity. In any given 11-year cycle, the positive spot may lead the negative spot in the northern hemisphere, while the opposite is true in the southern hemisphere. In the next successive 11-year cycle, the order of the spots is reversed, as shown in Figure 7.19. The general magnetic field of the sun also shows this alternating nature.

Rotation of the Sun

It would be an interesting experiment, if you could provide yourself with a safe method of observing the sun, to watch sunspots move across

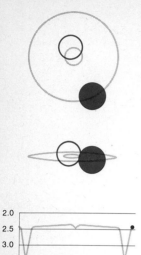

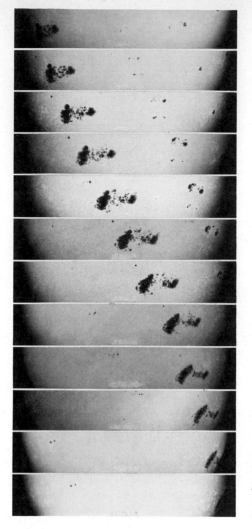

Figure 7.20 *The sunspots show the rotation of the sun, photographed over twelve successive days. North is at right. (Hale Observatories)*

the face of the sun, day after day. You would witness the rotation of the sun in approximately one month's time (see Figure 7.20). Careful study over several years, however, would reveal that the apparent period is not constant but varies with the latitude of the sunspot. When near the equator, a given sunspot group seems to require 27 days for one rotation. At successive latitudes north or south of the equator, longer periods are required, up to 33 days near the poles. This reminds us of the gaseous nature of the sun and specifically of its differential rates of rotation.

Some observers have speculated that it is the differential nature of the sun's rotation that creates its magnetic field and produces such strong fields at the sunspots.

Source of the Sun's Magnetism

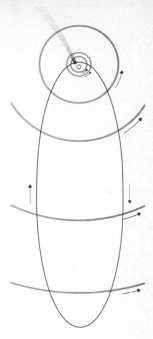

If the sun were standing still or rotating as a solid, one would expect its magnetic field lines to appear as in Figure 7.21 (a), with lines running north and south. Because the equator of the sun turns faster than higher latitudes (differential rotation), the magnetic lines tend to "wind up" as shown in Figure 7.21 (b–e), and the field is thereby strengthened. Experimentation with plasmas in the laboratory seems to confirm this possibility. Adjacent lines are thought to exert a buoyant force on each other, causing them to burst through the surface. At this eruption point a sunspot may occur, because the smooth flow of energy bypasses that region, leaving it cooler than its surroundings. While this theory of the magnetism of the sun represents some degree of speculation, note how it correlates with the remaining solar activities.

Plages

When the sun is photographed in the light of hydrogen-alpha or calcium II, a bright region often appears [see Figure 7.22 (b, c)]. The brightness of the region results from higher densities and temperatures

Figure 7.21 *Differential rotation of the sun may produce the magnetic fields associated with sunspots. This sequence shows how the faster rotation of the sun's equator may take north–south lines of magnetism and develop east–west lines and finally swell upward to form a loop over a sunspot.*

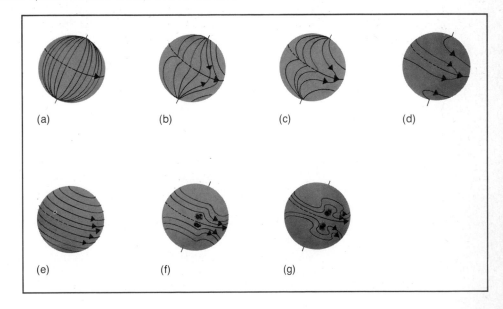

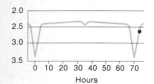

than exist in the surrounding chromosphere. Plages are thought to be due to the focusing of energy into that region by the strong magnetic fields which can be measured there. Further inspection of this region in white light [Figure 7.22 (a)] reveals that beneath the bright plage is a sunspot, dark owing to lack of energy. There appears to be a direct correlation between these facts. Furthermore, bright plages may appear before their corresponding sunspot and therefore herald its appearance. Plages often remain after the sunspot has disappeared.

Spicules

Near the *limb* (edge) of the sun, in the upper chromosphere, the spectroheliogram reveals spikelike columns of gas jetting out, much like the rising columns of gas in the photosphere. These *spicules* may show speeds of up to 32 km/sec and may sometimes reach to heights of 16,000 km (Figure 7.23). A given spicule lasts for only 10 to 15 minutes.

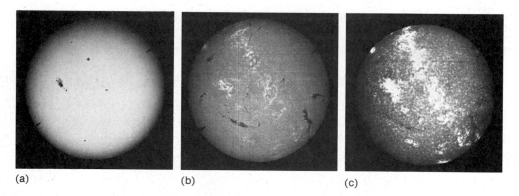

(a)   (b)   (c)

Figure 7.22   *The sun: (a) in ordinary light; (b) in the light of H$_\alpha$; (c) in the light of Ca II. (Hale Observatories)*

Prominences

For many years, activities of the solar atmosphere were observed only during a total solar eclipse, when the entire photosphere of the sun was covered by the moon. Gigantic protrusions were observed at that time. Now, by use of the spectroheliograph, these *prominences* may be studied at will. One kind of prominence may extend more than 320,000 km into the solar atmosphere and remain relatively fixed. This is known as the *quiescent* type. A second type, called a *loop prominence*, takes on the shape of an arch or loop and exhibits motion within the loop, suggesting the presence of a magnetic field (Figure 7.24).

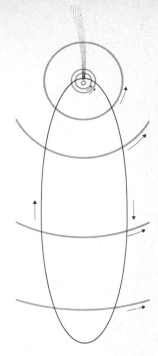

Figure 7.23 *Spicules in the sun's chromosphere photographed in red light of H_α at Big Bear Solar Observatory. (Hale Observatories)*

Figure 7.24 *A solar prominence that reached a height of 330,000 km, photographed in the light of calcium (K line). (Hale Observatories)*

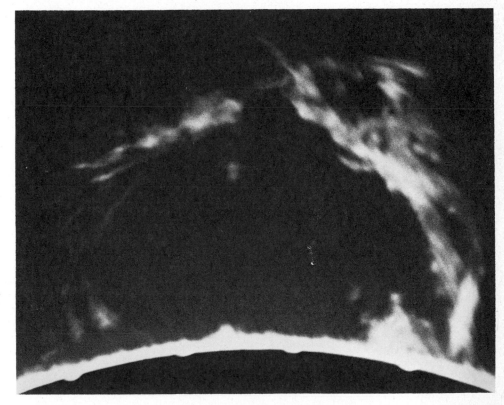

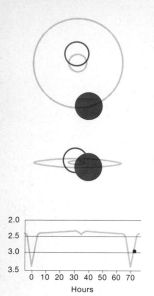

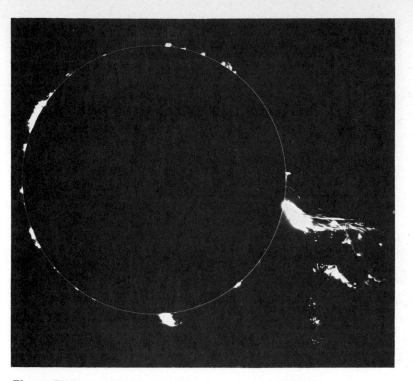

Figure 7.25 *Prominences visible on March 1, 1969. The very large eruptive prominence extends approximately 645,000 km above the surface of the sun. (Haleakala Observatory, Institute for Astronomy, University of Hawaii)*

A third type of prominence is the *eruptive* type, which may send material out thousands of kilometers into the corona of the sun at speeds up to 640 km/sec (Figure 7.25).

The dark filaments, as seen in H_α light, represent prominences viewed from above (Figure 7.26). They tend to form along the boundary between the regions of opposite magnetic polarity associated with sunspots.

The Solar Flare

The *solar flare*, the most dynamic activity associated with the surface or atmosphere of the sun, represents a tremendous release of energy in a very short time (Figure 7.27). The sudden brightening, usually in the vicinity of a sunspot, accompanies a violent outthrusting of material. Typically a flare occurs over an area 200,000 km in diameter, and the temperature associated with it may exceed 100 million degrees Kelvin. Flares often occur along the boundary between positively and negatively polarized regions of sunspot groups. It appears as though the

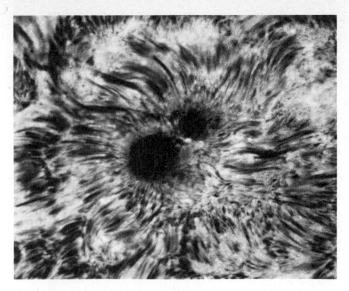

Figure 7.26 *A spectroheliogram showing dark filaments in the vicinity of a sunspot group. (Hale Observatories)*

magnetic fields had focused vast amounts of energy into a relatively small area.

Solar flares have a direct influence on the earth and on humans. The expulsion of high-energy particles from a major flare could kill astronauts if they were not protected by an atmosphere or by some artificial means. During a limited period a flare may radiate more X rays than the entire sun. The ionospheric layers of the earth's atmosphere are altered by such an outburst and may fail to reflect radio waves, thereby

Figure 7.27 *Solar flare. This sequence of spectroheliograms was taken over a period of 1 hr in the light of (a) H_α (hydrogen-alpha); (b) H_α plus 0.6 Å; (c) H_α plus 0.9 Å. Each sequence shows the development of the flare at a different level and also its association with a sunspot. (Aerospace Corporation; San Fernando Observatory)*

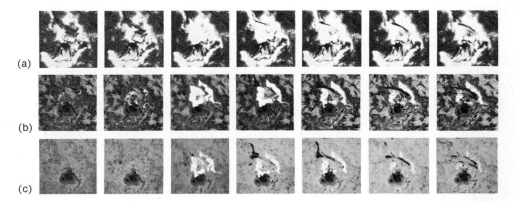

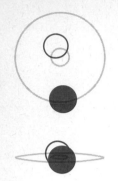

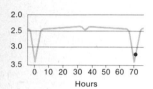

causing a communications blackout on the shortwave radio bands; the blackout may last for a few hours or up to several days. Aurora activity usually is greatly enhanced by the accelerated particles in the solar wind which are produced by the flare. About a day after the occurrence of a flare, the magnetic properties of the earth are usually disturbed, causing the needle of a compass to react in very strange ways.

Our Nearest Star

The sun is our nearest star, and we have been able to examine it quite closely (Figure 7.28). As we move out into space, to the other stars of our Galaxy, we ask the same questions about them: How far? How big? How massive? Do they rotate? What kind of atmospheres? Do they have spots like the sun? Do they have a magnetic field? What is their temperature? What kinds of energy do they radiate?

Figure 7.28   *The active sun.*

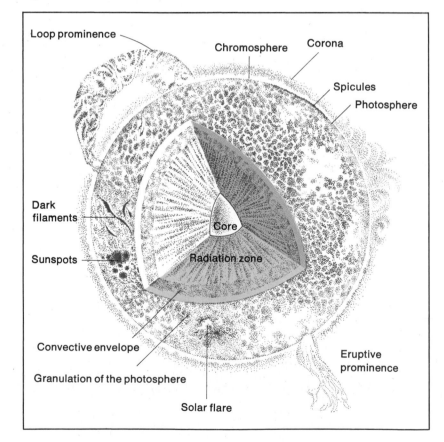

Questions

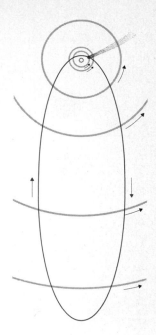

1. What basic characteristics identify the sun as a star rather than a planet?

2. What is the apparent diameter of the sun, as measured in degrees? How does this compare to the apparent diameter of the full moon?

3. List several means by which the rotation of the sun is measured.

4. Does the sun rotate as if it were solid? Explain.

5. What does the granulation of the photosphere seem to suggest?

6. State the minimum temperature necessary to sustain the process whereby hydrogen is converted to helium in the core of the sun.

7. If hydrogen is considered to be the "fuel" of the thermonuclear process in the core of the sun, then what occurrence within the process explains the tremendous release of energy that results?

8. How much material (mass) of the sun is converted into energy every second?

9. The dark line spectrum of the sun is produced at what level (or layer) of the sun?

10. List the things that an astronomer might deduce from the solar spectrum.

11. Of what does the solar wind consist?

12. The temperature of the corona of the sun has been measured by its ionization level and found to be approximately _____ °K.

13. Define the solar constant.

14. List the kinds of electromagnetic radiation that the sun produces.

15. The number and the location of sunspots seem to be repeated in a cycle of how many years?

16. When the spectral lines of the sun appear split, what condition is indicated?

17. When the sun is photographed in the light of hydrogen or of

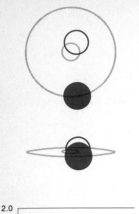

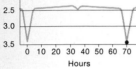

calcium, levels of the sun may be seen other than the photosphere. Explain what is seen in each of these lights.

18. The prominences of the sun are not seen in ordinary light. How may they be viewed?

19. What is the most dynamic activity associated with the sun?

20. How do solar activities affect the earth?

21. The normal spectrum of the sun is an absorption spectrum (dark lines). Why does the sun show an emission spectrum (bright lines) during a total solar eclipse?

22. Why is the chromosphere red?

23. The sun is now being photographed in the light of oxygen VI (ultraviolet) and in X rays. What do these photographs reveal by contrast with those taken in white light or in H_α (hydrogen-alpha) light?

24. What condition within a star or its atmosphere is necessary to ionize oxygen five times?

25. Supposing that the sun converts 5 million metric tons of mass into energy every second, find how much mass it has converted in 5 billion years.

Suggested Readings

BANDEEN, W. R., and MARANS, S. P. (eds.), *Possible relationships between solar activity and meteorological phenomena.* Washington, D.C.: Superintendent of Documents, Government Printing Office, NASA SP-366 1975.

EDDY, JOHN A., The case of the missing sunspots. *Scientific American* 236 (5), 80 (1977).

GINGERICH, O. (ed.), *Frontiers in astronomy* (Introductions to Chapters 8, 9, and 10). San Francisco: (W. H. Freeman & Company Publishers, 1970.

GOLDBERG, LEO, Ultraviolet astronomy. *Scientific American* 220 (6), 92–102 (1969).

GOSLING, J. T., and HUNDHAUSEN, A. J., Waves in the solar wind. *Scientific American* 236 (3), 36 (1977).

KUIPER, GERARD P. (ed.), *The sun.* Chicago: University of Chicago Press, 1953.

LIVINGSTON, W. C., Measuring solar photospheric magnetic fields. *Sky and Telescope* 43 (6), 344–349 (1972).

Noyes, Robert W., Ultraviolet studies of the solar atmosphere. *Annual Review of Astronomy and Astrophysics* 9, 209–236 (1971).

Pasachoff, Jay M., The fiery sun. *Natural History* 81 (5), 49–53 (1972).

Pasachoff, Jay M., The solar corona. *Scientific American* 229 (4), 68–79 (1973).

Zirin, Harold, *The solar atmosphere.* Waltham, Mass.: Blaisdell, 1966.

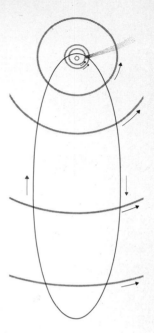

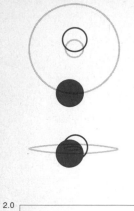

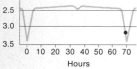

EIGHT

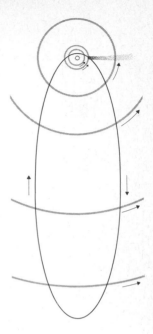

STARS IN GENERAL

In order to determine factors such as the direction, distance, velocity, size, temperature, and luminosity of the stars, the astronomer must rely entirely upon his ability to interpret the radiation that he receives from them. Let us first consider the method for determining the distance to nearby stars.

When it is necessary to measure an inaccessible distance on the earth, the surveyor uses a method of triangulation. For instance, suppose that we wish to know the distance from a dock (A) on the bank of a river to a dock (C) on an island in the middle of the river (Figure 8.1). By laying off a base line AB along the bank of the river, say, 100 m long, and then measuring the angles formed at A and B, one can solve for side AC and get the desired distance, using trigonometry.

For an object such as the moon, let us suppose that two observers are stationed at positions A and B on the earth (see Figure 8.2) and that they are in radio communication with each other. Observer B radios that he sees the moon in alignment with star D. At the same time, Observer A sees the moon in alignment with star C. Observer A can measure the angle between stars C and D, and since the

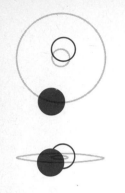

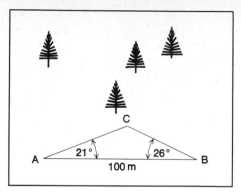

Figure 8.1   *Triangulation.*

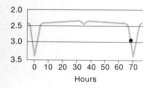

rays of light from star D are parallel* for both observers, we see that the angle between the stars is equal to the angle formed at the moon. This fact is expressed by the geometric theorem: *Alternate interior angles of parallel lines are equal.* Since the base line AE, the radius of the earth, is a known distance, it is possible to calculate the distance from A or B to the moon. This method, called *geocentric parallax*, is of only limited usefulness, owing to its relatively short base line. However, the usefulness of the idea may be extended by using a much longer base line—the radius of the earth's orbit (see Figure 8.3). Suppose that when the earth is in position A the nearby star S aligns with the more distant

---

*We may assume that the two lines pointing to star D (see Figures 8.2 and 8.3), even if separated by 93 million miles, may be treated as parallel, because the star itself is so far away—there is no measurable deviation from parallelism.

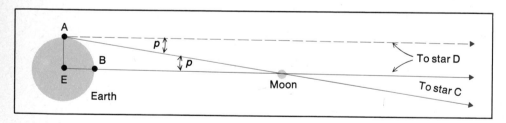

Figure 8.2   *Geocentric parallax.*

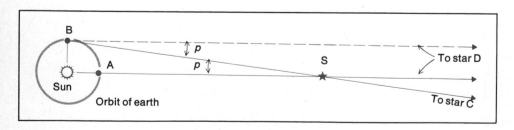

Figure 8.3   *Heliocentric parallax.*

star D, but that three months later, when the earth is at position B, star S aligns with star C. By measuring the angle between stars C and D, the astronomer then knows the angle formed at the star S. This angle (p), called the *heliocentric parallax* of the star, is always less than 1 second ($\frac{1}{3600}$ degree) in size. Triangles that have such a small angle can be solved by the following method. Consider a very large circle with center at star S and with radius equal to the distance r (see Figure 8.4). The base line (BF) is a small part of the large circle and compares to the entire circumference of the circle as the angle of parallax (p) compares to the entire circle of 360°, giving

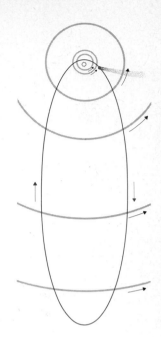

$$\frac{BF}{2\pi r} = \frac{p°}{360°}$$

Solving for r, we get

$$r = \frac{360°(BF)}{2\pi p°}$$

Changing units so as to use p in seconds and clearing the factor 2 yields

$$r = \frac{206{,}265(BF)}{p''}$$

Since BF is equal to 1 A.U., we can now find r in astronomical units:

$$r = \frac{206{,}265 \text{ A.U.}}{p''}$$

Figure 8.4 *Properties of small angles.*

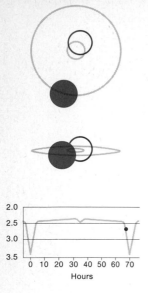

If we let 206,265 A.U. equal 1 parsec, we get a much simpler expression:

$$r = \frac{1}{p''} \text{parsec}$$

A *parsec* is simply the distance at which a star must be situated in order to exhibit 1 second of parallax, using 1 A.U. as a base line. The term "parsec" is a contraction of the two words *parallax* and *second*. Further calculations show that 1 parsec is equal to 3.26 light-years. Notice that the greater the distance (r) to the star, the smaller will be its parallax angle (p). Our nearest star, *Proxima Centauri*, has a parallax of 0.75 second, hence a distance of $1/0.75 = 1.33$ parsecs, or 4.3 light-years. For stars as distant as 100 pc (parsecs) the parallax angle is only 0.01″ (1/36,000°). The measurement of an angle this small is generally fraught with errors, which are large enough to nullify the effectiveness of the method beyond approximately 50 pc. In a subsequent section of this chapter ("Distance to Moving Clusters") we will study a method which is more accurate, but first we will need to look at several concepts regarding the motion of stars.

Motions of Stars

Stars show very little evidence of motion, and yet most are traveling at phenomenal speeds. These motions are not readily apparent because the stars are so far from us. However, if we could watch the stars in the Big Dipper over a period of 100,000 years, we would realize that they are moving in various different directions. Figure 8.5 shows the approximate location of the stars 100,000 years in the future. The small arrows indicate the direction of motion. (The left-hand flip pages beginning on page 284 reveal this proper motion of the stars in the Big Dipper.)

Proper Motion

The *proper motion* of a star—the rate at which its position in the sky changes—is measured in terms of seconds of arc per year. Many stars

Figure 8.5   *Motions of stars in the Big Dipper.*

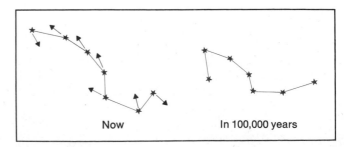

| Now | In 100,000 years |

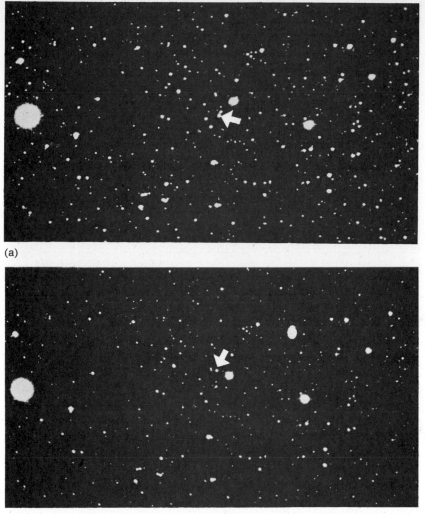

(a)

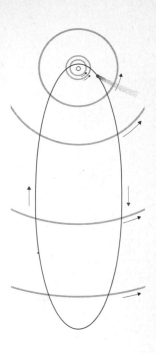

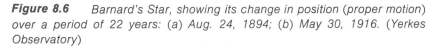

(b)

Figure 8.6 *Barnard's Star, showing its change in position (proper motion) over a period of 22 years: (a) Aug. 24, 1894; (b) May 30, 1916. (Yerkes Observatory)*

exhibit such small changes that they must be photographed over many years in order to measure that change. A star that shows a shift of 2 seconds of arc in a period of 10 years has a proper motion of 2 seconds divided by 10 years, or 0.2 second of arc/year. Barnard's Star has the largest observed proper motion, equal to 10.25 seconds of arc/year (Figure 8.6). This star shows relatively large proper motion because it is the second closest star to the sun. While slightly more than 300 stars have proper motions of at least 1.0 second of arc/year, the average star seen with the naked eye has a proper motion of less than that. If the distance to a star is known, its proper motion can be translated into a velocity that

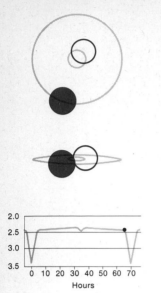

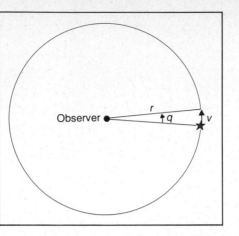

Figure 8.7
Tangential velocity (v).

is at right angles to the observer's line of sight. This is called the *tangential velocity* of the star. Suppose that a star A has a proper motion of q seconds of arc/year and is located at a distance of r pc. Its tangential velocity (v) may be computed as follows. Visualize a circle of radius r, with the observer at the center of the circle and star A on the circumference (see Figure 8.7).

The length of the velocity vector (v) compares to the entire circumference of the circle as the angle of proper motion (q) compares to the entire circle of angles (360°), measured in seconds of arc (360° = 360 × 60 × 60 seconds):

$$\frac{v}{2\pi r} = \frac{q}{360 \times 60 \times 60}$$

Solving for v:

$$v = \frac{2\pi \, rq}{360 \times 60 \times 60} \quad \text{(pc/yr)}$$

Converting units to kilometers per second (not shown):

$$v = 4.74rq$$

For example, the tangential velocity of Barnard's Star, given $q = 10.25$ second/yr and $r = 1.83$ pc, is

$$v = 4.74(10.25)(1.83)$$

$$= 88.91 \text{ km/sec}$$

Space Velocity

Suppose that star A is traveling in a direction AB, as shown in Figure 8.8. Its tangential velocity is shown by AC and its radial velocity by AD.

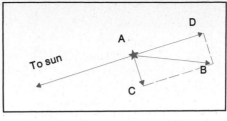

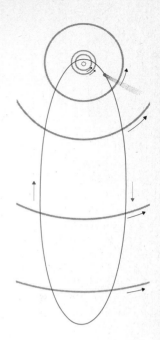

Figure 8.8 *The motion of a star relative to the sun.*

Recall from our discussion of the Doppler effect (page 83) that the radial velocity may be computed directly from the shift in a star's spectrum due to its motion. We may then define the space velocity of star A by

$$(AB)^2 = (AC)^2 + (AD)^2$$

a relationship which is true for any right triangle. For example, if the tangential velocity of a given star is 60 km/sec and its radial velocity is 80 km/sec, then its space velocity is

$$(AB)^2 = (60)^2 + (80)^2$$

$$(AB)^2 = 3600 + 6400$$

$$(AB)^2 = 10,000$$

$$AB = 100 \text{ km/sec}$$

In our discussion thus far we have ignored the motion of the sun and of our solar system, and so the space velocity of a star represents its motion with respect to the sun and not with respect to any absolute standard. The sun is moving at the rate of approximately 240 km/sec around the center of the Galaxy, and the Galaxy is moving with respect to all other galaxies (see Figure 8.9); how, then, do we find an absolute standard? It seems that we must be content to express the motion of any object as it relates to some stated frame of reference.

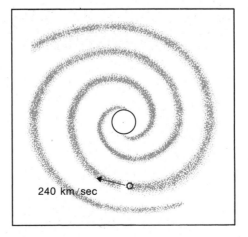

240 km/sec

Figure 8.9 *The sun's motion within the Milky Way galaxy.*

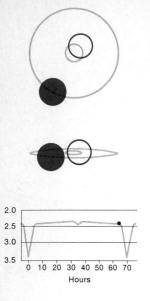

2.0
2.5
3.0
3.5
0 10 20 30 40 50 60 70
Hours

Peculiar Velocity

If we selected a group of stars, say, those within 100 parsecs of the sun, we could average out their velocities and arrive at a velocity that described the motion of this entire group of stars in general. To illustrate this idea, consider yourself driving on a freeway. Some cars would pass you from time to time and likewise you would pass some cars. Some might also be changing lanes at times. Nevertheless, it would be possible to describe the general motion of the entire group of cars within one mile of yourself. For instance, suppose a helicopter flew overhead and the pilot adjusted his speed and direction in order to fly along with your group of cars. His speed and direction might then be used to describe the speed and direction of the entire group. In this same manner, the local group of stars (within 100 parsecs of the sun) has a general speed and direction that may be called the *local standard of rest*. When we take into account the sun's own motion and specify a star's velocity with respect to the local standard of rest, we are referring to its *peculiar velocity*. Any star that is moving with the same speed and direction as the local group has a peculiar velocity of zero.

We are now prepared to look at one of the most fundamental methods of determining the distance to stars.

Distance to Moving Clusters

Clusters of stars are gravitationally bound to each other, and so they move through space together. For this reason we may expect all members of a given cluster to show the same peculiar velocity, and we can identify members by this fact. Consider, for instance, the Hyades cluster (in Taurus). This cluster, which is moving away from us, is spread out over a sufficient angle in the sky so that the proper motions of the individual stars appear to converge to a single point (see Figure 8.10)—like railroad tracks receding into the distance.

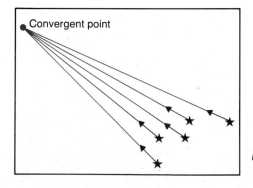

Convergent point

Figure 8.10 *Convergent point of a moving cluster.*

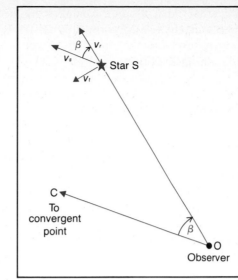

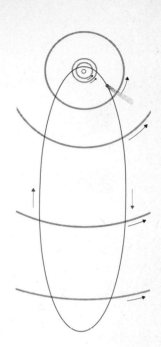

Figure 8.11 *The moving-cluster method.*

Because the convergent point of such a cluster can be observed, it follows that the angle between a given member of the cluster and that convergent point can also be measured. Note in Figure 8.11 that the line OC is parallel to the space velocity (v_s) of the star and therefore the angle (β) between the convergent point (C) and the star (S) is equal to the angle between the space velocity (v_s) and the radial velocity (v_r) of the star.

We have discussed previously (page 83) the fact that the radial velocity of a star may be measured directly by observing the Doppler effect in the star's spectrum. When both the angle β and the radial velocity are known, mathematicians can solve for the tangential velocity (v_t) as follows:

$$v_t = v_r \tan \beta$$

Recall that on page 304 we derived a relationship between tangential velocity (v_t), proper motion (q), and distance (r):

$$v_t = 4.74rq$$

Solving for r:

$$r = \frac{v_t}{4.74q}$$

Because we have calculated v_t above and have observed the star's proper motion, we now can compute the distance to the star. This method of moving clusters, because it is fraught with smaller observational errors

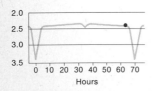

Hours

than that of trigonometric parallax, has become the fundamental method by which the H-R diagram (discussed later in this chapter) is calibrated, and it serves as the foundation stone upon which all other distance determinations rest.

Brightness of Stars

In the second century B.C. Hipparchus classified the brightest stars as first magnitude and the faintest naked-eye stars as sixth magnitude, ranking other stars as second, third, fourth, or fifth magnitude. He produced the first real catalogue of stars, showing both position and brightness. Of course, he was forced to rely upon his own ability to judge brightness without the aid of any instrument. Modern methods for measuring stellar brightness—the techniques of *photometry*—are much more precise.

Figure 8.12   *The apparent magnitude of a star may be judged from the size of the image it makes on a photographic plate. (Hale Observatories)*

Photography has played a very important role in this aspect of astronomy. The apparent magnitude of a star may be judged by the size of the image it makes on a photographic plate (Figure 8.12). The brighter the star, the larger the image on the film. A photographic film has a distinct advantage over the eye, for an image will continue to build up over time, causing a darker image to be formed as the exposure time is increased (Figure 8.13). The "image" formed on our eye fades and must be continually reinforced, so there is no buildup.

A more precise method, however, is that of photoelectric photometry. You are probably familiar with an exposure meter used in photography. When light falls on a sensitive surface at the front of the meter, a small electrical current is produced. This current then causes a needle to move, indicating the intensity of light falling on the sensitive surface. With a similar but much more sensitive device, the astronomer may bring the light of a star through the telescope and cause it to fall on the sensitive surface of a *photomultiplier* (Figure 8.14). The small current that the light produces in the photomultiplier is amplified to the point

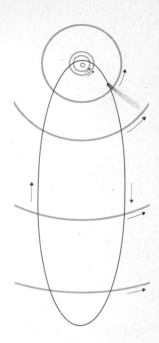

Figure 8.13   *Region of Orion Nebula, showing effect of aperture and time exposure. (a) Small aperture and/or short time exposure. (b) Large aperture and/or long time exposure. (Lick Observatory)*

(a)   (b)

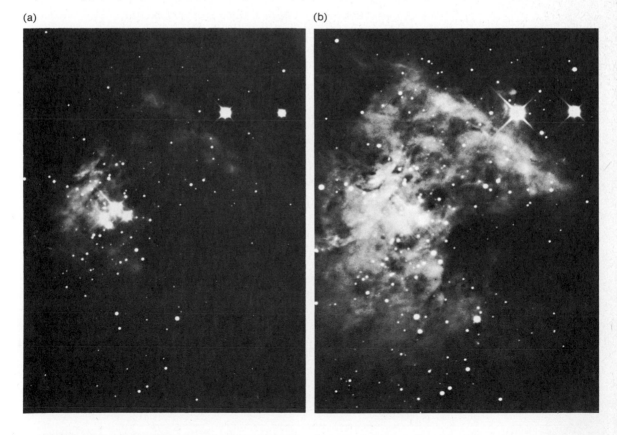

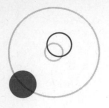

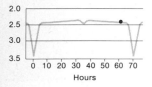

where it may be recorded. This method provides an extremely accurate means for ranking stars by brightness.

The Scale of Brightness

In an effort to preserve the familiar scale of Hipparchus, the modern astronomer has set forth the following scheme. The typical first magnitude star is 100 times as bright as the typical sixth magnitude star, so we seek a multiplying factor for each of the five steps between first and sixth magnitudes which will produce a 100-fold increase in brightness. Letting n be that factor, we write

$$n^5 = 100$$

$$n = \sqrt[5]{100} = 2.512 \simeq 2.5$$

Thus, if each step in the magnitude scale is 2.5 times as bright as the previous step, five such steps equate to the 100-fold increase:

Figure 8.14 *A photo multiplier tube showing (a) a cut-away view and (b) a functional diagram. The complex physical and electronic nature of this tube makes it sensitive to very small fluctuations in light intensity. (EMR Division of Weston Instruments, Inc.)*

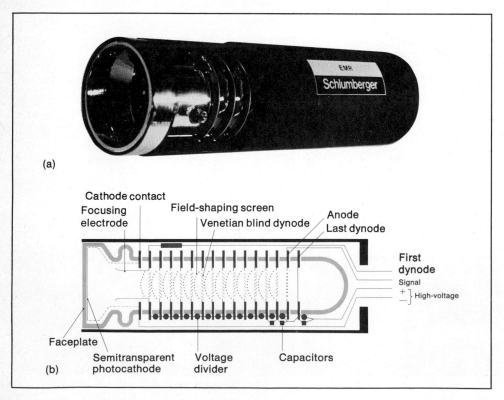

Table 8.1 *Light Ratios for Various Magnitude Changes*

MAGNITUDE CHANGE	LIGHT RATIO
0.5	1.6
1.0	2.5
2.0	6.3
3.0	15.9
4.0	39.8
5.0	100.0
6.0	251.2
10.0	10,000.0
20.0	100,000,000.0
25.0	10,000,000,000.0

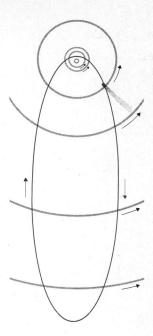

$$(2.5)\ (2.5)\ (2.5)\ (2.5)\ (2.5) \cong 100$$

Table 8.1 shows light ratios for given changes in magnitude. To say that a star is three magnitudes brighter than another is to say that it is 15.9 times as bright.

$$(2.5)\ (2.5)\ (2.5) \cong 15.9$$

Why should Table 8.1 include the larger values? The calibration of magnitudes of objects today shows two basic changes from the day of Hipparchus. Having defined the scale mathematically and using accurate detectors, modern astronomers recognize that some stars which Hipparchus would have called first magnitude are significantly brighter than others he placed in that same category. For instance, Sirius is approximately nine times as bright as Aldebaran, so the scale of brightness was extended upward to 0, −1, −2, and so on. The apparent magnitude of Sirius is −1.4. But some objects, such as certain planets, the moon, and the sun, are still brighter, hence the extension of the magnitude scale to include these, as shown in Table 8.2. This table also shows an extension in the other direction to include dimmer objects unknown to Hipparchus, which we see only by using optical aids. Even an average pair of binoculars will open up stars of magnitudes +7, +8, and +9 to our view. The larger telescopes with very sensitive detectors permit our recording of stars down to +25.

A magnitude change such as 17, not found in Table 8.1, may be calculated as follows. Since a change of 5 mag corresponds to a factor of 100 in light ratio, and of 2 mag corresponds to a factor of 6.3, then a total of 17 mag steps corresponds to a light ratio of 6,300,000:1:

Magnitude change: $17 = 5 + 5 + 5 + 2$

Light ratio: $(100) \times (100) \times (100) \times (6.3) = 6,300,000$

$$= 6.3 \times 10^6$$

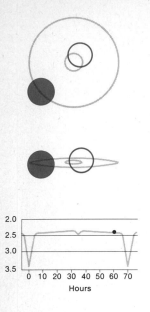

2.0
2.5
3.0
3.5

0 10 20 30 40 50 60 70

Hours

Table 8.2 *Apparent Magnitudes of Familiar Objects*

OBJECT	MAGNITUDE
Sun	−26.5
Moon (full)	−12.5
Venus (when brightest)	−4.4
Jupiter (when brightest)	−2.5
Sirius	−1.4
Rigel	+0.0
Aldebaran	+1.0
Polaris	+2.0
Naked-eye limit	+6.0
Binocular limit (average)	+9.0
15-cm telescope limit	+13.0
5-m telescope limit:	
visual	+20.0
photographic	+23.5

Note that whereas the magnitude changes add, their corresponding light ratios multiply.

Absolute Magnitude

Because stars are situated at widely different distances from us, it would be a serious error to assume that apparent magnitudes indicate their actual luminosity. We have all experienced the fact that as we move away from a source of light, its brightness seems to diminish. If we could accurately measure its brightness, we would find that when we were twice as far from the lamp, its brightness would be only one-fourth what it was from our starting position. The reason is that the energy which would fall on a screen at distance d (see Figure 8.15) would fall on a screen four times the size of the first when placed at distance $2d$. We then say that light falls off with the square of the distance, and that distance is a critical element in determining the apparent, as compared to the actual, brightness of a star. In order that all stars may be compared on an equal basis, let us mentally place them all at a distance of 10 parsecs and then determine the magnitude they would appear to have at that distance. We define this as their *absolute magnitude*. If our sun were placed at a distance of 10 parsecs, it would be very dim, just visible to the naked eye. Rigel, on the other hand, which is at a distance of 250 parsecs, if brought in to 10 parsecs would brighten to −7 mag. Rigel would then appear 625 times as bright as it does now.

Apparent magnitude, absolute magnitude, and distance are related by the proportion

$$\frac{L(10)}{L(r)} = \left(\frac{r}{10}\right)^2$$

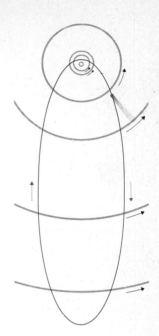

where $L(10)$ is the brightness of a given star at 10 parsecs and $L(r)$ is its brightness at r, its distance in parsecs. Suppose that a star situated at 1000 parsecs is brought in to 10 parsecs; what will be its change in magnitude? Letting $r = 1000$ in the equation above, we obtain the following brightness ratio:

$$\frac{L(10)}{L(1000)} = \left(\frac{1000}{10}\right)^2 = \left(\frac{100}{1}\right)^2 = \frac{10,000}{1}$$

The star would appear 10,000 times brighter. From Table 8.1 we see that this ratio of brightness corresponds to a change of 10 mag; thus if the apparent magnitude of the star was $+7$, its absolute magnitude will be $7 - 10 = -3$. (The absolute magnitudes of many stars are shown in Appendix 8.)

By similar calculations we may determine that a star situated at a distance of 16 parsecs will appear 1 mag dimmer than its absolute magnitude, that a star located at 25 parsecs will appear 2 mag dimmer than its absolute magnitude, and so on. This concept is summarized and extended in Table 8.3.

Figure 8.15   *In traveling twice the distance, the light covers four times as much area. Therefore we would expect its intensity to be only one-fourth.*

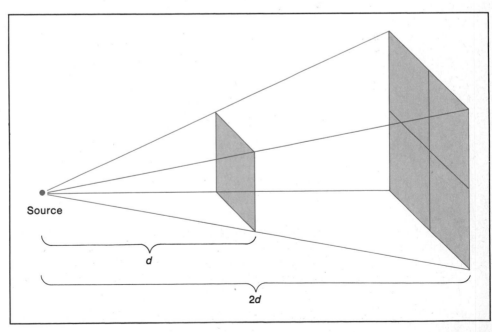

Table 8.3 *Difference in Apparent and Absolute Magnitudes at Various Distances*

Distances (parsecs)	16	25	40	63	100	160
Difference betweeen absolute and apparent magnitudes	1	2	3	4	5	6

Temperature of Stars

Stars radiate energy in a broad range of wavelengths. If we consider the total energy emitted in all wavelengths, we call this the star's *bolometric* magnitude. However, the astronomer is usually able to measure the magnitude of a star within only a limited range of wavelengths. If he uses a yellow-sensitive film, he gets a visual magnitude (m_v). This most closely resembles the magnitude as seen by the human eye. If a blue-sensitive film is used, the astronomer records a photographic magnitude (m_p). For most stars, m_v and m_p are different, and this difference is called the *color index* (C.I.) of a star, defined by

$$\text{C.I.} = m_p - m_v$$

The color of a star may be determined by this difference, as shown in Table 8.4.

We see that the color of a star is principally a function of its surface temperature, ranging from the very hot blue-white stars, with a surface temperature of approximately 50,000°K, to the very cool red stars, with a temperature of approximately 3000°K. Invisible stars exist with temperatures as low as 1000°K, radiating only infrared wavelengths.

As the light from distant stars travels through the dust clouds of space, it is reddened; therefore its original color may not be evident. We must depend upon more reliable means of assessing a star's surface

Table 8.4 *The Temperature-Color Relationship of Stars*

COLOR INDEX	COLOR	SURFACE TEMPERATURE
−0.6	Blue-white	50,000°K
−0.3	Blue	20,000°K
0.0	White	10,000°K
+0.2	White	7,000°K
+0.6	Yellow	6,000°K
+0.8	Orange	5,000°K
+1.5	Red	3,500°K

temperature. The spectra of stars give us a better indication, for some lines show up very clearly at certain temperatures and not at others (Figure 8.16). Stars were originally classified according to the strength of the hydrogen lines in their spectra. Class A has the strongest lines, Class B the next strongest, and so on to Class O. Thus we see that the idea of spectral class had an orderly beginning. When these same classes were rearranged according to surface temperature, however, the order was jumbled: O, B, A, F, G, K, M. This order may be remembered rather easily by the following mnemonic device. Setting a word to each letter class, we get: "Oh Be A Fine Girl, Kiss Me." Each spectral class is further subdivided into ten subdivisions as follows: B0, B1, B2, B3, B4, B5, B6, B7, B8, B9, A0, A1, A2, and so on. How may these classes be recognized from their spectra? At the extremely high temperature typical of the O-type star, the energetic collisions that occur among the atoms often knock electrons entirely free, thus producing ions with their own characteristic spectra. At the other extreme—the lower temperatures—atoms may unite to form compounds such as titanium oxide, as evidenced in the spectrum of an M-type star. Table 8.5 gives examples of stars in each class, together with their typical spectra.

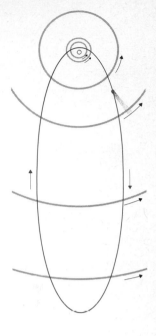

Hertzsprung-Russell Diagram

Early in the twentieth century a very important study concerning the relationship between the surface temperature of a star and its lumi-

Figure 8.16   *Examples of different spectral types. (Hale Observatories)*

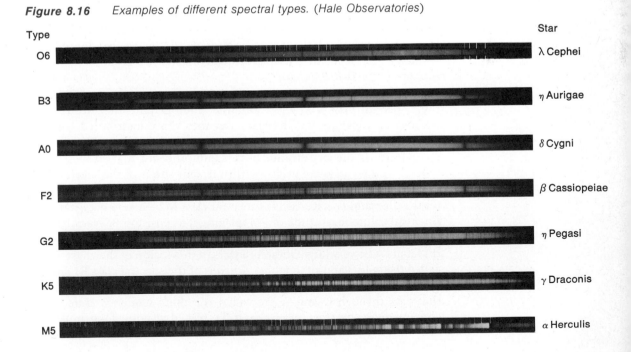

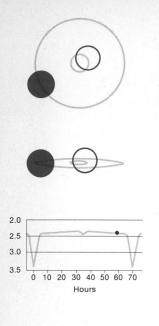

Table 8.5   *Stellar Spectra*

CLASS	EXAMPLE	SPECTRA	SURFACE TEMPERATURE
05	None visible to the naked eye	Ionized helium, nitrogen, oxygen	50,000°K
B3	Achernar	H, He strong	15,000°K
A1	Sirius	H lines at maximum Ca lines weak	11,000°K
A3	Fomalhaut	H lines strong Ca lines stronger Metals weak	9,000°K
F0	Canopus	H lines weak Ca lines strong	7,600°K
F5	Procyon	Ca lines very strong Neutral metals	6,600°K
G0	Capella Sun	Ca lines at maximum Iron lines strong	6,000°K
K2	Arcturus	H lines weak Molecular bands present	5,000°K
M2	Betelgeuse	Neutral lines strong TiO molecules present	3,500°K

nosity was made by two astronomers, Ejner Hertzsprung of Denmark and Henry N. Russell of the United States, working independently. The results of their work are shown graphically in Figure 8.17, known, in their honor, as the *H-R diagram*.

Stars of known distance have been plotted on the chart according to their spectral class and their absolute photographic magnitude. It is obvious at a glance that the stars are not evenly distributed over the entire chart, but rather seem to be grouped in several areas. We shall try to understand the significance of this grouping, but first let us see how several familiar stars are placed within it. Our sun is of the spectral class G2 and has an absolute magnitude of +5, hence it is placed at point A on the diagram. *Betelgeuse* is of spectral class M2—a very cool star—but has an absolute magnitude of −4, placing it at position B. We might ask ourselves how a cool star could possibly be so luminous. The answer is that it must be very large: therefore stars in this region of the chart will be called the *giants* and *supergiants*. *Spica* is of the spectral class B1 and has an absolute magnitude of −3, so it is placed at point C. The band of stars that pass through C and A are called the *main-sequence stars*. What characteristics must a star possess to be placed at point D? It must be a

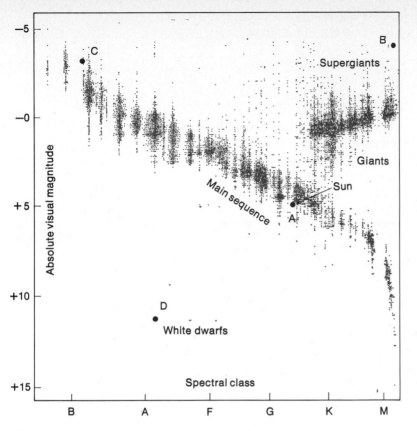

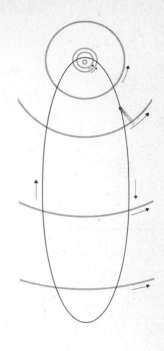

Figure 8.17 *The H-R diagram. The points represent approximately 6700 stars of known absolute magnitude. (Yerkes Observatory)*

very hot (white) star and yet rather dim, hence small. Stars in this region of the chart are called *white dwarfs.*

The H-R diagram shown here represents only a very limited number of luminous stars relatively close to the sun and should not be interpreted as an accurate picture of the distribution of all stars. Furthermore, it should not be inferred from the location of the sun on this diagram that it is a below-average star. If Figure 8.17 included all stars in a given region of space, we would recognize that less than 5 percent of them are brighter than the sun; thus the sun is far above average among the stars. In order to plot the additional stars, the H-R diagram would have to be extended downward and to the right, because most of the unplotted stars are both cooler and dimmer than those shown. Such stars are now being revealed by infrared techniques. Some familiar stars are plotted in Figure 8.18.

A further subdivision of the giants and main-sequence stars may be made as follows (represented graphically in Figure 8.19):

Ia Most luminous supergiants
Ib Less luminous supergiants
II Bright giants
III Normal giants
IV Subgiants
V Main-sequence stars

Class Ia supergiants, because they are so large relative to their mass, will have extremely low pressure and low density. The spectral lines of such a star will be very narrow (sharply defined). Each successive class represents stars of increasing pressure and density; this is shown by broader spectral lines. Accordingly, a star of unknown distance may be placed in its proper class and its distance inferred from its apparent magnitude. This method of determining distance is called *spectroscopic parallax.*

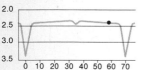

Figure 8.18   *Familiar stars plotted on the H-R diagram.*

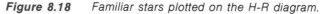
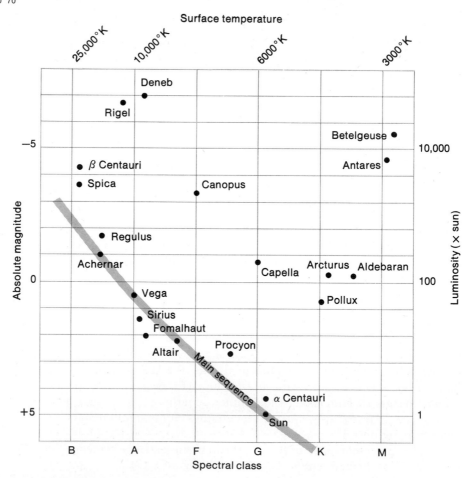

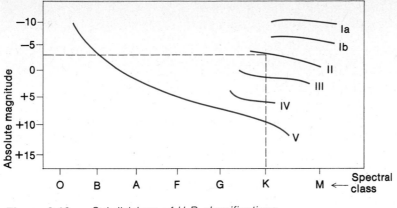

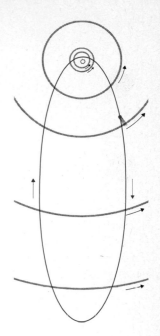

Figure 8.19   *Subdivisions of H-R classifications.*

Spectroscopic Parallax

We have discussed the method of heliocentric parallax, but we are also aware of its very definite limitation to nearby stars. The method of spectroscopic parallax is limited only by our ability to see the star (so that we can measure its apparent magnitude) and to photograph its spectrum. Because we will be using the H-R diagram, its proper calibration is essential. The moving-cluster method of calibration used here is discussed on page 306. Let us take a step-by-step approach to the method of spectroscopic parallax.

STEP 1. Place the given star in its proper pressure category (Ia, Ib, II, III, IV, V) on the H-R diagram. To do this, you must look at its spectrum to see the degree to which its lines have been broadened by pressure: Ia-type stars have the narrowest lines, owing to lowest pressure; types Ib, II, III, IV, and V have progressively broader spectral lines because of their reduced size and corresponding higher pressure.

STEP 2. Classify the star by spectral type, again using the spectrum as discussed earlier in this chapter.

STEP 3. Run a vertical line upward on the chart corresponding to the spectral classification, extending it until it intersects the proper pressure category line. For example, take a star of K0 type which is also in Category II, and plot it as shown in Figure 8.19.

STEP 4. From the point of intersection draw a horizontal line to the left-hand side of the chart and read out the corresponding absolute magnitude, which in this case is −3.

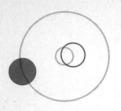

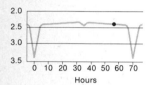

STEP 5. Observe the apparent magnitude of the star, using the photographic method explained earlier, i.e., +7.

STEP 6. By comparing the absolute brightness of a star to its apparent brightness, determine its distance:

$$\frac{L(10)}{L(r)} = \left(\frac{r}{10}\right)^2$$

$L(10)/L(r)$ is the ratio of absolute brightness to apparent brightness. The difference in apparent and absolute magnitude is 10 [7 − (−3) = 10]. This corresponds to a light ratio of 10,000:1 (see Table 8.1, page 311).

$$\frac{L(10)}{L(r)} = \left(\frac{r}{10}\right)^2$$

$$\frac{10,000}{1} = \left(\frac{r}{10}\right)^2$$

$$10,000 = \frac{r^2}{100}$$

$$r^2 = 1,000,000$$

$$r = 1000 \text{ pc}$$

Sizes of Stars

We have been discussing the variety of sizes within the stellar population, ranging from the dwarfs to the supergiants. Most stars are at such a great distance that they appear as mere points of light in even the largest telescopes; however, as we observed in connection with the H-R diagram, a cool star that is also bright (in absolute magnitude) must be a large star. Let us see how these factors are related.

Suppose that the spectrum of a star indicates a surface temperature of 3000°K, just half that of the sun, and that the absolute magnitude of the star is zero, 5 mag brighter than the sun. A difference of 5 mag indicates that the given star is 100 times as luminous as the sun. It is also known that a star whose temperature is one-half that of the sun only radiates one-sixteenth as much energy over each square centimeter of its surface. Taking these two factors into consideration, we see that the given star must have 1600 (16 × 100) times as much surface area as the sun. Since the surface area of a sphere is proportional to the square of the radius, we may compute the radius by taking the square root of 1600, which is 40. Hence, the radius of the given star is 40 times that of the sun. Among the largest known stars is Betelgeuse, in the constellation of Orion, having a diameter 750 times that of the sun—a diameter which is greater than the diameter of Mars' orbit around the sun.

The Great Variety of Stars

We have seen that stars vary greatly in size, surface temperature, color, and brightness; however, we have not yet perceived the complete range of this variety because we have been considering only stars which emit visible light. Our entire discussion has centered around an extremely narrow portion of the electromagnetic spectrum. It may surprise you to discover that, of the number of stars in the Milky Way galaxy, more than half cannot be detected using visible light; we must turn to other wavelengths and other tools for detection of these objects.

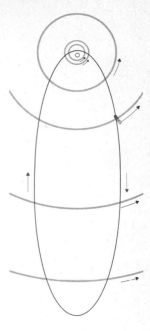

Figure 8.20   *The radiation curves for stars of different temperatures.*

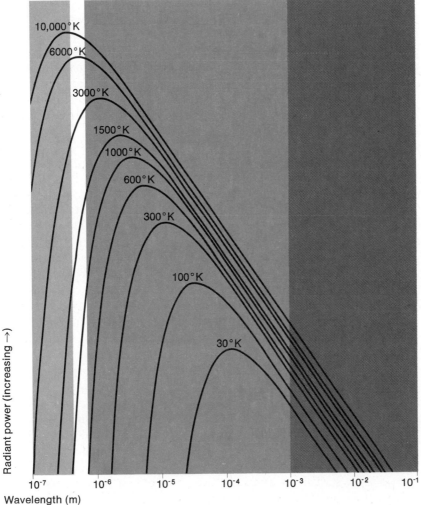

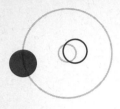

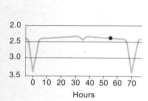

First, let us consider a fundamental relationship between the surface temperature of a heated object and the wavelengths that it radiates. The curves in Figure 8.20 show that as an object is heated, it radiates a broader range of wavelengths. The wavelength at which the maximum radiation occurs is related to the surface temperature by a simple mathematical statement known as Wien's law:

$$\text{wavelength}_{\text{max}} = \frac{3000}{T°}$$

where wavelength is measured in microns (abbreviated μm 1μm $= 10^{-6}$ m $= 1$ millionth of a meter); T is the surface temperature in degrees Kelvin. For example, the sun has a surface temperature of 6000°K and it radiates its peak energy at a wavelength of 0.5 μm, because $3000/6000 = 0.5 \mu$m. Note that 0.5 μm is the same as 5×10^{-7} m—near the center of the range of wavelengths called visible light. On the other hand, a star whose surface temperature is only 1500°K will radiate its maximum energy at 2 μm, which falls in the infrared range. Over 50 percent of the energy radiated at 2 μm will penetrate the earth's atmosphere, making this one of the better "windows" through which the astronomer may survey the infrared sky. The atmosphere is opaque to most of the infrared spectrum, and so Orbiting Astronomical Observatories are essential to the total infrared picture.

The Infrared Sky

A map of the infrared sky would look so entirely different from a map of the visible sky that we could not recognize familiar constellations. Compare the map of bright infrared objects in Figure 8.21 with the same region, namely that of Orion, shown for visible stars (see Appendix 12, *winter map*). Out of the many thousands of bright infrared sources identified, only a few can be seen with the naked eye; thus infrared astronomy has provided a wealth of new information, but also new questions. What kind of objects are being detected? What kind of objects would have temperatures in the range of 300° to 400°K and thereby radiate primarily in infrared? (it is interesting to note that our own body temperature, 311°K, produces peak infrared radiation of approximately 10 μm.) These are relatively cool objects, some of which may be *proto-stars*—stars in the making. Stars are thought to condense out of huge clouds of gas and dust, and one of the manifestations of such a condensation is heat. A condensation as cool as 100°K would radiate at 30 μm. Many such sources are large in angular diameter, which is consistent with the idea that they are clouds of gas.

Visible starlight is significantly scattered as it passes through the

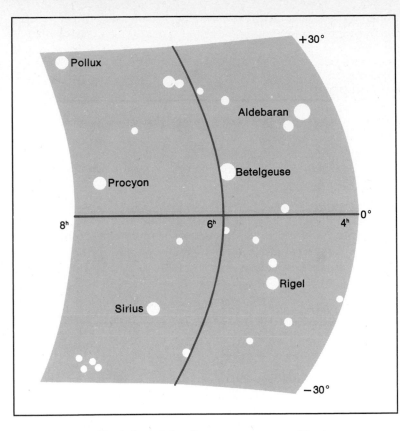

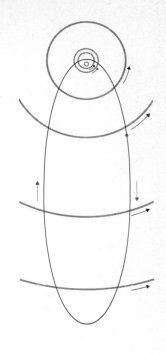

Figure 8.21 *The infrared sky. If our eyes were sensitive to infrared wavelengths alone, we would recognize only a few familiar stars. Compare this infrared star map with the visible winter sky in Appendix 12.*

dust concentrated near the center of our Galaxy, and this produces such severe loss of light that objects intrinsically bright in visible light may not be visible to us at all. On the other hand, infrared energy is not significantly scattered by that same interstellar medium; therefore, bright infrared objects whose line of sight passes through the center of our Galaxy are readily detectable. This ability to penetrate gas and dust was the critical factor in the recent discovery of two new galaxies (Maffei I and Maffei II), for their line of sight passes near the center of the Milky Way.

Stars are also believed to cool off in their old age, and so old stars represent still another possible class of infrared objects. Several observers have speculated that a star may throw off a shell of gas and/or dust, and then the core will heat that shell to some temperature less than 1500°K.

Some infrared sources may radiate energy at levels exceeding that of the sun by a factor of 1000 or more, yet their rate of production per

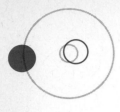

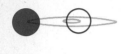

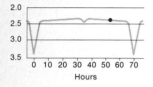

square meter of surface area is low. We must conclude, therefore, that such sources are extremely large.

X-ray Astronomy

At almost the other extreme from the cool, infrared object we see the very hot, X-ray object. By Wien's law an object having a surface temperature of 30 million degrees Kelvin would peak its radiation at a wavelength of 0.0001 μm (1 Å), in the X-ray portion of the spectrum. Phenomena other than high-temperature effects, however, produce X rays. Thus we cannot limit our discussion to hot stars alone.

X rays have been detected from what is thought to be an extremely small but dense rotating star (called a *pulsar*) in the Crab Nebula. Here the process of X-ray production is tied to a rapidly rotating object which accelerates electrons to speeds near that of light and sends them hurtling through a magnetic field. The Crab pulsar can be detected "beeping" on and off at radio wavelengths and also pulsating in X-ray wavelengths. Astronomers could never have developed the total picture of this and similar objects without inspecting their characteristics over a wide range of wavelengths.

One of the most striking speculations of the past few years, involving X-ray astronomy, has concerned the "black hole." Very massive stars are thought to collapse in their old age and to develop very high surface gravities. Under such conditions, material may be drawn from a nearby star toward the black hole. Some of this material is thought to go into orbit about the very dense core, possibly becoming so hot and turbulent as to generate X rays. Astronomers are quite confident that the X rays received from Cygnus X-1 represent the presence of a black hole. Black holes will be discussed more fully in Chapter 12.

Detection of X rays typifies what has been happening throughout all of astronomy in recent years—the opening of virtually the entire spectrum to our "view."

Questions

1. How long is the base line used in the method of heliocentric parallax?

2. The method of heliocentric parallax can be used to determine the distances to stars up to _____ parsecs.

3. Find the distance to a star whose heliocentric parallax is 0.50 second.

4. Which motion of stars appears to change the shape of the Big Dipper constellation?

5. The radial velocity of a star is indicated by what kind of observation?

6. Define the term *space velocity*. Why cannot this be called the true velocity of a star?

7. The sun moves within the Milky Way galaxy at a rate of _____ km/sec, owing to the rotation of the Galaxy.

8. What do we know about a star whose peculiar velocity is zero?

9. The computation of the tangential velocity of a star depends upon our knowledge of what two properties of the star's location and/or motion?

10. What was Hipparchus's system for categorizing stars by brightness?

11. List three means by which the apparent brightness of stars may be judged today.

12. A change of 1 in magnitude corresponds to a change of _____ in luminosity.

13. A star of −3 mag appears _____ times as bright as one of +2 mag.

14. True or false: Venus sometimes appears brighter than the brightest star.

15. True or false: The use of time exposures on film has extended the limiting magnitude of the Palomar 5-m telescope by several magnitudes.

16. The true brightness of two stars may best be compared by consideration of which of the following: (a) their apparent magnitude, (b) their absolute magnitude, (c) neither of these?

17. How is the apparent brightness of a star related to its absolute brightness and its distance from the observer?

18. Describe at least two ways by which the surface temperature of stars may be measured.

19. What is the general range of surface temperatures of stars? Can the surface temperature of a star be judged in even the roughest fashion by the naked eye? Explain.

20. What relationship is demonstrated by the fact that the majority of stars fall along the main sequence on the H-R diagram?

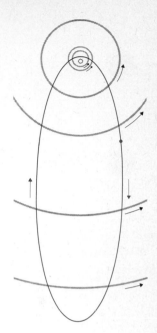

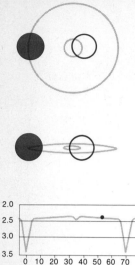

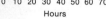

21. What advantage does the method of spectroscopic parallax have over that of heliocentric parallax for determining the distance to a star?

22. What is the range in star sizes as compared to the sun?

23. A 100-watt light bulb has a certain apparent brightness if viewed from a distance of 10 m. What will be the change in its apparent brightness if viewed from 20 m? from 50 m?

24. An object as cool as 300°K will radiate its maximum energy in what part of the spectrum?

25. The wavelength at which a star emits its maximum energy is inversely proportional to its surface temperature. If a star whose temperature is like that of the sun (6000°K) has its maximum output at 0.5 μm, find the wavelength for maximum output of a star whose temperature is 12,000°K. How will its color compare to that of the sun?

26. What increase in luminosity would accompany a change in temperature of a star from 3000°K to 6000°K, (a) assuming the diameter of the star remains the same, (b) assuming the diameter of the star is halved?

27. The sun appears to have a magnitude of −26.5, as seen from earth. What must be its magnitude as seen from Pluto, 40 times as far away?

Suggested Readings

ALLER, L. H., *Atoms, stars and nebulae.* Cambridge, Mass.: Harvard University Press. 1971.

GIACCONI, RICCARDO, X-ray astronomy. *The Physics Teacher 11* (3), 135–143 (1973).

JASTROW, ROBERT, *Red giants and white dwarfs.* New York: Signet (paper). 1969.

MIHALAS, DIMITRI, Interpreting early-type stellar spectra. *Sky and Telescope 46* (2), 79–83 (1973).

NEUGEBAUER, G., and BECKLIN, ERIC E., The brightest infrared sources. *Scientific American 228* (4), 28–40 (1973).

SCIENTIFIC AMERICAN, *New frontiers in astronomy* (readings). San Francisco: W. H. Freeman & Company Publishers, 1975.

SMITH, E. P., and JACOBS, K. C., *Introductory astronomy and astrophysics*. Philadelphia: Saunders, 1973.

STRAND, KAJ A. (ed.), *Basic astronomical data*. Chicago: University of Chicago Press, 1965.

STRUVE, OTTO, LYNDS, BEVERLY, and PILLANS, HELEN, *Elementary astronomy*. New York: Oxford University Press, 1959.

WEYMANN, RAY J., Stellar winds. *Scientific American* 239 (2), 44 (1978).

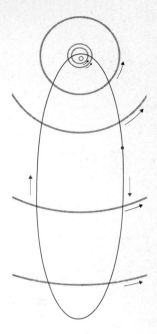

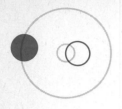

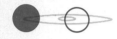

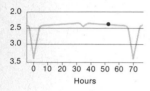

NINE

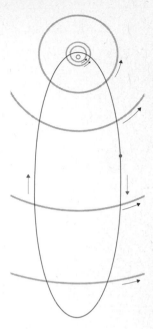

MULTIPLE STAR SYSTEMS

The casual observer would hardly suspect that more than half of the stars he sees with his naked eye are in reality multiple systems, containing two, three, or more stars. The astronomer, on the other hand, is well aware that when two stars appear close to each other, there are only two basic explanations: they may actually be close to each other, or they may be remote from each other and just happen to line up from the observer's point of view. Sir William Herschel, a famous observer of *binary* (double) *stars*, held to the theory that such stars differed greatly in distance from the earth, because one star of the pair was often much dimmer than the other. In trying to prove his theory, he found himself to be wrong, for he discovered that most binary stars are in each other's gravitational field—that is, they are close enough and massive enough to influence each other's motion directly. They literally revolve around some common point between them. In proving his original theory wrong, Herschel had made a very important discovery, and he continued his observations, cataloguing some 800 double stars by the year 1820.

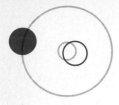

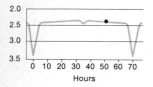

Hours

Optical and Visual Binaries

On the rare occasion that two stars appear close together from the observer's point of view but are really distant from each other, they are called *optical binaries*. This is not a true binary system.

A much more common class of binary stars are both in each other's gravitational field and visible as doubles in a telescope. They are called *visual binaries* (Figure 9.1).

Spectroscopic Binaries

If the components of a true binary system are so close together that they cannot be *resolved* (separated) by a telescope, they may still be identified as a binary system by their spectrum. Consider a pair of binary stars, A and B, in mutual revolution around a point that is the center of the system, as in Figure 9.2. When the stars are positioned as in Figures 9.2 (a) and (c), both have the same Doppler shift in their spectral lines. However, when they are positioned as in Figures 9.2 (b) or (d), one star is moving in a direction toward the observer and the other star is moving

Figure 9.1   *Stars may appear close together because (a) they are actually close or (b) they merely line up, from the observer's point of view.*

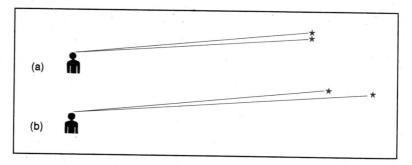

Figure 9.2   *Spectroscopic determination of a binary system.*

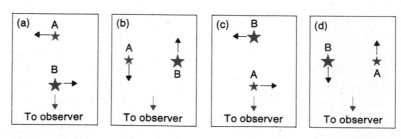

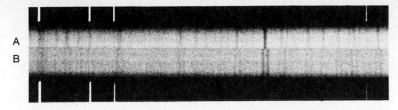

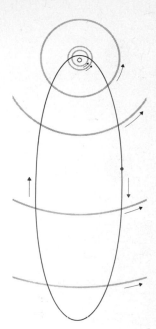

Figure 9.3   *The double lines (B) as seen in the spectrum of a spectroscopic binary star system. The top band (A) shows no separation, since the two stars are in alignment with the observer. (Hale Observatories)*

away; hence the spectral lines of one star will be slightly different from those of the other, producing double lines (Figure 9.3).

The periodic separation of the spectral lines, then, signifies a binary system—more specifically, a *spectroscopic binary system*. The spectrum may reveal the binary nature of a system by exhibiting lines that are characteristic of both a cool and a hot star.

You can easily identify a system of stars that illustrates both the visual and the spectroscopic types. You can see with your naked eye the two most obvious members of the system, Mizar and Alcor, in the handle of the Big Dipper. Several minutes of arc separate these stars (Figure 9.4). Now, if you look at Mizar through a small telescope, you will discover that Mizar is itself a visual binary; its brighter component is *Mizar A* and its dimmer component *Mizar B*. Next, if you were to analyze this system spectroscopically, you would discover that Alcor, Mizar A, and Mizar B are each spectroscopic binaries. Thus the system is composed of six stars.

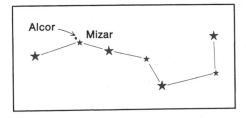

Figure 9.4   *The Big Dipper*

Astrometric Binaries

The dimmer component of a binary system may be so dim as to go unnoticed by any of the methods just described, and yet its presence could be detected by the gravitational effect that it has on the brighter star, causing it to move along a wavelike path in the sky. Such a system is called an *astrometric binary system*.

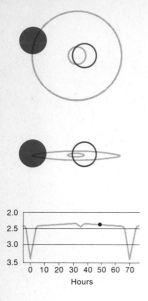

Eclipsing Binaries

Some binary systems are oriented in such a way that one star passes in front of the other; that is, we see the system edge-on to its plane or orbit. These are known as *eclipsing binary* systems. As one star is eclipsed by the other, we expect some variation in the total light reaching us from that system. If we plot the variations in light received as these eclipses occur, we may discover a number of facts from such a light curve. First, consider a binary system composed of a large star and a small one, as in Figure 9.5, where the small star is the hotter of the two. When the small hot star is in position 1 or 3, we see the combined light output of both stars. When it is in position 2, we see an annular eclipse, with only slightly less light being received, whereas when it is in position 4, we see a total eclipse of the hotter star, so that the least amount of light is received. If the orbit of this system were tipped slightly more in our direction, then we might observe only partial eclipses, with a resultant change in the light curve, as in Figure 9.6. The left-hand flip pages beginning on page 410 depict an eclipsing binary system in motion and also the light curve being received.

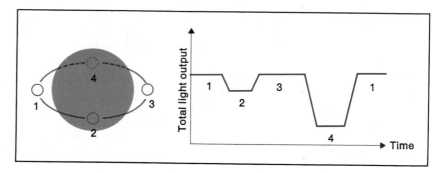

Figure 9.5   The light curve of a binary system.

Figure 9.6   Partially eclipsing binaries.

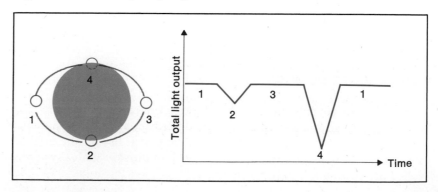

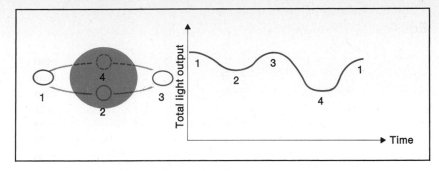

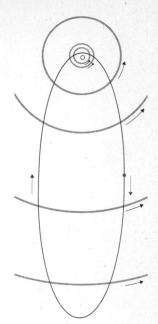

Figure 9.7   A binary system, distorted by gravitation.

Figure 9.7 shows the light curve for a pair of stars so close together that they distort each other's shape by their mutual attraction.

Whereas we have first proposed a model of a given system and then deduced the light curve, in actual practice the astronomer first observes the light curve and then tries to build a model to fit that curve.

The light curve also gives an indication of the size of the stars in relation to their orbital size. A smaller star is very rapidly eclipsed, making the sides of the curve very steep. A larger star requires more time for an eclipse and thus produces a less steep curve. This is one of the few ways astronomers have of determining the size of stars.

Close (Contact) Binaries

Some binary stars are in such close relationship to each other that they actually exchange material (mass). In order to understand some of the ramifications of this phenomenon, let us briefly preview Chapter 12. Stars are not static throughout their lifetime, but they change (evolve), sometimes growing very large, sometimes shrinking to a very small size (of the order of the size of the earth, or smaller). Let us suppose that two stars compose a binary pair and that one star has three solar masses whereas the other has only one solar mass. More massive stars typically go through their life cycle faster than the less massive stars, hence the 3-solar mass star may swell to become a red giant, while the 1-solar-mass star is still of the "main-sequence" variety. The outer limit of the 3-solar-mass star may come so close to the 1-solar-mass star as to experience a stronger gravitational attraction to it and, as a result, transfer some of its mass to that star. Two things will result: the mass of the two stars will be redistributed, changing the evolutionary rate of each star, and the deposition of material onto the low-mass star will generate X rays. Therefore, close binaries may be recognized by a characteristic pattern of X-ray radiation.

Further evolution in the more massive star may produce a highly condensed star, one that will pull material back from the low-mass star.

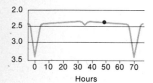

Hours

Thus the exchange of material can complicate the evolutionary pattern of the two stars involved.

Determination of the Mass of Binary Systems

A binary system lends itself very well to the determination of mass, both of the system and of the individual members. Three basic factors of the system are related: the mass of each star (M_1 and M_2), the separation of the stars (r), and their period of revolution (p). This relationship, first expressed by Kepler and later refined by Newton, is stated thus:

$$M_1 + M_2 = \frac{r^3}{p^2}$$

where M_1 and M_2 are expressed in solar masses (masses as compared to the sun's mass), p in years, and r in astronomical units. Since p and r may be observed for a number of binary systems, the sum of their masses ($M_1 + M_2$) may be determined. Suppose that two stars were orbiting each other at an average separation of 5 A.U. and that their period of revolution was 10 years; then

$$M_1 + M_2 = \frac{(5)^3}{(10)^2} = \frac{125}{100} = 1.25 \text{ solar masses}$$

Since this gives us only the total mass, we must next try to determine how the mass of the system is distributed between the two stars. If the stars were equal in mass, they would orbit around a point midway between the two, called the *barycenter* of the system. However, if one star is three times as massive as the other, we might expect the barycenter to be nearer the larger mass (Figure 9.8). These factors are related by the proportion:

$$\frac{M_1}{M_2} = \frac{x_2}{x_1}$$

By observing the displacement of the stars in a binary system, as compared with several nearby stars, it is possible to locate the system's barycenter, after which the distribution of mass be calculated.

Figure 9.8   *The barycenter of a binary system.*

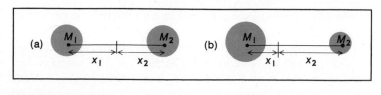

Consider the case of Sirius, a binary system with its A component the brightest star we see, and its tiny B component the first white dwarf ever discovered. As this system moves through space, its barycenter marks a uniform path, indicated by the dashed line in Figure 9.9. The two components, on the other hand, oscillate in varying degrees to either side of this path.

The approximate period of this system has been observed to be 50 years, and the separation of its members, 20 A.U. The total mass of the system is computed to be 3.2 solar masses:

$$M_1 + M_2 = \frac{(20)^3}{(50)^2} = \frac{8000}{2500} = 3.2 \text{ solar masses}$$

Sirius B is observed to orbit 2.2 times as far from the barycenter as does Sirius A; therefore the combined mass must be distributed in the ratio of 1:2.2. The mass of Sirius A is approximately 2.2 times that of the sun, and the mass of Sirius B is equal to that of the sun. The left-hand flip pages beginning on page 182 depict the earth and moon as a binary system. These objects may also be visualized as two stars in a binary system.

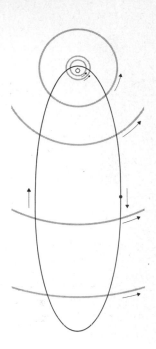

Mass-Luminosity Relationship

One of the most significant results of our being able to determine the masses of binary stars becomes obvious when we plot the masses of these stars against their absolute magnitudes (Figure 9.10).

(a)

(b)

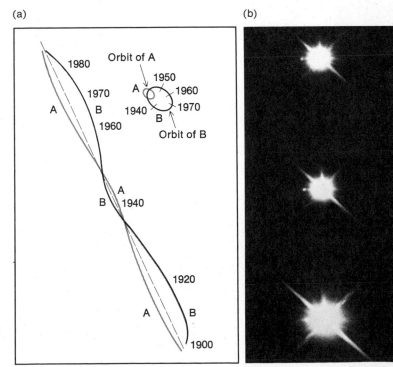

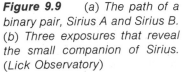

Figure 9.9 (a) *The path of a binary pair, Sirius A and Sirius B.* (b) *Three exposures that reveal the small companion of Sirius.* (*Lick Observatory*)

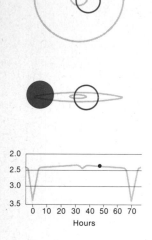

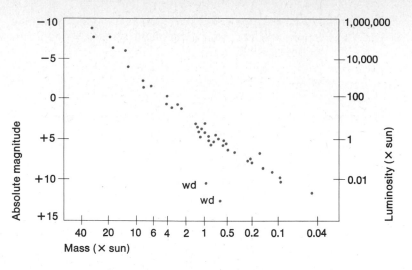

Figure 9.10   The mass–luminosity relationship.

It is evident that most of these stars fall within a narrow band running from upper left to lower right. Stars low in mass are also low in absolute magnitude. High-mass stars are high in absolute magnitude. This is called the *mass-luminosity* relationship. It is likely that all main-sequence stars satisfy this relationship. On the other hand, white dwarfs and red giants generally do not show this relationship. Figure 9.10 shows several white dwarfs that do not fit within the narrow band. (Note the similarity between this graph and the H-R diagram.)

Once the mass-luminosity relationship had been established, it became a tool of the astronomer; he could estimate the masses of main-sequence stars by calculating their absolute magnitudes, even though they were not part of a binary system.

Origin of Binaries

Binary systems are so numerous that any theory concerning the origin of stars must account for these special associations. Theories concerning the origin of binary systems run somewhat parallel to those of the solar system.

One theory suggests that two stars became entrapped within each other's gravitational field as a result of passing very close to each other. The chance of such an encounter is so small, however, that this theory is hard put to explain the multiplicity of binary systems.

A second theory suggests that a rapidly rotating star broke up into two or more parts. We might expect a rotating star to throw off some material from around its equator, but it is difficult to think of a star being torn in half by such a motion.

It is conceivable that, as a cloud of hydrogen gas began to condense to form a star, more than one center developed, eventually producing two or three stars within the same contraction. Such stars would possess a motion in common with the gas cloud out of which they formed. Thus, the existence of binary systems suggests a common origin.

Clusters

No matter what part of the sky is observed, if seeing conditions are fair to good, even a pair of binoculars will usually reveal from a dozen to hundreds of stars within its field of view. Certain regions, however, seem to "come alive" with stars when they are scanned. One such region is that of the *Pleiades cluster*, a group of stars visible to the naked eye but doubly spectacular in binoculars or in a low-power telescope. Hundreds of stars may be seen by using such an instrument, and yet we must look further to determine whether this is more than merely an accidental association. The answer to our question will be found in the motions of these stars. While each star differs slightly in speed and direction, there is a general trend toward the same direction and the same speed; hence they are a true *cluster*.

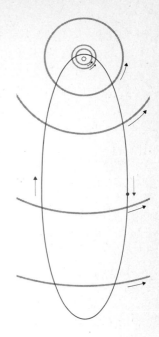

Origin of Clusters

It is possible that most stars were formed in clusters. Some have lost their identity with a cluster while others have retained it. If the stars of an identifiable cluster did have a common origin, we may assume that they are all at about the same distance from us, so that their apparent magnitude is proportional to their absolute magnitude. In other words, a star that appears to be brightest among the group *is* in fact brightest. Recalling the apparent relationship between the luminosity and the mass of a star, we see that the more massive members of a cluster ought also to appear the brightest. When clusters are observed with these relationships in mind, it appears that the more massive members of a cluster tend toward the central region. Perhaps clusters were formed with this concentration of mass around the center; on the other hand, the more massive stars may gradually be moving to this central position.

Cluster Types

Typical of the *open cluster* are the Pleiades (Figure 9.11), the double open cluster in Persei (Figure 9.12), and the cluster in Cancer (M67) as seen on page 328. Also known as *galactic clusters*, because they occur primarily in the disk of our Galaxy, open clusters usually contain young, hot (O- and B-type) stars numbering in the hundreds. They form a rather

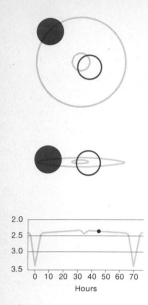

Figure 9.11 *The Pleiades* (*Seven Sisters*), a typical open cluster in Taurus. (*Lick Observatory*)

loosely bound system that may disperse in a matter of a few tens of millions to a few billions of years, owing to the gravitational forces within the Galaxy as a whole. Thus, by the time such stars age, they are already separated by distances which no longer suggest a common origin. Almost 1000 open clusters are known in our Galaxy.

The *globular cluster* forms the second class. A cluster of this type usually contains thousands of stars, which appear so tightly packed that the central regions cannot be resolved into individual stars, even with the largest telescopes (Figure 9.13). Just how close are these stars in the center of a globular? Even in the most dense portion there are probably no more than 64 stars per cubic light-year (Figure 9.14). Within this distribution the average distance between stars would be such that light would require 3 months to travel from one star to the next, a distance of 2.4 trillion kilometers. By comparison, if we took a cubic light-year at random in our Galaxy, our chances of finding even one star within that space would be only 1 in 64.

Globulars seem to form a halo around the nucleus of the Galaxy (Figure 9.15). They do not participate directly in the revolution of the Galaxy but rather seem to have elliptical orbits, moving in and out of the central region of the Galaxy in a rather random fashion. Their orbits and relationship to the Galaxy are somewhat similar to the orbits and relationship of comets to the solar system.* The distance to these objects may be determined by the variable stars that they contain, called the RR Lyrae, or cluster-type variables (see Chapter 10).

X-ray Bursters

A very interesting phenomenon is unfolding which appears to be related to globular clusters. Sudden bursts of X-rays and gamma rays seem

*The right-hand flip pages beginning on page 345 demonstrate the motion of the globulars in the halo of the galaxy.

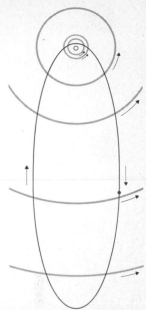

Figure 9.12 *The double open cluster in Persei. (Lick Observatory)*

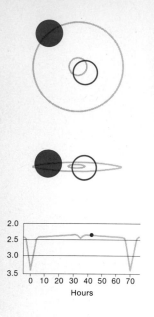

Figure 9.13 *The globular star cluster in Hercules (M13). (Hale Observatories)*

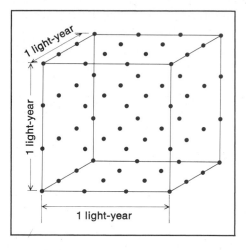

Figure 9.14 *The density of stars near the center of a globular cluster.*

to come from certain of these clusters. The discovery of these "bursters," as they are now dubbed, is a fascinating story that is typical of today's astronomy.

The space age ushered in all sorts of new techniques by which astronomers could use the entire electromagnetic spectrum in studying the universe. We have discussed the X-ray satellite "Uhuru" in Chapter 2, page 95. Another X-ray satellite is the ANS (Astronomical Netherlands

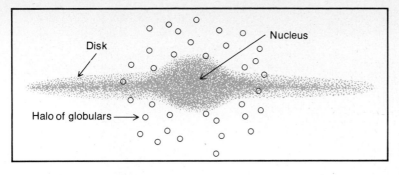

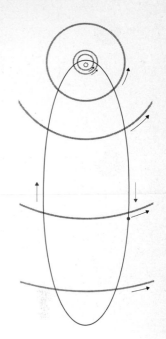

Figure 9.15   The Milky Way galaxy, showing the halo, the
nucleus, and the disk.

Satellite), built and operated by the Dutch but launched and partially
tracked by the United States. It was while scanning the computer output
of this latter satellite for September 25, 1975, that an American astronomer, Jonathan Grindlay, noticed a sudden increase in the X-ray count.
Upon obtaining a second by second record for this period (see Figure
9.16), he was astounded by the sudden tenfold increase in a period of two
seconds. Could it be a quirk in the complicated system of detection and
transmission? No—for the same kind of record was found in Uhuru data.
Furthermore, a satellite called Vela, which had been launched in 1973 to
detect gamma rays as an indication of violation of the nuclear test-ban
treaty, revealed a corresponding burst at even higher energy levels. An
inspection of earlier records of Vela showed that it had detected sudden
bursts of energy almost as soon as it had been launched. Shall we then
give Vela credit for first discovering "bursters"? No, we still reserve that
honor for the human element of recognition, which Grindlay would
share with several independent groups who made simultaneous discoveries. Without the human function of recognition the next logical
function, that of possible explanation, would never follow.

The "burster" recorded in Figure 9.16 is associated with the globular

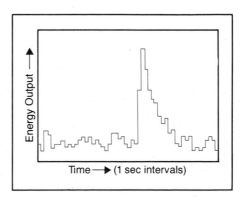

Figure 9.16   A typical X-ray
burst, as recorded in NGC6624.

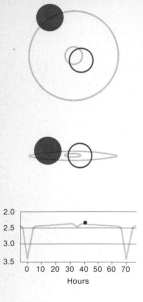

Figure 9.17   *NGC6624, an X-ray burster.*

cluster NGC 6624, a known source of X-rays (Figure 9.17). Since its discovery, at least 20 other bursters have been found. Some of these are associated with globular clusters, and some may be associated with other objects. The 10-second burst represented by this innocent-looking graph probably represents more X-ray energy than the sun emits in an entire year, so it is certainly worthy of explanation.

Since the models astronomers have devised to explain the "burster" all utilize very dense stars (which we have not discussed), let us begin by visualizing a star which has twice as much material (mass) as the sun, but has shrunk to a diameter of 10 to 20 km. The density of such a star, called a neutron star, must be equivalent to the density of the neutron itself (3×10^{14} g/cm³). Further, let us suppose that this star is only one member of a binary pair of stars and that, because of its ultracompact nature, it tends to gravitationally attract material from its mate. The flow of gases onto its surface would generate a background level of X-ray radiation, but this would not account for the bursts. Stars typically have a magnetic field, and this would tend to be strengthened by contraction. Such a field could hinder the flow of gases. When a sufficient mass of gas had collected, however, the field would collapse, and a sudden outburst of X-rays would occur as gas again fell on the star's surface. As evidenced by a number of "bursters," the whole process must then be repeated but within a variable interval of time. It seems that variability is characteristic of most X-ray radiation.

Grindlay's own model for bursters entails an even more massive and denser star called a black hole. Such a star is capable of pulling gas and/or objects toward its surface from greater distances than the neutron star. Gas falling toward its surface would experience frictional heating and would generate X rays. The surrounding gases would be heated by such radiation. Such heating could trigger an expansion—an outflow of gases which would at least temporarily halt the flow toward the surface. This deterrent could not be sustained very long, however, and a sudden collapse would trigger another outburst of X rays.

A third model might involve a black hole in a chance passage through a gas cloud. The sudden infall of gas would certainly produce a gigantic outpouring of X rays, without involving the globular cluster.

If you had asked astronomers what they expected to find with X-ray satellites, probably no one would have conceived of bursters. The unexpected is certainly a vital part of the fascination of the science today.

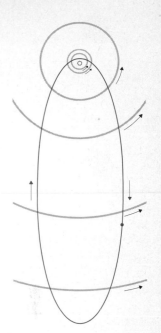

Associations

Star types O and B are not distributed at random, but rather appear to be associated in groups. Almost 100 such *associations* are known in the Milky Way galaxy, and one such is situated in the region of the Orion Nebula. These high-luminosity stars are thought to be very young, lying close to the flattened disk portion of the Galaxy. We will see in Chapter 13 how these very loose, short-lived associations mark the spiral arms of the Milky Way much as bright lights mark the streets of a city seen at night from the air.

Questions

1. Most of the pairs we see as double stars are (a) two stars that are many light-years apart but just happen to line up from our point of view, (b) two stars that are close enough together to control each other's motion. Which is true?

2. How may a true binary system be recognized, even if the two stars cannot be separated by the largest telescope?

3. In the case of eclipsing binary stars, what properties of such a system can be determined from its light curve?

4. What is the most important characteristic of stars that can be determined in a binary system?

5. What does the location of the barycenter of a binary system tell the astronomer?

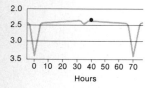

6. When the mass of a star is plotted against its absolute magnitude, what becomes evident from the graph?

7. What does the fact that many stars of a cluster are moving in the same direction and at approximately the same speed suggest about their origin?

8. What is the most obvious difference between the open-type and the globular-type clusters?

9. Within the Milky Way galaxy, where are the globular clusters found?

10. Describe the motion of the globulars in relation to the Milky Way galaxy.

11. (a) Find the mass of each star in a binary system if their total mass is 10 solar masses and their respective distances from the barycenter are 2 A.U. and 3 A.U. (b) Find their period.

12. Find the total mass in a binary system which has a separation of 3 A.U. and a period of 3 years.

13. What change would accompany a reduction of the distance between two binary stars if some unknown force shoved them closer together?

14. What assumption must be made about a star or star system if both the lines of ionized helium and the lines of molecules show up in the same spectrum?

Suggested Readings

ABT, HELMUT A., The companion of sunlike stars, *Scientific American* 236 (4), 96 (1977).

BAADE, WALTER, *Evolution of stars and galaxies* (Chapters 9, 10, and 12). Cambridge, Mass.: Harvard University Press, 1963.

CLARK, GEORGE W., X-ray stars in globular clusters. *Scientific American* 237 (4), 42 (1977).

EGGEN, O. J., Masses of visual binary stars. *Annual Review of Astronomy and Astrophysics* 5, 105–138 (1967).

IBEN, ICKO, JR., Globular-cluster stars. *Scientific American* 223 (1), 26–39 (1970).

KING, IVAN R., Dynamics of star clusters. *Sky and Telescope* 41 (3), 139–143 (1971).

PACZYNSKI, B., Evolutionary processes in close binary systems. *Annual Review of Astronomy and Astrophysics* 9, 183–208 (1971).

POPPER, DANIEL M., Determination of masses of eclipsing binary stars. *Annual Review of Astronomy and Astrophysics* 5, 85–104 (1967).

STRONG, IAN B., and KLEBESADEL, RAY W., Cosmic gamma-ray bursts, *Scientific American* 235 (4), 66 (1976).

Motion of globular clusters

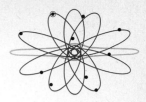

Rotation of the Milky Way

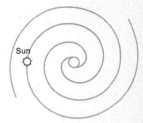

Sun

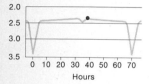

TEN

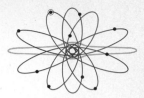

VARIABLE STARS

We have already seen a type of star system—the eclipsing binaries—that *appears* to vary in brightness. We shall now be concerned with stars that *actually* vary in light output. The existence of such stars was recognized by Hipparchus in the first century B.C., and today thousands of variables have been catalogued.

The Cepheids, perhaps the most important class of variable stars, are named after a particular star, δ *Cephei*, the delta star in the constellation *Cepheus* (Figure 10.1). This notation is an example of a system that was initiated by Johann Bayer in his catalogue of stars published in 1603. With few exceptions, he assigned the Greek letter α (alpha) to the brightest star in a given constellation, β (beta) to the second brightest, γ (gamma) to the third, δ (delta) to the fourth, ϵ (epsilon) to the fifth, and so on.

The variable nature of δ Cephei was discovered by John Goodricke, a young British astronomer, in 1784. The light curve of δ Cephei, shown in Figure 10.2, indicates its period of 5.4 days.

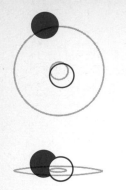

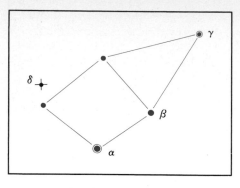

Figure 10.1   The constellation
of Cepheus.

Cepheid Variables

We can see from Figure 10.2 that the light of δ Cephei builds up rather rapidly from a magnitude of +4.3 to +3.6, a difference of 0.7 mag, which corresponds to a doubling of luminosity. This particular star appears twice as bright when at its maximum as at its minimum (Figure 10.3). Cepheids in general have periods of variation ranging between 1 and 50 days, and their light output varies over a range of 0.1 to 2 mag. Polaris, the North Star, is an example of a Cepheid that has only a small variation of 0.1 mag in a period of 4 days, not detectable by the naked eye. On the other hand, variations in δ Cephei are quite evident with the aid of binoculars.

We might wonder why certain stars fluctuate in their light output while others appear to be rather stable. What is happening to the star itself to cause these changes? The spectrograph will help us answer this question, for when a variable star is subjected to spectral analysis, it becomes very evident that the star is pulsating—growing larger for a time and then smaller again. Let us assume, for example, that a particular star is moving away from us and therefore has a red shift in its spectral lines due to its radial velocity. As this star grows, its spectral lines will be shifted slightly from their normal red shift position back toward the blue, since the star's motion due to expansion is basically toward us. On the

Figure 10.2   The light curve of δ Cephei.

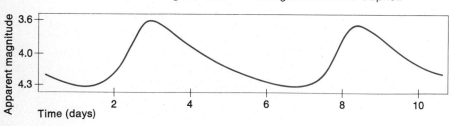

other hand, when the star is decreasing in size, both this motion and its radial velocity will increase the red shift. If we plot the rate of change in size, along with the light curve, we find a distinct relationship between the two. The actual size of the star at any given time is shown in Figure 10.4 (bottom graph).

If a star were to behave simply like a "ball of gas," we might expect the time of its maximum brightness to correspond to its most condensed state, when the star would also be hottest, owing to its contraction; however, it is evident that the point of greatest light output follows shortly after this time of greatest contraction. The core of the star may actually put out most energy at the time of greatest contraction, but this effect does not reach the outer layers until later. Can we explain the variation in light by the change in size alone? δ Cephei has an average diameter of 40 million kilometers, with its expanded diameter being about 2.4 million kilometers greater than its contracted diameter—a change of only about 6 percent. This in itself could not account for the doubling in luminosity; however, the spectrum of the star reveals that accompanying a change of diameter is also a change of temperature. A small change in temperature creates a much more drastic change in energy output than does a corresponding change in diameter, and for this reason the variation in light output is primarily a function of temperature.

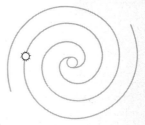

Figure 10.3 *A double exposure with slight displacement, showing the variable star WW Cygni at maximum and minimum brightness. The variable is located near the center of the photograph. (Hale Observatories)*

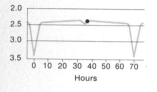

Cepheids: Indicators of Distance

In 1912 Henrietta Leavitt of Harvard College Observatory became interested in variables that had been discovered in the Small Magellanic Cloud (Figure 10.5). Traveling to the southern station of Harvard Observatory in South Africa, she took many photographs of this Small Cloud near the south polar region. From these photographs she made a remarkable discovery: there was a definite relationship between the apparent magnitude of the Cepheids and their period of variation. The longer-period Cepheids were also the most luminous. A plot of the stars that she studied is shown in Figure 10.6.

The very fact that these stars appear to lie along a diagonal line on this plot instead of being distributed randomly over the whole chart shows the relationship between the period and the apparent magnitude of these stars. The longer the period, the brighter the star's appearance. Does it necessarily follow that, because of this relationship, there is also a relationship between a star's period and its absolute magnitude? Miss Leavitt reasoned that because all the stars in the Small Magellanic Cloud belong to the same group, they are all approximately the same distance from the observer. This supposition was later proven correct. When a

Figure 10.4 *Graphs of a Cepheid variable. (a) Light curve. (b) Rate of change in size. (c) Change in size.*

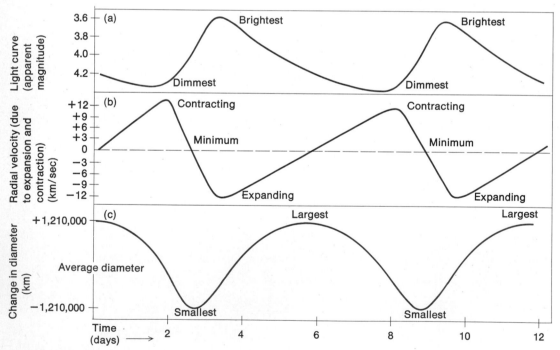

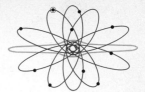

Figure 10.5 *The Small Magellanic Cloud. (Mount Stromlo and Siding Spring Observatories, The Australian National University)*

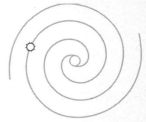

Figure 10.6 *A plot of Cepheids in the Small Magellanic Cloud.*

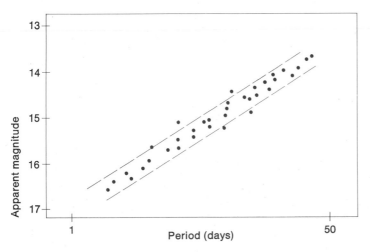

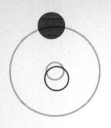

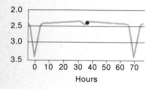

group of stars are found to lie at approximately the same distance from the observer, their absolute magnitudes compare in precisely the same way as their apparent magnitudes. It is only necessary to find the distance to a few Cepheid-type variables in order to make this conversion; however, at that time no Cepheids had been found within the limits of the method of heliocentric parallax.

Harlow Shapley and others joined Leavitt in the search for Cepheids in our own neighborhood of stars, and by using a method of statistical parallax they found the distances to a small number of Cepheids. Cepheids of a 1-day period were found to have an absolute magnitude of approximately zero, while those of a 50-day period were set at -4 in absolute magnitude; thus the latter were relatively luminous stars ranging from 100 times to 4000 times the luminosity of the sun. The Small Magellanic Cloud was recognized as being a very distant object, later to be established as another galaxy outside our own at a distance of more than 150,000 light-years.

Naming of Variables

If a variable found in a given constellation already has a Greek-letter designation, as in the case of δ Cephei, it is retained. If there is no previous designation, variables are given a prefixed letter in the order of their discovery within a given constellation according to the following scheme: R, S, T, . . . , Z; RR, RS, RT, . . . , RZ; SS, ST, SU, . . . , SZ; and so until ZZ is reached; then AA, AB, AC, . . . , AZ; BB, BC, BD, . . . , BZ; CC, CD, . . . , CZ; through QZ, with the letter J omitted. Up to this point, 334 different designations are possible. If additional designations are needed in any given constellation, the prefix V is used—V335, V336, and so on—always followed by the name of the constellation. Examples include RR Lyrae and V335 Herculis. The scheme by which variables are named has evolved over a period of time and now represents a blend of several schemes.

RR Lyrae Stars

A class of variables was found with periods ranging from a few hours to 1 day. These are called the RR Lyrae type, and each group member is estimated to have an absolute magnitude of zero; thus the group seems to be a continuation of the Cepheid type. Using this supposition, Walter Baade of Mt. Wilson and Palomar Observatories in 1952 conducted a search for variables in the Andromeda galaxy. He was able to photograph a variety of Cepheids in the galaxy, including those of 1-day period; however, he could find no star of RR Lyrae type, even though such a star should have been of equal magnitude. Look at Figure 10.7:

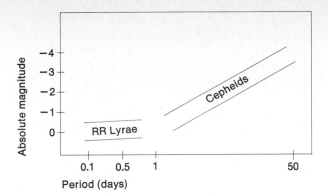

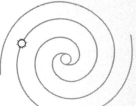

Figure 10.7   *A plot of RR Lyrae and Cepheids.*

notice that both the 1-day Cepheid and the RR Lyrae stars were thought to be of zero absolute magnitude.

The fact that Baade could not find RR Lyrae stars in the Andromeda galaxy made him suspicious of Shapley's calibration of the absolute magnitudes of the Cepheids. Further study revealed that there are actually two types of Cepheid variable stars: Type-I Cepheids, which are found in the disk of spiral galaxies like our own; and Type-II Cepheids, found in globular clusters. Baade made numerous observations using the 5-m telescope at Mt. Palomar, and he calibrated the 1-day, Type-I Cepheids as having an absolute magnitude of -1.5, which was 1.5 mag brighter than the Type-II Cepheids Shapley had observed. This was equivalent to a fourfold change in luminosity. Since the old Cepheid calibration had been used to determine the distance to nearby galaxies, and then the nearby galaxies had been used to estimate distances to other galaxies (and ultimately the size of the universe), a fourfold increase in luminosity estimates produced a doubling of all distance estimates. In other words, when Baade realized that the Type-I Cepheids he had been observing in the disk of the Andromeda galaxy were really four times as bright, he was forced to conclude that they were twice as far away. Remember that the apparent brightness of any source falls off with the square of its distance.

After the period-luminosity relationship of both Type-I and Type-II Cepheids was determined with greater accuracy, these stars came to play a vital role in measurement of distances to galaxies. Let us see how this has been accomplished. Suppose that the astronomer isolates a Type-I Cepheid in the disk of a spiral galaxy and observes its period to be 10 days; its absolute magnitude can be read off as -4 (Figure 10.8). Now suppose that same star has an average apparent magnitude of $+21$. Representing its luminosity as L, the astronomer may use the following proportion to determine its distance (r):

$$\frac{L(10)}{L(r)} = \left(\frac{r}{10}\right)^2$$

Apparent magnitude $+21$
Absolute magnitude $-\ 4$
Difference $\overline{25}$

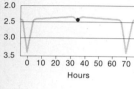

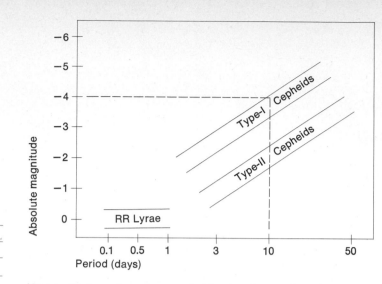

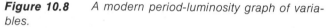

Figure 10.8 *A modern period-luminosity graph of variables.*

A change of 25 mag is equivalent to a change in luminosity of 10^{10} (see Table 8.1); therefore,

$$\frac{10^{10}}{1} = \frac{r^2}{10^2} \quad \text{and} \quad r^2 = 10^{12}$$

Thus,

$$r = 10^6 = 1 \text{ million parsec} = 3.26 \times 10^6 \text{ light-years}$$
$$= 3.26 \text{ million light-years}$$

Irregular Variables

A variable star may be irregular in both its period and the amplitude of its variations. The light curve of Mira Ceti is shown in Figure 10.9. We see that its magnitude at maximum is between 3 and 5 and at minimum between 8 and 10. Its period averages about 330 days; however, this does not mean that any single maximum is predictable. What kind of changes in a star can produce a 5-mag (hundredfold) change? Mira is an M-type star with a surface temperature, at minimum light, equal to about 2000°K and rising to only about 2600°K at maximum. Such an increase in temperature can produce only about a threefold increase in total light output, not the hundredfold increase observed. The second factor that may be considered is a change in size. Mira Ceti is one of the relatively few stars whose diameter we may compute by direct observations using

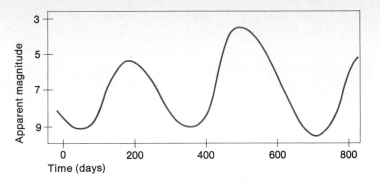

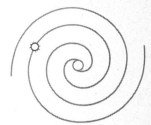

Figure 10.9 *The light curve of Mira Ceti, an irregular variable.*

an interferometer. Its diameter is approximately 300 times that of the sun, and during pulsations it increases to 360 solar diameters, an increase of 20 percent. This in turn produces almost a 50 percent increase in surface area, yet this in itself is not sufficient to explain the total variation in light output. While the final explanation is not certain, it is felt that the atmosphere of the star must also undergo drastic changes, and it may hold additional clues concerning the large light variations.

Betelgeuse is still another example of an irregular variable with a variation of one magnitude in a period of 5 to 6 years. Its diameter varies between 500 and 750 solar diameters. The T Tauri stars, still another group of irregular variables, are characterized by their very rapid variations. Stars of this type are usually found in young clusters, often with a certain amount of uncondensed gas (nebulosity) in the star's vicinity. This suggests that the T Tauri stars are probably very young, still contracting, and not yet stabilized as main-sequence stars. T Tauri stars can also be identified by a strong H_α (hydrogen-alpha) emission line in their spectra that is not characteristic of older pulsating stars.

Recurrent Novae

To witness the "birth" of a *nova* (new) star would indeed be a spectacular event in a person's life (Figure 10.10). Ancient observers thought that this was happening when from time to time a star would suddenly appear where no star had been visible before. Far from being a new star, the nova is probably a star nearing its old age. It is a star that suddenly brightens without warning, increasing in light output perhaps a thousandfold—an occurrence which may be repeated. A typical prenova is a small blue or white dwarf star of about the same magnitude as the sun (+5). When it brightens, its outer layers expand very rapidly, 1600 km/sec, and its magnitude may reach −7; that is, a change of

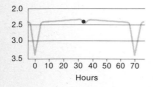

12 mag, or a 60,000-fold increase in luminosity. The expanding shell of gas, which may be detected spectroscopically, gives a clue to its distance, for if both its rate of linear expansion and its rate of change in angular diameter can be observed, then its distance may be determined. Figure 10.11 graphically depicts the light curve of a typical nova, showing its variability even at minimum.

The eruption may take place in only a few hours, then require several weeks, months, or years to subside. The nova usually returns to its prior luminosity or only slightly less than that, indicating that the explosion was a superficial one, involving only its upper atmosphere. Probably less than one ten-thousandth of its mass is lost. Nova Aquilae 1918, for instance, increased in brightness 13 mag (from $+5$ to -8), and a shell of gas was thrown off at a velocity of 1700 km/sec. At a distance of about 1200 light-years, this shell appeared to increase in diameter about 1 second of arc per year.

By what mechanism can a star expel a shell of gas and then return to its former brightness? Suppose that a hot dwarf star is a member of a binary pair in which the other member is expanding to a giant stage. As

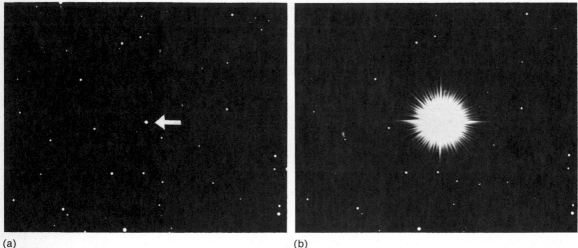

(a)

(b)

Figure 10.10   *Nova Herculis 1934, showing the large change in brightness between (a) March 10, 1935, and (b) May 6, 1935. (Lick Observatory)*

the star expands, some portion of its outer layer approaches the dwarf star close enough to experience its stronger gravitational force; as a result, some of the material of the cool giant will be transferred to the dwarf. Such a renewal of hydrogen-rich gases pouring onto the dwarf, which has used most of its hydrogen in thermonuclear reactions, may trigger a renewed reaction. Thus, a shell of gas would become heated and would then expand to produce the nova phenomenon. In this model, the dwarf

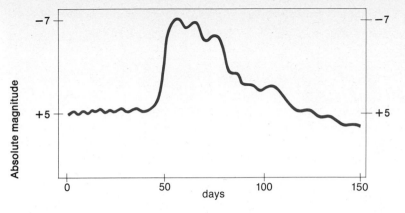

Figure 10.11 *The light curve of a typical nova.*

star may then return to its former dim state until another such reaction occurs. Some observers believe that a binary association is essential for nova-type reactions to occur. By contrast quite a different model is presented for the supernovae.

Supernovae

The name *supernova* implies a big nova, and the light curve of a supernova may superficially resemble that of a nova—but that is where the similarity ends. The record of supernovae suggests that they are very rare occurrences: only one has been recorded in the past four centuries in the Milky Way galaxy (in 1604). As you will see in our study of evolving stars (Chapter 12), at a certain point in time the core of a massive star may suddenly collapse, creating an implosion (something like an explosion except that the motion is inward). This collapse infuses new material into the central core of the star and provides added heat to trigger violent nuclear reactions (fusion). The result, from merely a visual point of view, is a sudden brightening, often to the point of outshining an entire galaxy in which the star resides. Such a star and such a reaction are called a *supernova*. It would not be unusual for such a star to brighten to 100 million times its normal brightness, a change of 20 magnitudes, then gradually subside to something less than its original magnitude. In the process of the implosion, a shock wave is thought to move outward from the star driving off a significant portion of the star's mass (several solar masses may be lost) at high velocities (up to 10,000 km/sec) and to produce a visual image like that of the Crab Nebula (Figure 10.12). The star which exploded to form this nebula was recorded by the Chinese in A.D. 1054 as a star (a supernova) which could be seen in the daytime. Far

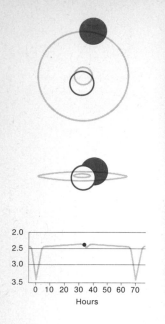

(a)   (b)

Figure 10.12 *Supernova in NGC 5236 (a) before and (b) after brightening in 1972. (Hale Observatories)*

from being a new star, as the word nova implies, the star which produces a supernova reaction is in its later stage of evolution.

R Coronae Borealis Stars

A class of variable stars having a light curve that seems almost upside-down from those that we studied is the R Coronae Borealis type (Figure 10.13). Not having a predictable period, these stars suddenly drop several magnitudes and then slowly return to normal brightness. When at

Figure 10.13 *The light curve of R Coronae Borealis.*

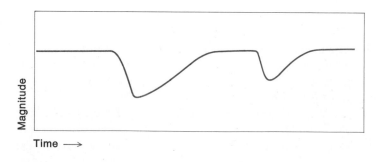

a minimum, some reveal a small shell of gas surrounding the star; the spectrum of this shell is surprisingly similar to that of a planetary nebula. Furthermore, the distribution of these stars is quite similar to that of the planetary nebulae. We cannot be sure of this relationship, but the R Coronae Borealis stars form an interesting group and constitute at least a possible predecessor to the planetary nebulae (to be considered in Chapter 12).

Flare Stars

The final class of variable stars to be discussed is exemplified by UV Ceti, a flare star. We have discussed flares that occur on the sun, and this may be the same kind of phenomenon which causes certain stars to brighten periodically. A flare on the sun is a localized hot spot. For stars that are in reality very bright, it is difficult to spot a flare, since the proportionate increase in light output is small. On the other hand, a flare on an intrinsically faint star, say, an M-type dwarf, is much more apparent. Typically, the flaring process creates a very rapid variation in light output, lasting only minutes. If this phenomenon in a star is similar to the localized flare on the sun, it represents a tremendous release of energy in a very short period, hence a real explosion—perhaps much more violent than that we observe on the sun.

We have studied a number of special types of stars, some of which surely represent different stages in the life cycle of stars in general. In Chapter 12 we shall try to discover an evolutionary progression for these various types.

Questions

1. How much variation in brightness is shown by δ Cephei?

2. True or false: Of all the different types of variable stars, the Cepheids can be distinguished by their light curve.

3. What physical change appears to produce the variations of brightness in the Cepheids?

4. What relationship makes the Cepheids a valuable distance indicator?

5. What characteristic of the RR Lyrae stars makes them good distance indicators?

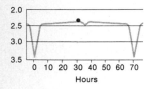

6. The discovery that there are really two types of Cepheids (Type I and Type II) led to a reevaluation of the size of the known universe. Explain this reevaluation.

7. What circumstances are thought to have produced the Crab Nebula?

8. What evidence suggests that a planetary nebula bears little or no relationship to a nova?

9. There is some indication that stars may evolve through several variable stages. Is the T Tauri group thought to be composed of young or old stars? What evidence supports your answer?

10. What distinguished the light curve of the R Coronae Borealis stars from that of other variable types?

11. Why is it thought possible for novae to recur over and over, whereas a supernova probably represents a one-time event for a star?

12. RR Lyrae stars are classified as Type-II variables. Explain the meaning of this classification and tell where the RR Lyrae stars are usually found.

13. The period-luminosity graph of Figure 10.8 tells us that a 1-day, Type-II Cepheid has an absolute magnitude of -1. Suppose this same star had an apparent magnitude of $+14$; find its distance.

14. True of false: A flare star can be recognized as showing a variation in its light output only if it is a rather cool star.

Suggested Readings

GLASBY, J. S., *Variable stars.* Cambridge, Mass.: Harvard University Press, 1969.

JONES, KENNETH G., *Messier's nebulae and star clusters.* London: Faber and Faber, 1968. (A pictorial guide to interesting objects which can be seen with the naked eye or with a small telescope.)

KIRSHNER, ROBERT P., Supernovas in other galaxies. *Scientific American* 235 (6), 88 (1976).

SMAK, JOZEF, The long-period variable stars. *Annual Review of Astronomy and Astrophysics 4*, 19–34 (1966).

SMITH, E. P., and JACOBS, K. C., *Introductory astronomy and astrophysics* (Chapter 16). Philadelphia: Saunders, 1973.

STEPHENSON, F. R. and CLARK, D. H., Historical supernovas. *Scientific American* 234 (6), 100 (1976).

STRUVE, OTTO, *The universe* (Chapter 5). Cambridge, Mass.: M.I.T. Press, 1962.

WARNER, BRIAN, and NATHER, R. E., High-speed photometry of cataclysmic variables. *Sky and Telescope* 43 (2), 82–85 (1972).

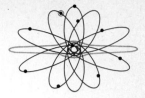

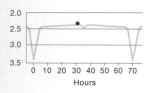

ELEVEN

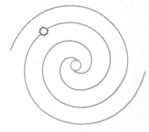

INTERSTELLAR MEDIA

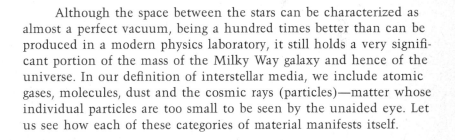

Although the space between the stars can be characterized as almost a perfect vacuum, being a hundred times better than can be produced in a modern physics laboratory, it still holds a very significant portion of the mass of the Milky Way galaxy and hence of the universe. In our definition of interstellar media, we include atomic gases, molecules, dust and the cosmic rays (particles)—matter whose individual particles are too small to be seen by the unaided eye. Let us see how each of these categories of material manifests itself.

Interstellar Gases

The most obvious visible manifestation of interstellar gases is that of the *emission nebula.* For example, the Orion Nebula must have been seen by early observers, for it is visible to the naked eye as a fuzzy path beneath the left-hand star in the belt of Orion, as shown in Figure 11.1. This constellation can be found in the central portion of the winter sky map in Appendix 13 (page 513). The black-and-white photograph, taken by the 3-m telescope of Lick Observatory, reveals many of the details of the nebular structure and gives an indication of the four very hot stars imbedded in its central

363

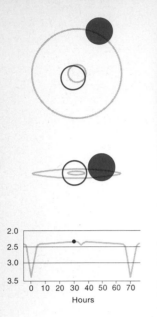

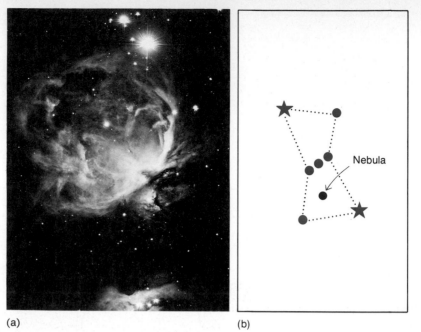

(a)   (b)

Figure 11.1   (a) *The Orion Nebula, NGC 1976. (Lick Observatory) (b) The nebula identified in the constellation of Orion.*

region—called the trapezium. The atoms which compose this nebula are excited by the radiation of these four stars, especially their very energetic ultraviolet radiation. Upward electron transitions result from this stimulation; in fact the atoms in many cases are ionized by the ultraviolet radiation—we refer to this as *photoionization.* Upon recombination of free electrons with the ionized nucleus, downward transitions result in the production of light and radio waves, as explained in Chapter 2, page 82. The Orion Nebula is also shown in Color Plate 11. The reddish color originates from the bright red line of the hydrogen spectrum, as shown in Color Plate 1. The spectrogram of this nebula, and in fact of most emission nebulae, clearly shows the predominance of hydrogen gas (more than 75 percent) over every other element. Because the hydrogen atoms are ionized in virtually every emission nebula, such regions are called *H II regions.* Typically these regions have relatively higher densities (greater than 100 atoms/cm^3) and high temperatures (10,000°K). Approximately half of the interstellar gases may be found in such concentrations (Figure 11.2), yet these same concentrations fill only a very small portion of interstellar space (less than 1 percent). The remaining portion of gases may be described as generally distributed throughout the galaxy.

Because such gas is typically very cool, and the hydrogen is in a neutral state, called H I, the electrons are in their lowest energy level and

Figure 11.2 *The Lagoon Nebula in Sagittarius, NGC 6523 (M8). (Lick Observatory)*

cannot emit light. Fortunately another atomic mechanism is at work by which the presence of neutral hydrogen may be known.

In addition to any other motions the proton and electron of a hydrogen atom may have, each of these members is spinning on its own axis. Normally the orientation of their spin axes is parallel and pointed in the same direction. On occasion, the spin axis of the electron will flip into the opposite direction, and in that process a small amount of energy will be radiated away in the form of a 21-cm radio signal. The beauty of this phenomenon lies in the fact that no stars need be nearby to excite this action; hence the distribution of hydrogen can be discovered by its own radiation. To do this the radio astronomer tunes the radio telescope to receive only the 21-cm signal, rejecting all other wavelengths, even as you might tune in a certain radio station, rejecting all others. Then, by pointing the telescope in various directions, the astronomer may plot the distribution of neutral hydrogen gas in the Galaxy. The concentration of hydrogen can be judged by the strength of the radio signal, and the motion of various regions can be judged by the Doppler effect as it acts to lengthen or shorten the characteristic wavelength. All of these factors have been taken into consideration in producing the plot you see as Figure 11.3.

Hydrogen gas may reveal itself in one further way—by the absorption process of upward electron transitions. Again, because H I regions are cool and electrons are in their lowest state, any light from a bright star which passed through the region would create upward electron transitions from the first energy level. The absorption lines which result would

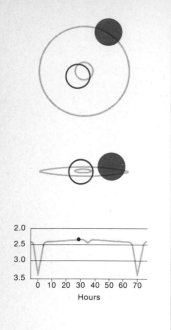

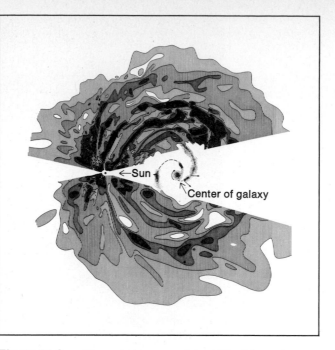

Figure 11.3 *The spiral nature of the Milky Way galaxy as constructed from 21-cm, neutral-hydrogen observations made at Leiden Observatory, Netherlands, and at Radio Physics Laboratory, Sydney, Australia.*

all lie within the ultraviolet portion of the spectrum, which is obscured by the earth's atmosphere. We have now surmounted that problem by orbiting the ultraviolet observatory (satellite) called Copernicus. This fine tool produces ultraviolet spectrograms, as illustrated in Chapter 2, page 94. It has not only confirmed the work of radio and optical astronomers but has generally broadened our understanding of the universe. While our emphasis has been upon hydrogen gas, we should note that elements such as helium, calcium, sodium, oxygen, nitrogen, and carbon have also been identified in the interstellar media.

Interstellar Dust

Interstellar dust consists of solid particles of appreciable size (about 0.00005000 cm) as compared to the diameter of the hydrogen atom (about 0.00000004 cm). The composition of these dust grains probably includes many of the elements (and compounds) found in meteorites and in certain ordinary rocks—carbonates, silicates, and water being most likely. Some observers have suggested magnetite, a natural magnetic mineral that would help to explain the apparent alignment of these grains with magnetic fields, as we shall see shortly.

One of the ways that dust exhibits itself is in the *reflection nebula*.

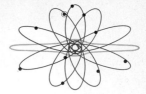

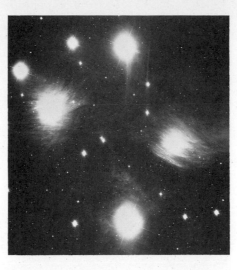

Figure 11.4 *The nebulosity surrounding the Pleiades is thought to be primarily dust. (Hale Observatories)*

For example, the Pleiades are enveloped in a nebulosity that might be mistaken for an emission nebula (Figure 11.4); however, the spectrum of the nebulous material is almost the same as that of the stars that it surrounds. This indicates that the nebulous material is merely reflecting the light and so must be of the nature of dust. Gas molecules are not large enough to reflect light, nor can they effectively absorb light over the

Figure 11.5 *The Horsehead Nebula in Orion. The dark "horsehead" figure represents a dust cloud that obscures the light from behind. (Hale Observatories)*

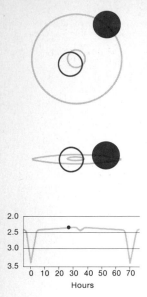

Figure 11.6 *The North American Nebula in Cygnus. (Lick Observatory)*

continuous spectrum. Gas clouds absorb a limited number of wave-lengths to produce an absorption spectrum, but this does not appreciably dim the light from a source beyond the cloud. Dust clouds, on the other hand, do in fact appreciably dim such objects. The Horsehead Nebula is an example of such a cloud (Figure 11.5).

The North American Nebula is an emission nebula; however, the portion roughly corresponding to the Gulf of Mexico is an absorption nebula, basically consisting of cosmic dust (Figure 11.6).

The dark central regions of the Milky Way, near the constellation Sagittarius, were once thought to be "windows" through which the observer might look to the other side of the Galaxy (Figure 11.7). Now it is known that these regions are clouds of cosmic dust that obscure our view of almost all objects that fall behind them. Early observers called this region the "Zone of Avoidance," for no galaxies are evident in that direction. Modern astronomers are penetrating this zone by use of infrared techniques. Infrared wavelengths are not appreciably affected by dust, and recently new galaxies have been discovered in infrared photographs near the center of the Milky Way (see Chapter 14). The presence of obscuring clouds makes the sky look spotty in many regions, and what appears to be a lack of stars in a given area of the sky may only signal the

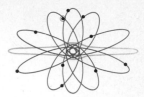

Figure 11.7   *The Milky Way near Sagittarius. (Lick Observatory)*

Figure 11.8   *Dark globules in the Lagoon Nebula (M8). (Lick Observatory)*

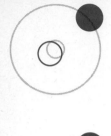

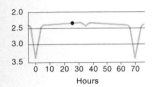

2.0
2.5
3.0
3.5

0 10 20 30 40 50 60 70
Hours

presence of such clouds. Concentrations of dust may appear as small black globules against the background of a glowing nebula (Figure 11.8). In Chapter 12 we will see that these globules are the sites of new star formation. Dust plays a significant role in this process.

The starlight which enters the telescope is often found to be polarized, and most astronomers agree that the most likely cause of this polarization is the presence of dust in the path of the light. Interstellar dust grains are believed to be irregular in shape, perhaps in the form of a long string of molecules. If their form is longer in one dimension (perhaps even needlelike), then their ability to polarize starlight would seem intuitively simple, especially if they existed in a magnetic field which would align the particles. They would act like the long narrow crystals which are laid down to form polarized sunglasses today. Whether this is the correct explanation, astronomers have measured the polarization of starlight in virtually every direction within the Milky Way galaxy and have built a map like that shown in Figure 11.9. Each line shows by its length the degree to which the light is polarized and shows by its direction the plane of polarization. While there is some randomness,

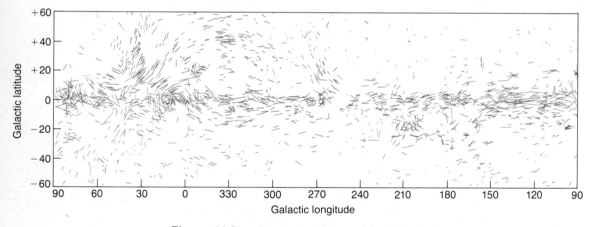

Figure 11.9 *The polarization of starlight in the Milky Way suggests the overall structure of its magnetic field.*

there appears to be an overall pattern which suggests a general magnetic field within the Galaxy. Furthermore, if the general distribution of interstellar gas were plotted on a similar map, we would find minor concentrations of that gas in the form of long filaments (and in some cases, loops) that tend to follow the pattern of polarization and hence the pattern of the general magnetic field of the Galaxy (Figures 11.10 and 11.11). Some observers suggest that expanding "shells" are very typical of the interstellar gas and that these produce shock waves through the media. Several shells may have interacted to form the complex filaments we see. Could a star explosion be sufficient to produce such shells? Many of these ideas are relatively new and are subject to continuing research; however, they do illustrate the interrelationships among gases, dust,

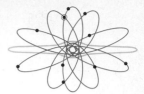

Figure 11.10 *The Veil Nebula, by its form, tends to suggest the presence of a shock wave moving through space (Lick Observatory).*

Figure 11.11 *Cygnus Loop, a shell of expanding gas thought to have been formed by the explosion of a star. (Hale Observatories)*

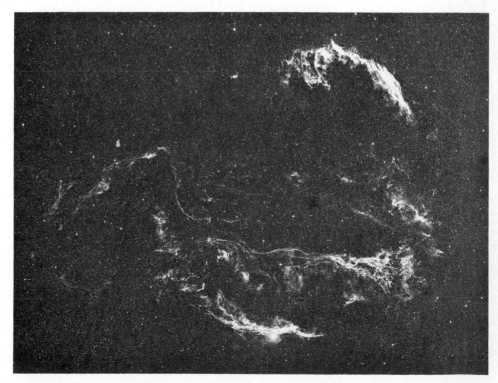

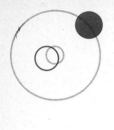

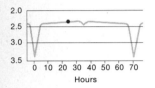

Hours

magnetic fields, and the ultimate structure of interstellar media. If dust were not imbedded in the gas, polarization would not have occurred.

Besides dimming the light of objects beyond these obscuring clouds, cosmic dust also reddens the light that does penetrate. This indicates that dust is also selective in that it absorbs and scatters blue light better than red (short wavelengths better than long wavelengths), thus leaving the light reddened.

Interstellar reddening affects our entire picture of the galaxy in regard to the distribution of stars. We judge the distance to a star by its apparent brightness; however, the presence of dust in the line of sight causes a star to appear dimmer than it would otherwise. A correction must be applied by saying that the star is really closer than it appears on the basis of brightness alone. The amount of correction depends upon the amount of reddening, and this can be judged by comparing the spectral class of the star as determined from its spectrum to its spectral class as determined from its apparent color. In other words, if a B-type star looks like an A-type, then the observer can compute the amount of reddening and make allowance for this factor in determining distance. Such striking effects of interstellar dust would seem to suggest that it must constitute a significant part of interstellar material; but, on the contrary, the total mass of dust represents only about 1 percent of the mass of the interstellar gas clouds.

Interstellar Molecules

The first hint that molecules exist in interstellar space came in 1937 with the recognition of certain spectral lines associated with the methylidyne radical, an atom of carbon linked to an atom of hydrogen (CH), its ion (CH+), and the cyanogen molecule (CN). The real breakthrough came with the advent of radio astronomy, which ultimately revealed some very complex molecules. One of the reasons radio telescopes are so effective in the study of molecules in space is that radio energy penetrates dust clouds that are opaque to visible light. It is within such dust clouds that the more complex molecules are found.

Molecules have certain inherent abilities to generate radio wavelengths which independent atoms do not have. When two or more atoms bond chemically to form a molecule, their bonds are not static, but rather the atoms vibrate in relation to each other. They also rotate end over end or in some other fashion. Any change in this rotation or vibration pattern represents a change of energy state, somewhat like an electron transition or an electron spin-flip. When the molecule changes from a higher to a lower state, an electromagnetic disturbance is generated, and typically the energy released is in the form of a radio signal. Fortunately, each molecule has its own characteristic wavelengths of radiation. These would correspond to the visible spectral lines with which we are familiar. Table 11.1 presents a partial listing of interstellar molecules and their charac-

Table 11.1 *Interstellar Molecules and Their Characteristic Wavelengths*

MOLECULE	FORMULA	CHARACTERISTIC WAVELENGTH
Methylidyne	CH	4300 Å
Methylidyne (ionized)	CH^+	3958 Å
Cyanogen radical	CN	3875 Å, 2.6 cm
Hydroxyl radical	OH	18.0, 6.3, 5.0, 317, 2.2 cm
Ammonia	NH_3	1.3, 1.2 cm
Water	H_2O	1.35 cm
Formaldehyde	H_2CO	6.6, 6.2, 2.1, 1.0, 0.2 cm
Carbon monoxide	CO	2.7 cm
Methyl alcohol	CH_3OH	35.9 cm
Hydrogen cyanide	HCN	3.4 mm
Cyanoacetylene	HC_3N	3.3 cm
Formic acid	HCOOH	18.3 cm
Silicon monoxide	SiO	2.3, 3.4 mm
Carbon monosulfide	CS	2.0 mm
Formamide	NH_2CHO	6.5 cm
Carbonyl sulfide	OCS	2.5 mm
Methyl cyanide	CH_3CH	2.7 mm
Isocyanic acid	HNCO	1.36 cm, 3.4 mm
Methylacetylene	CH_3CCH	3.5 mm
Acetaldehyde	CH_3CHO	28.1 cm
Thioformaldehyde	H_2CS	9.5 cm
Methanimine	CH_2NH	5.8 cm
Hydrogen sulfide	H_2S	1.8 mm

teristic wavelengths. In effect we are saying, "If you want to find the distribution of methyl alcohol in space, tune your radio telescope to 35.9 cm." When you measure a strong radio (noise) signal at that wavelength, you are detecting methyl alcohol along the line of sight of your telescope, for no other molecule emits that wavelength.

We should recognize that some of these molecules have been found only in a few locations and, even there, represent only trace amounts compared to the hydrogen present. For every 10,000 hydrogen atoms we might find four of oxygen, two of carbon, one of sulfur, and one of nitrogen. Such estimates are based on the strengths of their respective spectral lines, as judged by the peaks in a radio spectrogram. Figure 11.12 shows the profile of the spectral "line" of ammonia in radio emission, with its peak output at 1.27 cm (rounded to 1.3 cm in Table 11.1).

Certain molecules can be described as masers in interstellar space. Lasers and masers are very similar in their operation. Their names are acronyms, respectively, for "light amplification by stimulated emission of radiation" and "microwave amplification by stimulated emission of radiation." You are familiar with the various energy levels within an atom (as discussed in Chapter 2, page 82). There also exist energy levels

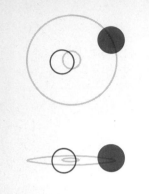

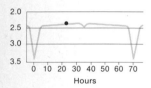

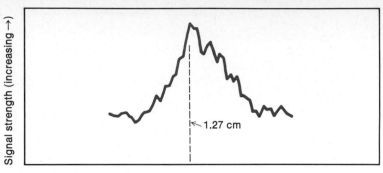

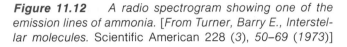

Figure 11.12   *A radio spectrogram showing one of the emission lines of ammonia. [From Turner, Barry E., Interstellar molecules.* Scientific American 228 (3), 50–69 (1973)]

within the molecule which we may think of in a similar way. In the case of a maser (or laser), molecules are "pumped up" (excited) in such a manner that many more atoms exist at a higher energy level than one would expect for a given temperature. This pumping action may result from collisions among the molecules, but more likely from radiation of a cosmic energy source. When a molecule returns, as it naturally does, to a lower energy state, then its characteristic wavelength is emitted and thereby the molecule is detected. The key ingredient of the maser (or laser) is a source of energy to keep the molecule "pumped up"—in an excited state, with a preference for a certain state. Then, in place of emitting a lot of different wavelengths, the molecule typically emits only one (or sometimes two). In the case of the laser, only a single wavelength of light is emitted, and the optical behavior of a single wavelength is quite different from that of white light (a mixture of many wavelengths). For instance, a beam of laser light can be sent to the moon with very little spreading, thus maintaining its energy in a very narrow beam.

The hydroxyl molecule (OH) is a particularly good maser, when stimulated, and several hundred such sources have been identified. Water (H_2O) masers are also found in many locations. Thus the list of places where we have found water, water ice, or water vapor in the Galaxy (and thus presumably throughout the universe) now includes not only earth, ice caps of Mars, rings of Saturn, frost of numerous moons—to name a few—but also interstellar space. In fact, as one looks at the molecules listed in Table 11.1, an interesting pattern evolves in one's mind. Five predominant elements compose these molecules: H, O, N, C, S. These same five elements appear to dominate the organic molecules of life as we know it. Some of the organic molecules listed in Table 11.1 are precursors of amino acids, which are precursors of the proteins essential to life. One might think that the entire universe was in some sense destined for life. We will look at this question more closely in Chapter 16.

Molecules do not have a general distribution throughout the Galaxy but have preferential locations. Typically they are found in cool, dense regions, where temperatures are just a few degrees above absolute zero (4° to 25°K) and densities are from 100 to 10,000 particles/cm³. These conditions remind us of the dark globules found in certain nebula (see Figure 11.2, page 365)—cool because of the presence of dust which obscures the radiant energy of starlight. The very cold state allows for slower collisions among the atoms, resulting in a higher probability of their sticking together (bonding) to form molecules. Dust is thought to play the additional role of providing a surface upon which atoms may attach themselves and eventually "get together" to form molecules. Thus dust plays a vital role in the formation of interstellar molecules. Other factors, such as ultraviolet radiation, tend to destroy (split up) molecules, so wherever we find molecules we must conclude that they are being formed faster than they are being destroyed in that location.

Cosmic Rays

A major component of the interstellar media consists of charged particles, called *cosmic rays*, which have been accelerated to almost the speed of light. These particles include the nuclei of almost every kind of atom, with a predominance of hydrogen nuclei (simple protons) and helium nuclei (alpha particles). Also included are electrons, positrons, and neutrinos. Cosmic rays represent a very energetic form of radiation. Their energy is basically due to velocity: a small mass traveling at very high velocity possesses high kinetic energy. Imagine an iron nucleus (with 56 protons and neutrons) traveling at 95 percent of the speed of light. Its kinetic energy compares to that of a tennis ball traveling 150 km/hr. If you could stop the iron nucleus (which is not actually possible), it would be like stopping the tennis ball.

Cosmic rays appear to approach the earth equally from all directions, and as they enter the earth's atmosphere, collisions occur that split various nuclei into smaller (secondary) parts, a process called *spallation*. From the surface of the earth we see largely these *secondary* particles. The rocket or satellite observatory gives us a better view of the *primary* components as they exist above the earth's atmosphere. Even these primary particles may have experienced some changes as they traveled through interstellar media for millions of years.

As we search for the source of cosmic rays, we naturally look to energetic (violent) processes to explain the tremendous velocities we see in these particles. We know that the sun ejects particles at high speeds, especially on the occasion of a flare, but such energies are too low to explain more than just the low-energy cosmic rays. Furthermore, the flow of cosmic rays is not preferential to the sun's direction. The supernova (an exploding star) has long been suspect, because the energy (and

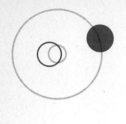

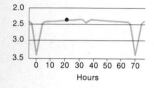

particles) released are sufficient to explain the energy in the cosmic rays, assuming that supernovae occur about once in 25 years per galaxy. If the supernovae represent the source, one is still hard pressed to explain why we do not sense more cosmic rays from one direction than another. What process could have smoothed out this radiation? Still another source lies within the nucleus of the Galaxy. Many very energetic processes are going on there. Charged particles accelerated by any of these processes would be influenced by the magnetic fields of the Milky Way and, in fact, most of the cosmic rays would be contained by that magnetic field. If this is true, we would not be inclined to look outside our own Galaxy for a source, assuming that most cosmic rays would be contained in their own galactic regions. On the other hand, some observers have suggested

Figure 11.13   *Cosmic ray abundances as compared to abundances in the solar system.*

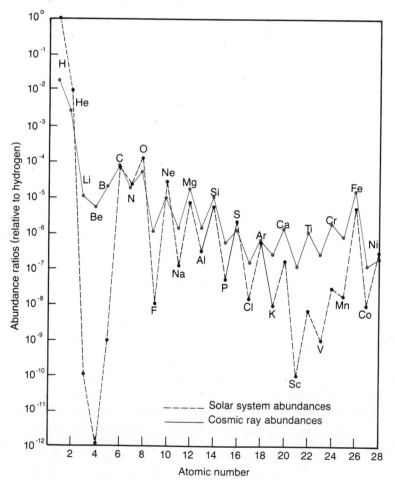

quasars as a source, for the process whereby they radiate energy equivalent to that of 1000 normal galaxies certainly should be adequate to accelerate particles as well. As a final suggestion, some observers believe that shock waves passing through the interstellar media may accelerate those media.

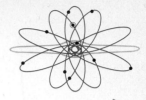

One way in which astronomers are testing their hypotheses is by analyzing the abundance of elements (and their isotopes) which compose the cosmic rays. The findings can then be compared to the abundance ratios in various sources. Figure 11.13 shows such a comparison between cosmic rays and solar system abundances. Note the similar pattern in most elements except lithium (Li), berylium (Be), and boron (B). Some observers would explain the higher abundance of these elements in the cosmic rays as due to the splitting of heavier elements such as carbon (C) and oxygen (O). The solar system abundance ratios are not significantly different from those of nucleosynthesis (the manufacture of heavier elements) in the cores of hot, massive stars. These same massive stars usually explode as supernovae at some stage in their life cycle; hence the evidence points toward the supernovae. Only with refinement of abundance and isotope ratios will questions of source be settled via this approach.

Noting Albert Einstein's statement that our perception of space is influenced by the distribution of matter that produces gravitational fields, we realize that it may also be influenced by the distribution of electrical charges—a distribution which produces electrical and magnetic fields.

Questions

1. By what process does an interstellar gas cloud emit its own light?

2. By what process may a nebula produce a radio spectrum?

3. How can the astronomer distinguish a reflection nebula from an emission nebula?

4. Describe the effects that interstellar dust clouds have upon light passing through them.

5. In what sense is the term "cosmic ray" a misnomer?

6. What are possible sources of cosmic rays?

7. What protection does the earth have from continual bombardment by cosmic rays?

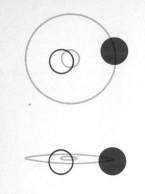

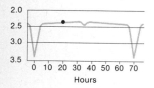

8. What did early observers mean by the term "Zone of Avoidance"?

9. How has the modern astronomer penetrated the "Zone of Avoidance" to discover new objects beyond the central bulge of the Milky Way galaxy?

10. How does neutral hydrogen generate a 21-cm radio signal?

11. Looking at Table 11.1, list the individual elements of which these compounds are formed.

12. Can you explain how molecules found in interstellar space may be distinguished from each other?

13. How is polarization related to the study of dust between the stars?

14. Although, like protons and electrons, cosmic "rays" are charged atomic particles, what distinguishes them from ordinary atomic particles?

Suggested Readings

BOK, BART J., The birth of stars. *Scientific American* 227 (2), 49–61 (1972).

CHAISSON, ERIC J., Gaseous nebulas. *Scientific American* 239 (6), 164 (1978).

DICKINSON, DALE F., Cosmic masers. *Scientific American* 238 (6), 90 (1978).

HEILES, CARL, The structure of the interstellar medium. *Scientific American* 238 (1), 74 (1978).

LINSLEY, JOHN, The highest energy cosmic rays. *Scientific American* 239 (1), 60 (1978).

LYNDS, B. T. (ed.), *Dark nebulae, globules and protostars.* Tucson: University of Arizona Press, 1971.

MIDDLEHURST, BARBARA M., and ALLER, LAWRENCE H. (eds.), *Nebulae and interstellar matter.* Chicago: University of Chicago Press, 1968.

MIDDLEHURST, BARBARA M., and ALLER, LAWRENCE H. (eds.), *Stars and stellar systems.* Chicago: University of Chicago Press, 1968.

SHU, FRANK H., Spiral structure, dust clouds and star formation. *American Scientist* 61 (5), 524–536 (1973).

SMITH, E. P., and JACOBS, K. C., *Introductory astronomy and astrophysics,* Chapter 17. Philadelphia: Saunders, 1973.

SOLOMAN, PHILIP M., Interstellar molecules. *Physics Today* 26 (3), 32–40 (1973).

TURNER, BARRY E., Interstellar molecules. *Scientific American* 228 (3), 50–69 (1973).

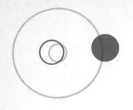

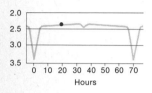

TWELVE

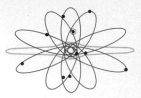

STELL/R EVOLUTION

Our study has revealed that stars vary greatly in size, mass, temperature, color, and luminosity. Does this great variety indicate that stars are basically different from one another, or does it indicate that stars are merely in different stages of a life cycle, throughout which their appearance drastically changes? While stars differ widely in their masses, and this does influence their life cycle, the fundamental reason for variety is that stars undergo change. Like humans, stars go through stages which could be termed pregnancy, birth, youth, adulthood, old age, and death. Only under the rarest circumstances, however, may an astronomer hope to witness a transition from one evolutionary stage to another, for the life cycle of an average star extends over billions of years; and so the astronomer must be content to observe stars in different stages and try to arrange them in the proper order.

The problem is somewhat like the one confronting an intelligent being from another planet who has only one day to spend on the earth, and who has no prior knowledge of the life cycle of humans. What approach might this observer use to develop a theory about humans' aging sequence? Obviously it could observe humans in all stages of development—but how could it determine which stage

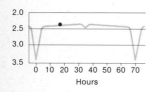

came first? It might judge the size of humans to indicate various ages, or it might consider vigor, metabolism, condition of skin, speech qualities, and so on. As a result of these observations, it might deduce the proper life cycle. In a very similar way, the astronomer observes various stars in different stages of their life cycle and attempts to place these in their proper order.

To complete this analogy, we must make two assumptions: 1) stars, like humans, have been and are being created continually and hence are of different ages; and 2) stars do change with time, however slowly. The pieces of our puzzle include the main-sequence stars, the dwarfs, the giants and supergiants, the pulsating variables, the novae, and so on—but which came first? Let us approach this problem from a theoretical point of view. We assume that a star is a ball of gas and that, under prescribed conditions, it should behave as a gas. What does this mean?

Gas Laws

Consider a certain number of gas particles (atoms or molecules) in a container with a plunger-type top. Such particles are always in motion. In fact, the temperature of a gas is a measure of the average velocity of these particles. The small arrows in Figure 12.1 indicate their direction of motion, and the length of the arrows indicates their speed. If the plunger

Figure 12.1 *Operation of the gas laws: (a) large volume and low temperature produces low pressure; (b) reduced volume and same temperature yields higher pressure; (c) increased temperature and same volume as in (a) yields higher pressure; (d) high temperature and expanding volume results in reduced pressure.*

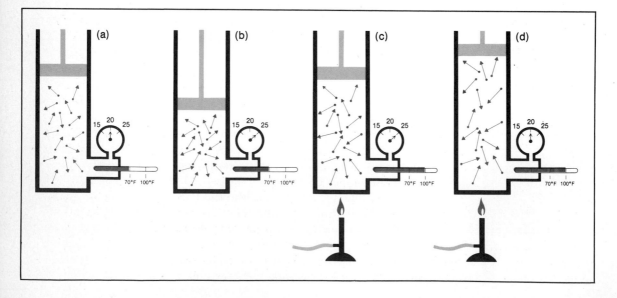

is lowered to reduce the volume, but the temperature is held constant [Figure 12.1(b)], then the pressure increases. You can visualize this increase in pressure by noticing that the particles would be closer together and hence collisions would be more frequent.

Next, let us compare Figure 12.1(a) with Figure 12.1(c), in which the gas has been heated. As the temperature rises, the particles move faster on the average (as indicated by longer arrows), and this increases the tendency for particles to collide, producing an increase in pressure. Alternately, if the plunger were lifted, as in Figure 12.1(d), the increase in temperature could produce an increase in volume, leaving pressure constant, When a star is heated, it will tend to expand as long as it behaves like a gas. But then, an expanding gas will typically experience a cooling effect. These reactions provide a safety valve in relation to the energy production in a star. As a star produces energy at an ever-increasing rate, it heats up, which causes it to expand, but expansion produces cooling, which tends to slow the rate of energy production and keep it in check.

Initial Energy Source

The very fact that the particles of a gas cloud are spread out over a given volume and that they attract one another gravitationally represents a form of energy called *potential energy*. If that cloud contracts, it is capable of converting some of this potential energy into *kinetic energy* (energy associated with motion) by speeding up the gas particles—heating the gas. Thus, as a gas cloud condenses, it not only gets more dense but also increases in temperature, owing to energy conversion. Then, as in the sun, if temperature and pressure are sufficiently high, other sources of energy (thermonuclear) will also become activated.

Model Building

The astrophysicist does not experiment with star formation and evolution in the same way as a chemist or physicist experiments in the laboratory, for he cannot get his hands on a star. Rather, the astrophysicist builds models based on his knowledge of the fundamental laws of nature: the law of gravity (see page 107), the laws of motion (page 39), and the gas laws (described above). The model of a star also includes a statement of factors such as mass, density, and temperature at various radii. To see how a star might evolve under a given set of circumstances requires mathematical computations entailing billions of separate steps, and this requirement stymied such an approach until the advent of the electronic computer. The modern computer can perform millions of arithmetic steps per second. When programmed to follow the basic laws of nature and instructed as to the initial conditions of mass, density, and

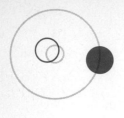

temperature, it can trace the evolution of a star and predict its state at any specific time. The theoretical model which evolves from the computer shows a star contracting under the force of gravity, heating, initiating a thermonuclear reaction like the one we saw in the sun, stablizing, then later growing into a huge but cooler star, and finally taking on one of several death forms. The question which naturally follows this simulation is, "Can we find evidence among the stars that these changes really take place?"

Formation of a Star

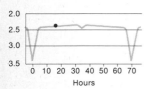

Where shall we look to find a star in the making? We know that gas and dust, the building blocks for stars, are scattered throughout interstellar space; however, the density of that material is only 0.1 to 10 particles/cm³ and its average temperature is 100°K ($-173°C$). Under these conditions the atoms would be moving so rapidly (in a random fashion) that there is little likelihood that gravity could arrest their motion and hold them together in a knot. A more likely situation would involve a denser concentration of matter and a cooler temperature. As you can see in Figures 12.2 and 12.3, small dark regions often appear silhouetted against bright nebulae. These *globules* are dark primarily because they contain not only a higher concentration of gas (10^4 to 10^6 atoms/cm³) but also dust particles, which effectively screen out the ultraviolet light of nearby stars. Temperatures in these dark globules average only 10°K (10° above absolute zero). These regions are also typified by the presence of molecules, verified by their radio emissions. We have already noted in Chapter 11 that dust particles are very logical locations for molecules to form.

The slow random motion which is naturally associated with low temperatures makes the formation of a star far more likely; however, to

Figure 12.2 *The dark globules in this nebula (IC 2944), in the constellation of Centaurus, suggest regions of cool gas and dust, perhaps regions of star formation. (Cerro Tololo Inter-American Observatory)*

Figure 12.3 *Dark globules in the Lagoon Nebula (M8). (Lick Observatory)*

say that a cloud of gas and dust will turn itself into a star by its own self-gravity may be a bit too simplistic. When small regions of such a nebula become unstable and start to form a knot of material, that knot experiences pressure forces from within that tend to destroy it. On the other hand, certain types of external forces may act on the globule to trigger star formation. For instance, the shock wave produced when certain stars explode as supernovae tends to sweep up material, creating pockets of higher density (see Figures 11.10 and 11.11, page 371). If enough mass of gas is compressed, then its own gravity may be sufficient to overcome internal pressures and continue to accrete (pull in) material from the surrounding nebula. Figure 12.4 shows the site of a supernova; you can see the shell of material thrown off by this stellar explosion. The expanding shell generates a shock wave which produces higher concentrations of material along its border. A new star is detected as an infrared source.

For years astronomers have recognized young stars that have begun to shine by their own internal energy, but of late they have succeeded in detecting stars earlier in their formation, perhaps 100,000 to 1 million years before they will be clearly visible in ordinary light. Assuming that the observers have chosen a dark globule, with molecular radio emissions, their first real clue to a star in the making comes via infrared wavelengths. Infrared radiation penetrates the dust which surrounds a forming star core. It is in fact this dust, heated as a result of the energy given up in the gravitational contraction of the core, that radiates the infrared energy.

As the core continues to heat, say to 500°K, some of the surrounding gases will be ionized, producing an H II atmospheric region. Such a region may be detected by a broad range of radio emissions. Bart Bok,

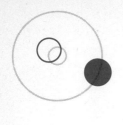

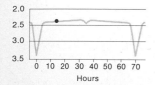

who has been identifying far less obvious dark regions (Figure 12.5) for years and associating them with possible star formation, is delighted with the confirmation that is now in progress. With specially sensitized photographic plates, he is beginning to record faint images penetrating the dust. This particular type of dark globule has been termed the Bok globule in his honor. The latest approach to the detection of dark globules involves a very sensitive detector called c.c.d. (charge coupled detector). This is another example of how electronic devices are extending the level of sensitivity beyond that which has been achieved photographically.

As the condensation knot grows and a temperature of about 1500°K is reached, the protostar becomes clearly visible on red-sensitive film. The photo record of condensations in the Orion Nebula actually shows growth over a period of only a few years (Figure 12.6). This may represent

Figure 12.4   *The Omega Nebula is the result of a supernova explosion. Astronomers suspect that the shock wave produced by such an event may have triggered star formation in the location indicated by an infrared object—a star in the making. (Lick Observatory)*

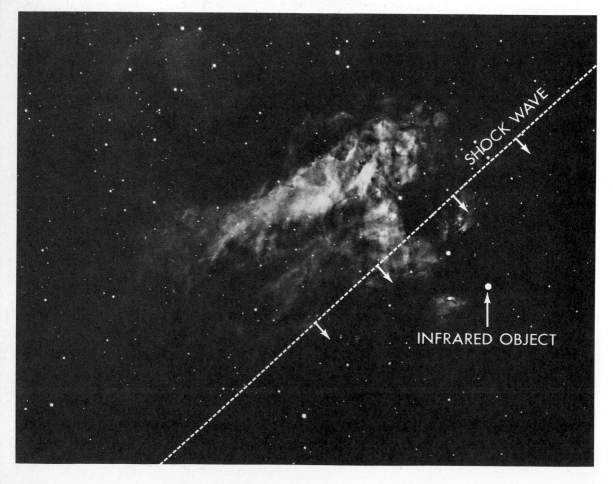

Figure 12.5 *An enlarged section of the nebula in Monoceros, showing dark regions. See Figure 12.7 for a view of the entire nebula. (Hale Observatories)*

Figure 12.6 *Herbig-Haro objects in Orion, taken (a) in 1947, (b) in 1954, and (c) in 1959, showing growing concentrations. (Lick Observatory)*

(a)   (b)   (c)

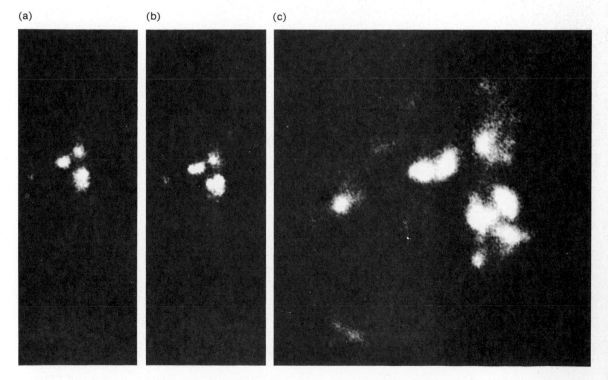

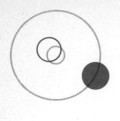

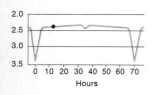

2.0
2.5
3.0
3.5

0 10 20 30 40 50 60 70
Hours

one of those rare examples of a crucial stage in stellar evolution occurring on a time scale which humans can witness. Of course, we are seeing only a very brief segment of a star pregnancy in these photos, for the average time needed for a star (such as the sun) to reach this state is of the order of a million years.

To this point the only force acting has been that of gravity, shrinking the cloud to smaller and smaller diameters. Where will this condensation be halted and by what force? The force which tends to counter that of gravity is pressure, and already we have noted that pressure is gradually increasing with the decrease in volume. Furthermore, the protostar is about to reach that significant point at which is central temperature has built up to 10 million degrees Kelvin and its central pressure has reached a billion earth atmospheres. The proton-proton cycle begins (see Chapter 7, page 269) with the release of an enormous amount of energy, and pressure is significantly increased.

A Star Is Born

Although a star begins to shine before the initiation of the proton-proton cycle, with such ignition we may truly say that a star has been born. However, it will continue to contract until the outward force of pressure exactly balances the inward force of gravity at each point within the star, a condition called *hydrostatic equilibrium*. This stabilized state probably does not occur as a simple halting of contraction but rather may be accompanied by a period in which the star pulsates in size and appears to vary in its light output. When astronomers observe certain regions of gas and dust, as in the Monoceros cluster (Figure 12.7), they see a type of variable star which we have already studied: the T Tauri type. These stars show rapid, irregular variations and may be recognized by their typical emission spectra, which are thought to originate in the envelope of gas and dust which still surrounds these young stars. By contrast most stars have absorption spectra. A plot of the T Tauri stars on the H-R diagram seems to suggest that they may represent an evolutionary stage just preceding the stable "main-sequence" stage, for most of them appear just to the right and above that stable group. In Figure 12.8 the T Tauri stars in the Monoceros cluster (NGC 2264) are shown by the "plus" symbol. As further evidence of the youthfulness of the T Tauri stars, they are always found in the denser regions of gas and dust.

As a means of visualizing the changes taking place as stars go through their life cycle, astronomers draw their evolutionary tracks on an H-R diagram. The tracks presented in Figure 12.9 are due to Margherita Hack. Tracks are shown for stars of differing masses. We have already demonstrated (see Chapter 9, page 336) that more massive stars appear higher on the main sequence. The numbers shown along the main-sequence line indicate the number of solar masses in each star.

As you view these evolutionary tracks in Figure 12.9, and others later in this chapter, keep in mind what the changes on the H-R diagram

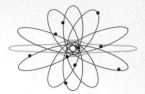

Figure 12.7 *Nebulosity seen in Monoceros. (Hale Observatories)*

mean. A vertical change upward means a brightening and downward means a dimming. A horizontal change to the left means an increase in surface temperature and to the right means a decrease. Thus an evolutionary track which slopes upward to the left indicates an increase in luminosity (a brightening) accompanied by an increase in surface temperature. This is what we might expect of a star that is heating up but not changing its size appreciably. What must be happening when a track is horizontal, moving to the left? The surface temperature is increasing, yet the luminosity remains unchanged. We conclude that the surface area must be decreasing at the same time.

Main-Sequence Stage

A protostar becomes a main-sequence star when the energy of hydrogen fusion becomes the main source of energy and when contraction ceases—a state of hydrostatic equilibrium. The star's size, surface

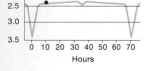

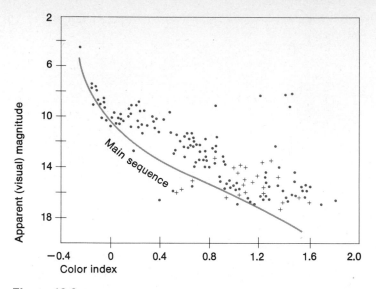

Figure 12.8   *A plot of the stars in NGC 2264.*

temperature, and luminosity now are basically determined for the major portion of its life cycle. Stars appear to spend at least 95 percent of their visible life on the main sequence. The evidence is that on an H-R plot of any large number of stars selected at random we find the majority of stars on the main sequence, and there is a direct correlation between the

Figure 12.9   *The evolutionary track of stars of different masses. (Diagram prepared by Margherita Hack)*

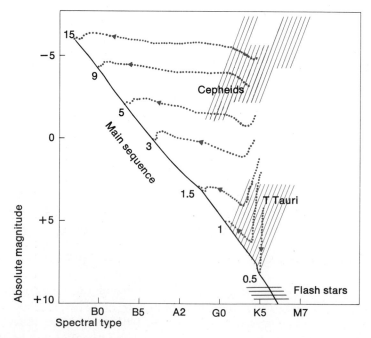

number of stars found in any stage and the length of time stars spend in that stage (Figure 12.10).

To clarify this, let us return to our analogy of the intelligent being from outer space who comes to the earth and tries to decipher the life cycle of humans in one day: it would find less than 1 percent of humankind in cribs, and it would therefore conclude that the crib stage is very short. It would find perhaps 3 percent to be of preschool age—another short period in human life. Then, if youth and adulthood are grouped into one stage, this would account for about 90 percent of the population, hence 90 percent of the human life cycle. The remaining 6 percent would be the elderly and the dying. The combined youth and adulthood stage would correspond to the main-sequence stage in a star.

The sun characterizes the main-sequence star: a stable (nonpulsating) star with a rather constant output of energy derived from the *proton-proton cycle.* Surface activities, such as sunspots and flares, represent only rather superficial variations in output. On the other hand, it is clear that not every star is like the sun even in its stable (main-sequence) state. Stars more massive than the sun become much hotter in

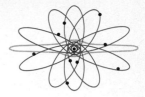

Figure 12.10 *The H-R diagram. The points represent approximately 6700 stars of known absolute magnitude. (Yerkes Observatory)*

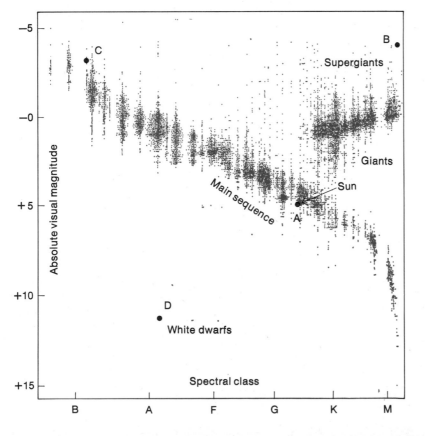

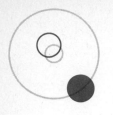

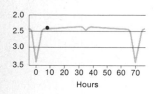

2.0
2.5
3.0
3.5

0 10 20 30 40 50 60 70
Hours

the core, and they experience additional energy-producing cycles; one such is called the *carbon cycle* or *CNO* (carbon, nitrogen, oxygen) *cycle*. In this cycle, carbon serves merely as a catalyst (and is neither created nor destroyed). Note that four hydrogen atoms ($_1^1$H) are used to make one helium atom ($_2^4$He), with the same conversion of mass to energy as in the proton-proton cycle. This cycle depends upon the presence of carbon (C) in a star, and nitrogen (N) and oxygen (O) are intermediates in the process; hence the (CNO) cycle:

$$_6^{12}C + {_1^1}H \longrightarrow {_7^{13}}N + \text{gamma ray}$$

$$_7^{13}N \longrightarrow {_6^{13}}C + \text{positron} + \text{neutrino}$$

$$_6^{13}C + {_1^1}H \longrightarrow {_7^{14}}N + \text{gamma ray}$$

$$_7^{14}N + {_1^1}H \longrightarrow {_8^{15}}O + \text{gamma ray}$$

$$_8^{15}O \longrightarrow {_7^{15}}N + \text{positron} + \text{neutrino}$$

$$_7^{15}N + {_1^1}H \longrightarrow {_6^{12}}C + {_2^4}He$$

Throughout the main-sequence stage the star is building a helium core, and this development finally leads to rather drastic changes. As the supply of hydrogen available in the core is depleted, the rate of energy production decreases. One might think that the star would be ready to die at this point. Rather than fading away, however, the star undergoes changes that give it new life. With diminished hydrogen burning, the source of internal support is drastically weakened, and so the helium core contracts under the weight of the overlying layers. The reduction in gravitational (potential) energy which this contraction represents shows up in the form of heat. The increased core temperature in turn excites the surrounding hydrogen shell to even more vigorous burning. The additional energy is expended in the expansion of the outer layers of the star, resulting in an increase in total luminosity which outranks anything the star has achieved before.

Let us look more closely at this process. First, an expansion of gases is generally followed by a cooling trend—a movement to the right on the H-R diagram. With such a cooling, say, from 6000° to 3000°K, the rate at which energy is radiated for each square meter of surface area drops to one-sixteenth its previous level, for the luminosity of a star is proportional to the fourth power of temperature [in our example, the temperature is decreased by one-half: $(\frac{1}{2})^4 = \frac{1}{16}$]. This drop in luminosity could be compensated for by a fourfold increase in size, since a star's luminosity is directly proportional to the square of any increase in radius. Thus, a star like the sun would need only a fourfold increase in radius in order to balance the loss of surface temperature. Observations of some large nearby stars, such as Betelgeuse, show diameters ranging up to 750 times that of the sun, and so it is not hard to see how the net output of an expanding star could increase despite its loss in surface temperature.

Furthermore, we can see such a star taking up new positions on the H-R diagram, moving to the right as it cools but also moving upward as its total output of energy increases. Because it is cool on the surface (red), and because it is very large, this kind of star has been dubbed a *red giant*.

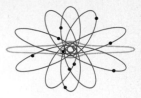

The Red Giant

Margherita Hack has plotted the evolution tracks for the more massive stars (Figure 12.11). Since the massive stars go through each step of their life cycle more rapidly, more observational evidence exists for these stars. This is, not very many stars of one solar mass have had time to evolve to the red-giant stage, but many stars of three or more solar masses have already reached it. Proof for these statements comes when we plot the stars of an old globular cluster onto the H-R diagram. Take, for instance, the globular cluster listed as M3 (NGC 5272). This cluster is composed of stars of varying masses but all of about the same age. We know that the more massive stars plot high on the main sequence, and it is obvious from Figure 12.12 that most of these stars have already evolved away from the main sequence to become red giants. There is a strong possibility that they have followed the tracks outlined in Figure 12.11; in fact, some may have evolved to still later forms, which we have not yet discussed. The stars of one solar mass or less evolve so slowly that even in

Figure 12.11   *Evolutionary track of the more massive stars. (Diagram prepared by Margherita Hack)*

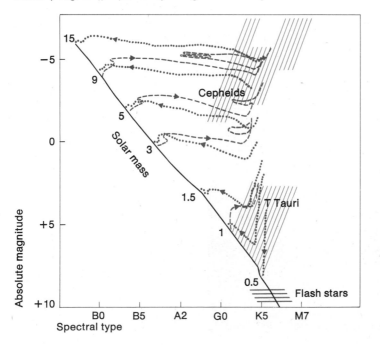

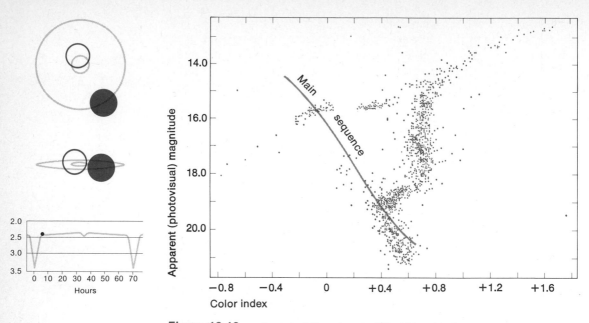

Figure 12.12 *A plot of the stars in M3. (After H. C. Arp, W. A. Baum, and A. R. Sandage)*

an old cluster like M3 (Figure 12.13) they still remain on the main sequence. A plot of younger open clusters in the disk of the galaxy (Figure 12.14) shows the same tendency for the more massive stars to evolve away from the main sequence first. With the exception of M67, these clusters are so young that not even their most massive stars have become red giants.

We must not forget that within the red giant lurks a core that is still increasing in temperature, owing to further contractions. When the core temperature rises to the order of 100 million degrees Kelvin, helium

Figure 12.13 *The globular cluster NGC 5272 (M3). (Lick Observatory)*

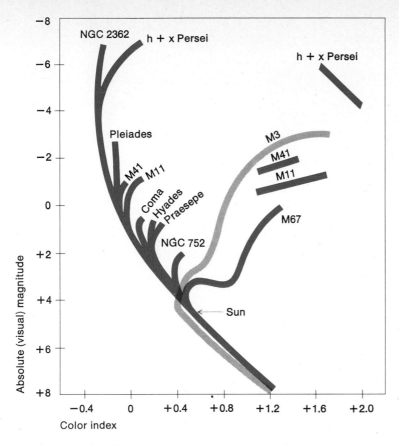

Figure 12.14 *The evolutionary tracks of several familiar galactic clusters superimposed on the track of M3.*

fusion takes place; that is, helium atoms begin to fuse, producing still heavier elements, such as beryllium (Be), carbon (C), oxygen (O):

$$\begin{aligned}
{}^{4}_{2}\text{He} + {}^{4}_{2}\text{He} &\longrightarrow {}^{8}_{4}\text{Be} + \text{gamma ray} \\
{}^{8}_{4}\text{Be} + {}^{4}_{2}\text{He} &\longrightarrow {}^{12}_{6}\text{C} + \text{gamma ray} \\
{}^{12}_{6}\text{C} + {}^{4}_{2}\text{He} &\longrightarrow {}^{16}_{8}\text{O}
\end{aligned}$$

Now the star has two sites (layers) of energy production (see Figure 12.15), a hydrogen-burning shell and a helium-burning core. This process eventually yields a carbon and oxygen core (see Figure 12.16).

At this point in our story, the star's mass appears to determine the next steps. First let us consider a star of approximately one solar mass. At some point in its contraction and heating, the gaseous core degenerates to a densely packed mass of protons and electrons, which take on the rigidity of a steel ball. If that core were like a gas, it would expand with the increased temperature and start to cool down, acting as a safety valve,

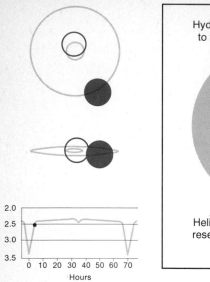

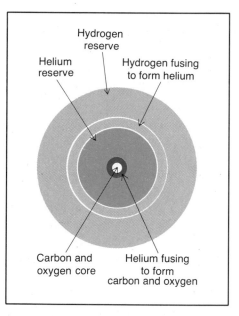

Figure 12.15   Sites of fusion within the core of an evolving star.

Figure 12.16   Sites of fusion within the core of an evolving star in a more advanced stage.

controlling the rate of energy production. Because of its dense nature, however, the core no longer responds like a gas, and without a safety valve the core continues to heat up until the thermonuclear process literally runs away with itself, producing a very explosive reaction called the *helium flash* (see Figure 12.17, point A). We don't literally see such a flash, for most of the energy is absorbed in the expansion of the core to restore its gaseous nature. Hydrogen burning in the shell which surrounds the core appears to be halted by the helium flash, and in fact there

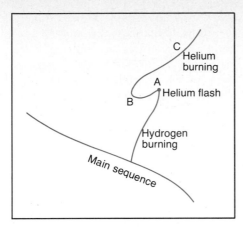

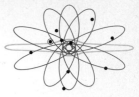

Figure 12.17   The evolutionary track of a star as it passes
from the hydrogen "burning"
stage to the helium "burning"
stage.

is a temporary decline in energy output (see Figure 12.17, point B). However, the core will once again contract under the influence of gravity, heat up, and then helium burning will become the primary source of stellar energy. In the more massive stars the degenerate core probably does not develop at this time, and so a helium flash is not typical. Helium burning is typical of such stars when a core temperature of 100 million degrees Kelvin is reached. Helium burning produces a carbon core, and with increasing temperatures carbon burning may occur, producing still heavier elements.

In the most massive stars this process is thought to continue until an iron core is produced, meanwhile at each fusion step converting mass to energy. But then the fusion process stops, for iron represents the most stable form in which protons and neutrons can exist. Iron will not fuse to form a heavier element without some external source of energy, because the nucleons of heavier elements are themselves more massive than those of iron and cannot yield energy as they are fused. At some point in these later developments, when cores heavier than helium are being formed, one of the most violent and spectacular reactions imaginable may take place. A series of events somewhat like those which produced a helium flash in less massive stars will occur in these more massive stars. By continued contraction, an incompressible (degenerate) core will form which will no longer respond as a gas and therefore does not possess the normal safety-valve characteristic. Uncontrolled heating results in a core temperature of the order of one billion degrees Kelvin, which, when combined with uncontrolled core collapse, produces a gigantic explosion called a supernova.

The Supernova

We have said that energy must be supplied in order to fuse elements heavier than iron. The supernova explosion supplies such energy, and

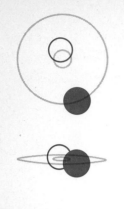

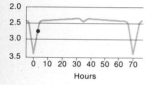

during the explosion elements heavier than iron are thought to be created by the rapid capture of nucleons (protons, alpha particles, and/or neutrons). For instance, if an iron atom (Fe) captured an alpha particle (helium-4), an atom of nickel (Ni) would be formed. Many observers believe this to be the principal way in which elements up through uranium are formed in the universe. At the same time the explosion of a supernova is the means whereby these elements may be dispersed to mix with the interstellar gases, enriching them for future generations of star formation. Our sun, its system of planets, and our own bodies are a product of star formation from such an enriched cloud of gas and dust. Had our sun been formed during the very early stages of the universe, when only hydrogen and helium were present, life as we know it would not have been possible on the earth.

A supernova may increase in brightness by a factor of one billion in a very short time (a few days); if it appears in one of the less luminous galaxies, it may temporarily outshine that entire galaxy. The remains of a supernova recorded in A.D. 1054 by the Chinese is still visible as the Crab Nebula, which continues to expand at the rate of more than 1300 km/sec. From our knowledge of this velocity, together with its apparent increase in size over the years, we can determine its distance at more than 4000 light-years (Figure 12.18). We have already noted that a mass several times that of the sun is required to create the conditions which produce supernovae, so it follows that not all stars will experience this event during their lifetime.

Figure 12.18 *The Crab Nebula, the remains of a supernova first seen in* A.D. *1054. (Hale Observatories)*

Dying Stars

We see that with the development of first a helium core and then a carbon core, higher and higher temperatures are required to ignite nuclear fusion. Eventually the star will simply not have sufficient gravitational force to ignite any further thermonuclear processes, and so this nuclear source of energy is lost.

The star still has a few sources of energy left: what remains of its gravitational force will produce limited heating through contraction, and its rotational energy may still contribute to production of radiation.

Its mass appears to be a very significant factor in determining the way a star dies. Stars of approximately one solar mass are thought to have a rather nonviolent demise. Their cores continue to be heated by the conversion of gravitational energy to kinetic energy as they continue to shrink in size. These cores become very dense. In fact, they become as dense as their atomic nature will allow without destroying the separate existence of protons and electrons. The modern physicist suggests that a star of less than 1.2 solar masses does not have sufficient mass to push its condensation beyond this state. Thus a star of one solar mass would shrink to the size of the earth but would have a density of 2×10^6 g/cm³ (about 34 tons/in.³).

In a star of this nature, heat from the core is conducted very readily to the surface, where the temperature ranges from 50,000° to 100,000°K.

The White Dwarf

The star is now hot (white) and relatively small, hence a *white dwarf*. This represents a drastic change in its nature, for its tenuous atmosphere (outer layers) has been shed—but by what mechanism?

Many white dwarfs have been observed, and some of these have large, slowly expanding shells of gas surrounding the hot core. These objects are called planetary nebulae, because when they were first discovered in a small telescope they resembled the disklike image of a planet. Modern views of two such "planetaries" are shown in Figure 12.19. Measurements reveal that a single planetary nebula would dwarf our entire solar system, being 20,000 to 200,000 A.U. in diameter. Often such a nebula appears somewhat like a ring, because our line of sight near the center passes through less of the shell's thickness, but in reality the shell is spherical. More than 1000 planetary nebulae have been catalogued, and their distribution in the Milky Way galaxy resembles a flattened sphere surrounding the nucleus, somewhat similar in distribution to the RR Lyrae stars. This suggests that planetary nebulae and the stars with which they seem to be associated are older objects, well along in their evolutionary development.

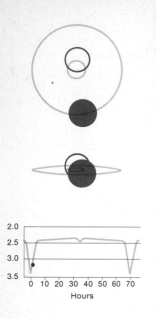

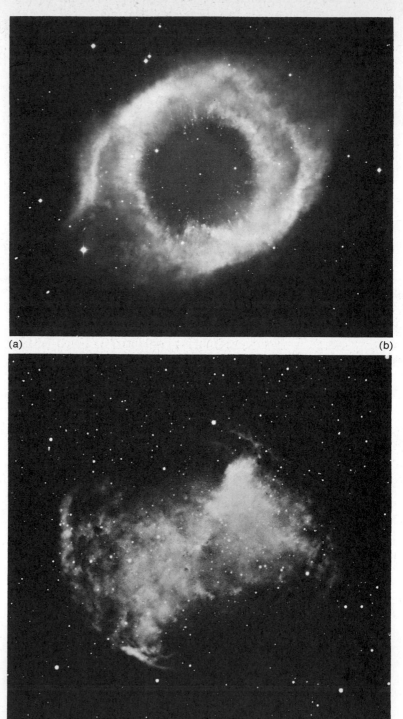

Figure 12.19 (a) The planetary nebula in Aquarius, NGC 7293. (b) The Dumbbell Nebula, a planetary nebula in Vulpecula. (a, Hale Observatories; b, Lick Observatory)

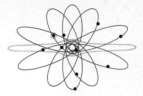

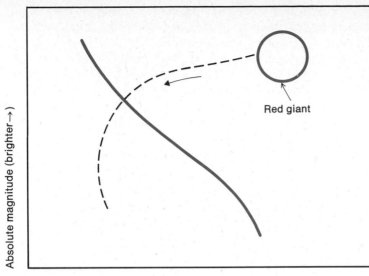

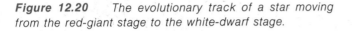

Figure 12.20 *The evolutionary track of a star moving from the red-giant stage to the white-dwarf stage.*

We suspect that at least some stars on their way to becoming white dwarfs have shed their outer layers this way, thus explaining how a red giant could turn into a white dwarf. The mechanism may be as follows. As the outer layers of the red giant star continue to expand and cool, protons (ions) recombine with free electrons to form neutral atoms. This involves the process of downward electron transitions (see page 82), producing photons—electromagnetic radiation. This radiant energy, in turn, is largely absorbed by the star's outer layers. The radiant energy is translated into kinetic energy, which can escape the gravitational attraction of the core only from these outer layers.

The evolutionary track of a star moving from the red-giant stage to the white dwarf is shown in Figure 12.20. The star is plotted further to the left on the H-R diagram as the hotter surface temperature of the core is revealed. But then, as the star shrinks in size, its luminosity decreases and it gradually cools, sending it downward to the right on the H-R diagram, into what will eventually be a yellow-, red-, and black-dwarf death. This final cooling process may require up to a trillion years.

The Neutron Star

Let us suppose that a star originally contained four solar masses and that half its mass was expelled in a supernova-type explosion, leaving a remnant core of two solar masses. If such a core were to contract to only 32 km in diameter, what would be its density? The answer is 200,000,000,000,000 g/cm³ (200 trillion g/cm³)! To conceive of the

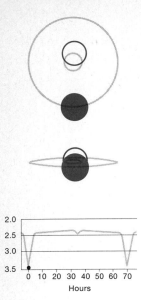

2.0
2.5
3.0
3.5
0 10 20 30 40 50 60 70
Hours

meaning of such a density, let us first visualize a greatly enlarged model of the ordinary hydrogen atom. Imagine a single golf ball placed at the center of a football field. This represents the single proton of the hydrogen atom. Now imagine a gnat flying around the outside of the huge stadium which surrounds the field. This represents the one electron. This model illustrates the fact that an ordinary atom is mostly empty space, and its mass is concentrated in a very small volume at the center. Furthermore, in our analogy, two atoms in an interstellar gas cloud would be separated like two stadiums in adjoining towns. But in the neutron star virtually all empty space has been taken up, and the gnats (electrons) have been forced inside the golf balls (protons), producing neutral golf balls (neutrons), which are packed so closely together as to touch. Should stars such as this exist, how would they be likely to manifest themselves? Perhaps as pulsars.

Pulsars

While investigating certain sources of radio interference, a small group of British astronomers at the University of Cambridge discovered a weak radio signal that appeared as pulses at very regular intervals. Speculation immediately suggested that intelligent beings might be trying to contact the earth, and the sources were soon dubbed LGMs (Little Green Men). But this possibility was soon eliminated, for many similar sources were found. Their extraterrestrial origin, proved by their moving with the stars, also eliminated the possibility that these signals originated in mechanical devices on earth, a constant source of interference for the radio astronomer. Because of their pulsing output they are called *pulsars*, the first being called CP 1919, meaning "Cambridge Pulsar located at 19 hr 19 min right ascension." The first nine pulsars to be discovered are listed in Table 12.1. By 1979 over 150 pulsars had been catalogued.

Table 12.1 *Pulsars[a]*

PULSAR	RIGHT ASCENSION	DECLINATION	PERIOD (SEC)	DISTANCE (PARSECS)
CP 1919	19^h 19^m 37^s	$+21°$ $47'$	1.337301	126
CP 0950	09^h 50^m 29^s	$+08°$ $11'$	0.253065	30
CP 1133	11^h 33^m 36^s	$+16°$ $08'$	1.187911	50
CP 0834	08^h 34^m 22^s	$+06°$ $07'$	1.273764	128
HP 1506	15^h 06^m 50^s	$+55°$ $41'$	0.739678	200
CP 0328	03^h 28^m 52^s	$+54°$ $23'$	0.714463	270
CP 0808	08^h 08^m 50^s	$+74°$ $42'$	1.292231	60
PSR 1749	17^h 49^m 49^s	$-28°$ $06'$	0.562645	500
PSR 2045	20^h 45^m 48^s	$-16°$ $28'$	1.961663	115

[a]CP indicates Cambridge Pulsar; HP, Harvard Pulsar; PSR, Pulsar; h is hours; m, minutes; s, seconds.

A search was made for a visible counterpart to these pulsars, but none was immediately found, and early speculation concerning the signals' origin took many forms. Most theories suggested a pulsating star, a rotating star, or a revolving system. Theoretical models were devised to account for the very rapid recurrence of the signal; most of these models were of very small, dense stars or systems.

Then, in 1969, a team of astronomers at Steward Observatory, University of Arizona, realized that a star clearly visible in the Crab Nebula was actually a pulsar. This was Pulsar NP 0532, with a period of about 0.0333 sec—one of the shortest known—hence the flashing of its optical counterpart was too rapid to be detected by the eye. A special rotating aperture was perfected that allowed astronomers at Lick Observatory to photograph this pulsar, showing that it is "on" part of the time and "off" part of the time (Figure 12.21). Pulsar NP 0532 is an 18-mag object, and its flashes are equivalent to 15 mag. Its light curve is indicated in Figure 12.22.

Observers now feel more confident that pulsars are high-density neutron stars, more massive than the sun and yet only 16 km to 32 km in diameter, their pulses being emitted in step with their rotation. This model of the pulsar is born out by the very nature of the radiation it emits—radiation that is strongly polarized and continuous in its spectral character. We are no longer seeing the kind of radiation produced by the thermonuclear processes of a normal star but rather a process character-

Figure 12.21   *Pulsar NP 0532 in the Crab Nebula, photographed using a special rotating disk, showing the pulsar "off" in the left-hand view and "on" in the right-hand view. This object blinks on and off approximately 30 times per second. (Lick Observatory)*

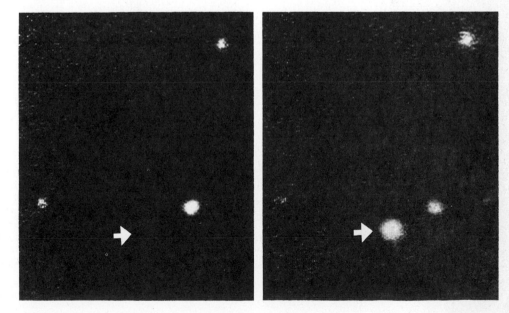

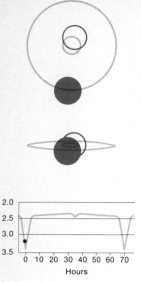

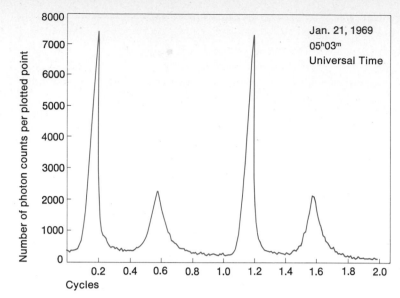

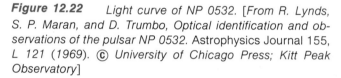

Figure 12.22 *Light curve of NP 0532. [From R. Lynds, S. P. Maran, and D. Trumbo, Optical identification and observations of the pulsar NP 0532.* Astrophysics Journal 155, *L 121 (1969).* © *University of Chicago Press; Kitt Peak Observatory]*

istic of a cold star, whose remaining energy lies in its rotation and its possession of a strong magnetic field. When electrons are accelerated in a magnetic field, they tend to spiral around the magnetic lines, and in so doing they emit the very kind of radiation seen in the pulsars. This form of radiation, called *synchrotron radiation,* is believed responsible for both the radio and light pulses we receive from the Crab and other pulsars. Further, the magnetic field tends to concentrate this radiation into two beams or cones of emission. If this model of a pulsar is correct, the presence of such a star will be sensed only if the observer's line of sight falls within one of these cones (Figure 12.23). Perhaps only 10 percent of all neutron stars that fit this model are aligned in such a way as to be sensed from the earth's position in space.

The periods of the pulsars seem to be increasing, which indicates that these objects are rotating more slowly as time passes, giving up some of their rotational energy to the acceleration of electrons; this in turn produces the radiation. Based on these observations, the duration of the stage of evolution represented by the pulsars is seen to be relatively short, in some cases as little as 1 million years. Since the primary source of energy for the pulsar is its rotation, when that rotation stops or the magnetic field collapses, the object will no longer radiate energy, and the last stage in the evolution of such a star has been reached. In several cases,

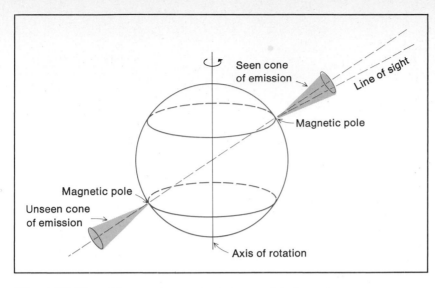
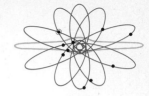

Figure 12.23   *The rotating neutron-star model of a pulsar.*

pulsars have been known to suddenly speed up. This sudden increase in rotational velocity may indicate that the star has contracted to a smaller volume, which would necessitate a rearrangement of the neutrons (somewhat like the collapse of a crystalline structure). This could be termed a "starquake." The star would respond to such a reduction of size by an increase in rotational velocity in order to conserve its angular momentum. A change of only 1 cm in the radius of a neutron star would be sufficient to explain a measurable change in the period of a pulsar.

Although pulsars show neither emission nor absorption lines in their spectra, their pulses do have a form of dispersion when observed in a variety of wavelengths. This phenomenon, called *velocity dispersion*, becomes a valuable tool for the study of the interstellar media that cause it. When a pulse of energy is created at the source, all wavelengths are generated simultaneously; however, in traversing the distance to the observer, the shorter wavelengths travel more rapidly and arrive ahead of the longer wavelengths. Figure 12.24 illustrates the velocity dispersion (due to arrival time delays) for a single pulse observed in five different wavelengths. The amount by which these signals are delayed is a direct measure of the number of free electrons (products of ionization) that lie along the line of sight to the pulsar. Then, to the degree that the density of free electrons can be estimated, the velocity dispersion becomes a distance indicator for the pulsars: the more free electrons, the greater the dispersion and the greater the distance. Furthermore, free electrons, under the influence of even a weak magnetic field, tend to polarize the energy of the pulse. Astronomers are using this fact to study the magnetic field of the Galaxy.

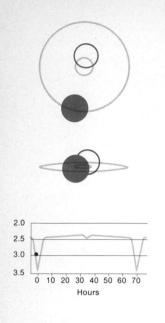

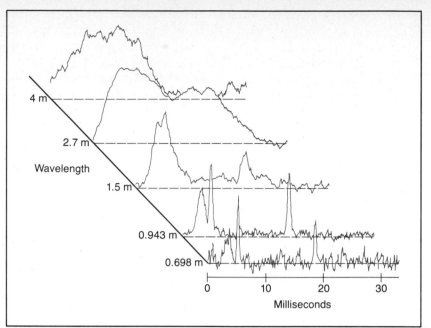

Figure 12.24 *The velocity dispersion for a single pulse, as observed in five different wavelengths. (Cornell University).*

The Black Hole

Some stars are many times as massive as stars we have discussed up to this point. The life cycle of such stars parallels closely those already traced to neutron stars. Because these more massive stars have even higher levels of gravitational force, however, temperatures may be developed in their carbon cores of the order of 1 billion degrees Kelvin, and elements as massive as iron may be produced before the core collapses. Under the extremely large gravitational forces associated with 10 to 20 solar masses, the collapse of such an iron core transcends anything discussed to this point. While a significant portion of the star's mass may be thrown off in the accompanying supernova explosion, the momentum created in the condensing core cannot be stopped by neutron forces—forces that dictate the stable size of a neutron star, and its inevitable fate as a *black hole.*

Albert Einstein predicted the possibility of such a collapse in his general theory of relativity (published in 1915), and Karl Schwarzschild worked out some of the consequences of this theory in 1916. Schwarzschild said that when a star collapsed to a certain size (dependent upon its mass only), the radiation from that star could no longer escape, owing to the enormous gravitational forces that would ensue. For example, if a star of 3 solar masses were to collapse to become a black hole, this

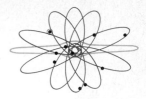

discontinuance of radiation would occur when the radius was reduced to 9 km. This does not mean that the collapse would stop at that point, for the collapse is thought to continue into a condition approaching infinite density and infinite tidal forces. This condition is called a *singularity*.

The Schwarzschild radius defines a sphere within which particles and radiation can only fall toward the singularity. In fact, if a particle were approaching a black hole, its radiation would cease as it passed through this sphere, and the event could no longer be witnessed (Figure 12.25). Before reaching this sphere, a particle would experience rapid acceleration and consequent heating by collision, perhaps to 1 billion degrees Kelvin, so that a portion of its radiation could be expected to fall in the X-ray spectrum.

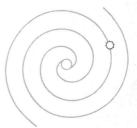

These speculations led astronomers to reexamine known X-ray sources and ponder the most likely circumstances under which material would be available to be sucked into black holes. One very natural suggestion is that of a binary star system in which one member is a black hole and the other member is a red giant. The surrounding envelope of the red giant might well extend close enough to the black hole so that material from that envelope would be drawn off and fall toward the black hole (see Figure 12.26). Instead of flowing directly onto the surface of the black hole, the gases would tend to form an orbiting disk around the ultradense object. Gas particles would orbit at very high velocities and become highly turbulent at certain elevations above the black hole. Frictional heating, which might raise the temperature of the gas to tens of

Figure 12.25 *As a particle passes within the sphere defined by the Schwarzschild radius, on its way toward the singularity, the particle's radiation ceases because it cannot escape the gravitational field of the black hole.*

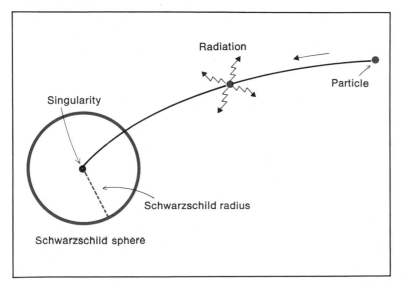

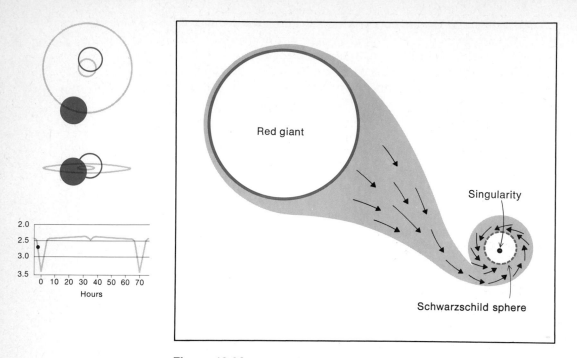

Figure 12.26 *Mass exchange in a pair of binary stars. The very strong gravitational field of the black hole attracts material from the outer portions of the expanded red giant.*

millions of degrees, would result in a condition ripe for X ray emission. Many astronomers believe that this explains what they are observing in Cygnus X-1, a source of X-rays in which the visible star behaves as though it were a member of a binary system, with the unseen member having a mass of more than 3 solar masses.

Rotating Black Holes

From our discussion to this point, it is difficult to imagine a black hole giving up some of its energy to the outside universe. However, a rotating black hole may do so. As a spinning ice skater pulls in her arms, she spins faster and faster, conserving angular momentum. Likewise a rotating star will spin faster as it collapses. As a consequence of this rotation, a deformed region called an *ergosphere* develops outside the black hole (Figure 12.27). A particle that enters the ergosphere might be split into two parts—one part entering the black hole, the other part being ejected with more mass-energy than the original particle had. The additional mass-energy is derived from the rotational energy of the black hole.

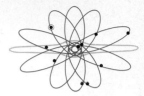

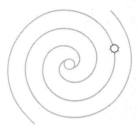

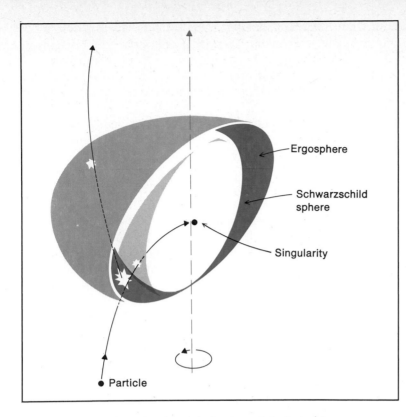

Figure 12.27 *Rotating black hole. A particle that enters the ergosphere (shaded) may split into two, one part passing through into the Schwarzschild sphere and the other being ejected back outside the ergosphere with more mass-energy than the orginal particle had.*

Exploding Black Holes

Almost from the outset, theorists have been willing to accept that nothing can escape the interior of a black hole. After all, if light cannot escape and if nothing can travel faster than light, then nothing can escape. This thought pattern persisted until theorists were attempting to fit together several of the most fundamental theories of all time: Einstein's special theory of relativity (1905), Einstein's general theory of relativity, (1915) and the quantum theory (1926). The general theory of relativity predicts the collapse of dense stars into what are now essentially confirmed as black holes. This same theory describes gravity in a manner vastly different from Newton's classical approach, but no one had been able to unite the concept of gravity with quantum mechanics. Quantum mechanics basically describes the behavior of particles and energies on

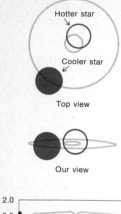

Top view

Our view

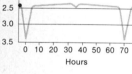

Hours

Light curve

the atomic and subatomic scale—the world of the extremely small. It deals with the uncertainty of our knowledge of location and momentum of subatomic particles, partially because we end up disturbing such particles in our inspection of them. Who would think that a theory concerning the microworld would have any connection with the macroworld of the black hole?

A young British theoretical astrophysicist, Steven W. Hawking, made one such connection. He reasoned that because the universe is lumpy today, it was probably lumpy when it was formed—in the Big Bang event. Hawking proposed that the extreme pressures associated with the birth of the universe could easily have created small black holes, with as little as 0.00001 g of mass and a Schwarzschild radius of 10^{-33} cm. However, a more typical dimension would include a mass of 10^{15} g (a billion tons) and a radius of 10^{-13} cm, the size of a proton. Certainly this represents the size scale to which quantum mechanics is usually applied.

Hawking solved the equations of general relativity for this 10^{15}-g black hole and to his unbelieving amazement his solution predicted the creation and emission of particles and radiation from such an object. In fact, it predicted the same spectrum of particles and radiation as would be seen emanating from a hot object at 120 billion degrees Kelvin. This seemed to violate two long-held principles concerning black holes: nothing left their core, and they had a temperature of virtually 0°K. The contradictions were only apparent, however, the same equations predicted 0.000,001°K for a 3-solar-mass black hole and the escape of virtually no particles. The breakthrough came with the recognition that small black holes are different.

If small black holes do exist and if they emit particles and radiation as predicted, then they will tend to waste away (lose mass). But this will only accelerate the process, for temperature will tend to rise, and eventually this trend will result in a gigantic explosion—the final reaction being equivalent to 10 million, million hydrogen bombs exploding simultaneously. Certainly such an event would seem easily observable; however, the distribution of these objects may make their detection more difficult. Their thermal character predicts emissions in the range of hard gamma rays (which have very short wavelengths), and at present large orbiting detectors are not available. The Space Shuttle program may remedy this in the near future. Hard gamma rays are also known to produce a reaction in the earth's atmosphere, which may be detected.

White Holes

As one might guess, white holes are the opposite of black holes, in that they emit energy and eject particles. We have just seen, however, that small black holes may fit this very description. Certainly there are many sites within the universe where unexplainable large amounts of energy are outpoured, such as quasars and the nuclei of galaxies. As we study these objects in the next chapter, let us remember the processes we have just discussed.

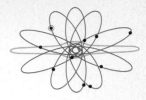

In this chapter we have seen that stars form and evolve in response to the fundamental force of gravity, together with the reactions which it engenders: condensation, high temperature, high pressure, production of energy by thermonuclear fusion, and the continual dissipation of that energy (cooling). Figure 12.28 summarizes this evolutionary cycle on the H-R diagram.

Questions

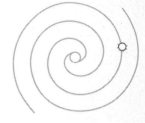

1. Describe a likely birthplace of stars and the process that is thought to bring them into being.

2. What evidence suggests that certain stars are relatively young?

3. Not all stars are thought to evolve at the same rate. Which kind appear to evolve most rapidly?

4. Of the following, which stage in evolution appears to be the longest and which the shortest: variable, T Tauri, white-dwarf, red-giant, main-sequence, or neutron star? State evidence for your answer.

Figure 12.28   *Proposed evolutionary life cycle of a star.*

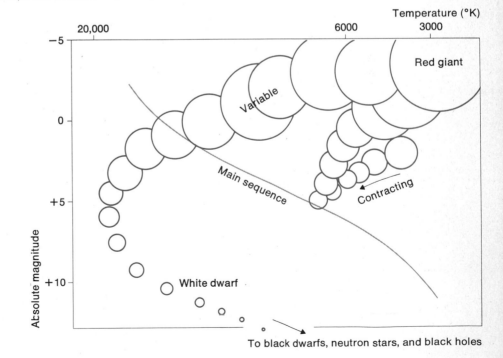

5. Describe a neutron star.

6. How does the astronomer sense that he is receiving energy from a pulsar, as compared to other types of stars?

7. What changes have already been observed in the pulsars?

8. When the evolutionary life cycle of a star is shown on an H-R diagram, a change in location on the diagram does not indicate a change in location in the sky. What does such a change on the diagram indicate?

9. If a star increases in temperature, what change must take place if its internal pressure remains substantially the same?

10. In a gas cloud, gas molecules and dust particles have a certain form of energy which may be converted to heat (kinetic energy) upon collapse of the cloud. Name this form of energy.

11. What is the role of dust in creating a condition in which a star is more likely to form?

12. Name at least two ways in which dust is revealed visually.

13. What role does carbon play in the carbon cycle in stars?

14. Interpret a horizontal movement to the left on an H-R diagram.

15. What evidence do astronomers have that stars spend a large percentage of their time in the main-sequence stage?

16. What is thought to trigger the evolution of a star from the main-sequence to the red-giant stage?

17. What evidence do we have that the T Tauri variables are associated with very young stars and not with old stars?

18. How is it possible for a star that is getting cooler also to get brighter?

19. When a star runs out of available hydrogen in the core, it should cool down, but instead it heats up significantly. Why does this occur?

20. How are the elements heavier than iron manufactured in stars?

21. Compare the densities of a white dwarf, a neutron star, and a black hole.

22. Explain what is meant by the terms *singularity* and *Schwarzschild radius* in relation to a black hole.

23. Why is it *not* thought possible for the sun ever to become a black hole?

24. Why is it *not* thought possible for the sun to fuse elements as heavy as iron?

25. How may a black hole be recognized?

Suggested Readings

ASIMOV, ISAAC, *The collapsing universe: The story of black holes.* New York: Walker and Co., 1977.

BOK, BART J., The birth of stars. *Scientific American* 227 (2), 48–61 (1972).

CHIU, H. Y., and MURIEL, A. (eds.), *Stellar evolution.* Cambridge, Mass.: M.I.T. Press, 1972.

DICKMAN, ROBERT L., Bok globules. *Scientific American* 236 (6), (1977).

GORENSTEIN, PAUL, and TUCKER, WALLACE, Supernova remnants. *Scientific American* 225 (1), 74–85 (1971).

HARRISON, B. K., THORNE, K. S., WAKANO, M., and WHEELER, J. A., *Gravitation theory and gravitational collapse.* Chicago: University of Chicago Press, 1965.

JASTRO, ROBERT, *Red giants and white dwarfs.* New York: Signet (paper), 1969.

MARAN, STEPHEN P., The gum nebula. *Scientific American* 225 (6), 20–29 (1971).

METZ, WILLIAM D., Discovery and the ubiquity of black holes. *Science* 195 (4275), 276–277 (1977).

MISNER, C. W., THORNE, K. S., and WHEELER, J. A., *Gravitation.* San Francisco: W. H. Freeman & Company Publishers, 1973.

PENROSE, ROGER, Black holes. *Scientific American* 226 (5), 38–54 (1972).

RUDERMAN, MALVIN A., Solid stars. *Scientific American* 224 (2), 24–31 (1971).

RUFFINI, REMO, and WHEELER, JOHN A., Introducing the black hole. *Physics Today* 24 (1), 30–41 (1971).

WERNER, M. W., BECKLIN, E. E., and NEUGEBAUER, G., Infrared studies of star formation. *Science* 197 (4305), 723–732 (1977).

ZEILIK, MICHAEL, The birth of massive stars. *Scientific American* 238 (4), 110 (1978).

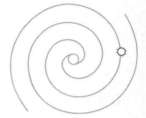

THIRTEEN

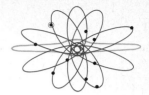

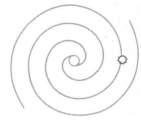

THE MILKY WAY

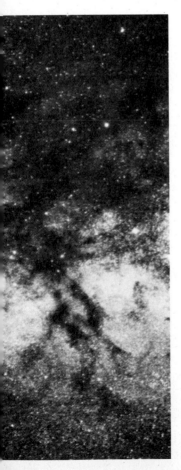

Humans' perception of the universe took a giant leap forward when they realized that the nearby collection of stars, of which the sun is one, is really only an example of a fundamental unit in the universe, the *galaxy*. By the late eighteenth century, Messier had catalogued more than 100 nebular objects without ever suspecting that some of them were collections of stars completely separated from our own collection, the Milky Way galaxy. By the early nineteenth century, more than 5000 nebulae had been identified, largely as a result of the work of Sir William Herschel and his son John. The latter published the *General Catalogue of Nebulae* in 1864. By 1908 almost 15,000 nebulae had been listed in a volume called the *New General Catalogue*, together with its supplement, the *Index Catalogues*. Objects in Messier's list also carry NGC numbers; for example, M31 is represented by NGC 224 in the later work.

The term *nebula* in those days referred to any object which appeared fuzzy in the telescopes of the time. Sir William Herschel did express his belief that certain objects that he found were collections of stars separated from our own; lacking proof, however, he did not pursue the idea. As late as 1912 Henrietta Leavitt, in her work with the Cepheids in the Magellanic Clouds, did not realize that these "clouds" were galaxies outside our own.

Figure 13.1   *The Andromeda galaxy, a spiral galaxy considered to be very much like the Milky Way galaxy. (Hale Observatories)*

With the installation of the 2.5-m telescope on top of Mt. Wilson in 1917, and the use of the photographic methods available by that time, individual stars were being resolved in the nearer galaxies (Figure 13.1), and novae were discovered in them. It was reasoned that if these novae were as bright as some that had been discovered in our own Galaxy, and yet appeared very dim to the observer, they must be at very great distances. However, the matter was not settled until 1924, when Edwin Hubble, working at the Mt. Wilson Observatory in California, discovered Cepheid variables in several nearby galaxies. From the periods of these stars their absolute magnitude may be computed, as explained in Chapter 10. They appear as such faint objects that they are undoubtedly outside our own Milky Way galaxy.

Figure 13.2   *The Milky Way from Sagittarius to Cassiopeia. (Hale Observatories)*

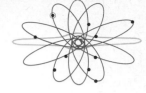

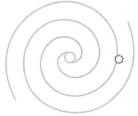

Figure 13.3   *Counting stars.*

By this time astronomers realized that galaxies exist in a wide variety of shapes. Just a glance into the first few pages of Chapter 14 will show spherical, elliptical, and flattened galaxies, none of which have arms, as well as the spiral galaxies, the barred spirals with arms, and the irregulars. It seems reasonable to suspect that our own Galaxy looks something like one of these types, but since we are situated within it, determining its shape is not so simple.

If the Milky Way galaxy were spherical, then we might expect a rather even distribution of stars within it. Let us perform a simple experiment to see if this is the case. Cut out a small cardboard frame with a square opening of 5 cm $\times$ 5 cm. Hold the frame at arm's length and count the stars you see within it. Repeat this for different parts of the sky, recording the location and number of stars counted on each occasion. It will become apparent that the distribution of stars is not uniform and that there are higher concentrations of stars in certain parts of the sky. Try the region near Sagittarius (Figures 13.2 and 13.3) or the one near Orion; in fact, try any area through which runs the fuzzy band of light we refer to as the Milky Way. What are we really looking at when we see this milky band on a clear dark night? A small telescope or even a pair of binoculars will begin to reveal that the band is composed of millions of stars. If we observe the Milky Way over a period of several months, we realize that it forms a complete circle in the sky (Figure 13.4). This suggests that we are part of a flattened system of stars.

Figure 13.4   *The Milky Way as shown by Martin and Tatjana Keskula. Sagittarius is in the center, and the two bright objects in the lower right portion are the Magellanic Clouds. (Lund Observatory, Sweden)*

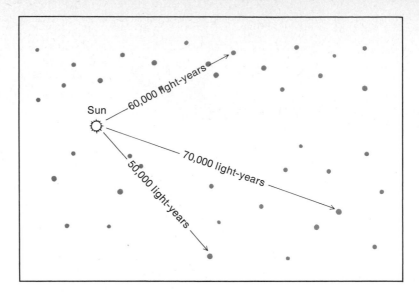

Figure 13.5   *Distribution of the globulars in relation to the sun.*

Where are we in the Milky Way system? The globular clusters will help us answer that question. Recall that globulars contain RR Lyrae-type variables with absolute magnitudes very close to zero. By comparing this absolute magnitude to the apparent magnitude observed for each, astronomers have determined their distances and plotted their position in relation to the sun (Figure 13.5).

Figure 13.6   *If the sun were located at the center of the Milky Way galaxy, the globulars would be thrown off center in relation to the Galaxy.*

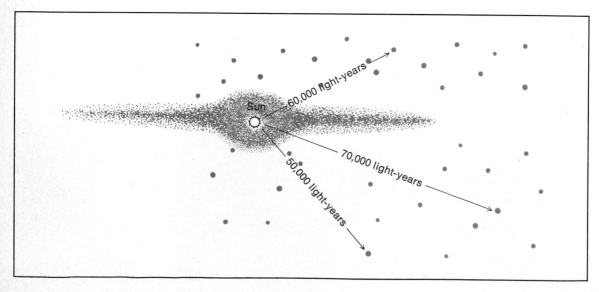

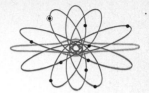

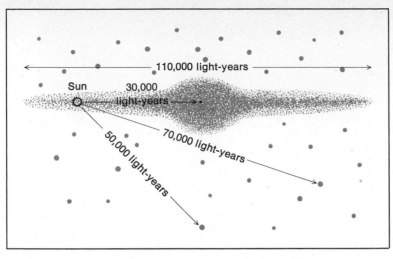

Figure 13.7   *The position of the sun in the Milky Way galaxy.*

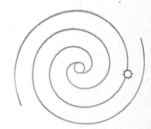

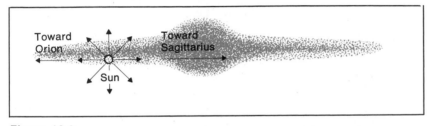

Figure 13.8   *The Milky Way from our own point of view (edge-on).*

Suppose we were to assume that the sun is at the center of the Milky Way galaxy; then the picture would look like Figure 13.6. However, it seems more likely that the globulars are distributed symmetrically with respect to the Galaxy (Figure 13.7). Thus, the sun and its planets are thought to be located in the flattened disklike portion of the Galaxy, approximately 30,000 light-years from its center, and the overall diameter of the Galaxy is placed at 110,000 light-years.

The center (or *nucleus*) of the Galaxy is further identified by the high concentration of stars toward the constellation of Sagittarius. When you view the Milky Way band near Orion, you are looking out through the disk toward the outer edge of the Galaxy (Figure 13.8). The Milky Way band is indicated on the star maps in Appendix 12.

Next we might ask whether the Milky Way galaxy has arms like the many spiral galaxies we can view, or whether its disk is more uniformly populated with stars. By observing other galaxies of the spiral type, astronomers have noticed that very young O- and B-type stars (hot stars) populate the arms and that they form a spiral pattern themselves. These hot stars are easily recognized as bright dots in the photograph of the Whirlpool galaxy (Figure 13.9). Furthermore, we would suspect that gas

Figure 13.9 *The Whirlpool galaxy, showing bright O- and B-type stars outlining its spiral arms. (Lick Observatory)*

and dust would accompany these bright young stars, because gas and dust are the building blocks of stars.

Turning this knowledge toward the Milky Way galaxy, astronomers have determined the distance to numerous O- and B-type stars, and a plot of these stars strongly suggests a spiral nature for our Galaxy (Figure 13.10). This finding was soon confirmed and expanded to a much larger portion of the Galaxy by radio astronomy. The presence of dust in the Milky Way galaxy obscures much of the starlight; however, the Galaxy is much more transparent to radio wavelengths. In fact radio signals characteristic of neutral hydrogen may be received from almost the entire Galaxy.

We have seen how hydrogen, when excited, produces certain characteristic wavelengths in the visible spectrum, such as the Hα (hydrogen-alpha) line produced by a downward transition of the electron. Neutral hydrogen also produces a characteristic radio "line" which has a wavelength of 21.11 cm and which is simply referred to as the 21-cm signal of hydrogen. The process whereby the hydrogen atom emits or absorbs this particular wavelength is associated with the spin of its nucleus and the spin of its electron. The electron may *align* its spin axis with that of the proton, in which case it has more energy than when it *opposes* the spin axis of the proton. Thus, when the electron flips over from an aligned orientation to an opposed orientation, it *emits* energy corresponding to the 21-cm radio signal. When it flips back to an aligned direction, it *absorbs* the same wavelength (Figure 13.11).

When an astronomer tunes his radio receiver to this 21-cm wavelength, much as we would tune in a local station on an ordinary radio, he can detect hydrogen in the Galaxy. The hydrogen clouds, however, are in

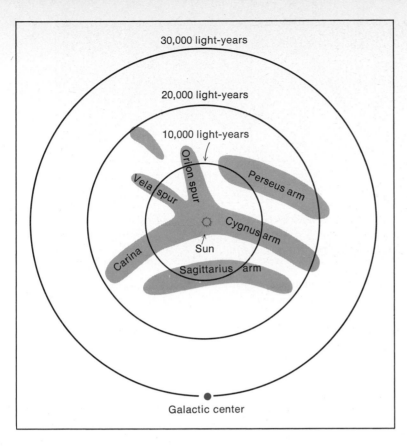

Figure 13.10 *The optical pattern of the Milky Way galaxy based on a plot of O- and B-type stars in the vicinity of the sun.*

Figure 13.11 *Electron spin reversal in the hydrogen atom. When the spin axes of the proton and the electron are aligned (a), the atom contains more energy than when they are opposed (b). When the electron flips from state (a) to state (b), it emits a 21-cm radio signal.*

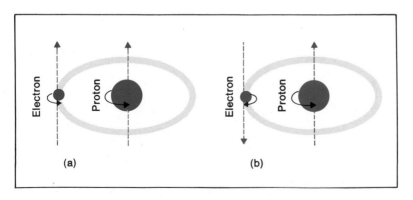

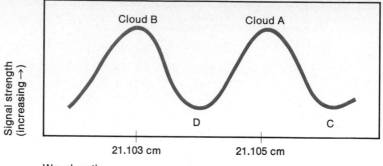

Figure 13.12 *A radio scan of two hydrogen clouds show-ing wavelengths which have been shortened because the clouds are moving relative to the sun.*

motion relative to the sun, as demonstrated by the fact that the typical 21.110-cm signal is not always detected by our receivers at exactly that wavelength but is Doppler-shifted to a shorter wavelength if the cloud is moving toward the sun. Figure 13.12 shows the wavelength of cloud A shortened by 0.005 cm and that of cloud B shortened by 0.007 cm, making it possible for us to compute their relative velocity toward the sun. This is a powerful tool in determining the rotation and structure of the Galaxy. If we assume that, like the planets, the hydrogen clouds which are closer to the center of the Galaxy move faster and those farther away move slower, then the Doppler shifts of these clouds will reveal their position. Based upon this assumption, the astronomer translates the radio scan of Figure 13.12 into the theoretical physical model shown in Figure 13.13, cloud B being closest to the center of the Galaxy because it showed the largest Doppler shift, and cloud A being farther from the center of the Galaxy because it showed a smaller Doppler shift.

Suppose the radio astronomer had scanned along a different line of sight, say, to the other side of galactic center, as in Figure 13.14. He would have found the wavelengths associated with hydrogen clouds in this direction lengthened, indicating that the materials which compose these arms are receding relative to the sun. Plotting their positions (E and F), he not only would find a spiral structure beginning to develop but would confirm that the Galaxy is turning in a clockwise direction (as viewed from the north, rela-

Figure 13.13 *A model of the Milky Way galaxy in which clouds A and B are assumed to be located in the arms, and points C and D represent spaces between the arms.*

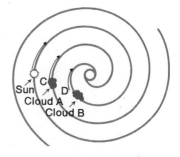

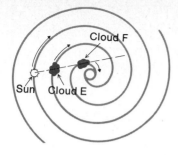

Figure 13.14 *When viewed along this line of sight, hydrogen clouds show a lengthening of their normal wavelength (indicating recession), thus confirming the clockwise rotation of the Milky Way galaxy (see text).*

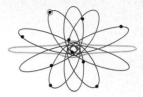

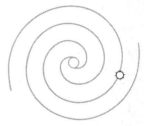

tive to the earth). Using this method, we can map the Galaxy's structure more completely, as shown in Figure 13.15. The darker regions represent higher concentrations of neutral hydrogen, and their pattern strongly suggests a spiral. Certain regions, viewed through the center of the Galaxy, are omitted, because the concentrations of stars and dust along those lines of sight prevent accurate observation.

Let us see now whether we can specify the Galaxy's rate of rotation, using the sun's motion as an indicator.

Velocity of the Sun Within the Galaxy

In Chapter 8 we considered the various reference systems within which we might describe the motions of stars, including space motions relative to the sun and peculiar motions relative to the local standard of rest. To describe the sun's motion within the Galaxy we need to relate that motion to a standard of rest for the Galaxy. The globular clusters do not seem to participate in the rotation of the Galaxy but rather have a motion that crosses the galactic plane at right angles. The in-and-out motion of these objects, as shown in the right-hand flip pages beginning on page 345, helps to establish a zero reference. This reference may also be established by galaxies outside our own. In other words, we may view the sun's motion from a distant point in space—for instance, from a point on the northerly side of the galactic plane. We find that the sun is revolving clockwise in a nearly circular orbit around the center of the Galaxy and has a velocity of approximately 250 km/sec, making one revolution in about 200 million years. These two facts, when considered together, give one an idea of the immense size of the Milky Way galaxy.

Mass of the Galaxy

While the actual determination of the mass of the Milky Way galaxy is beyond our immediate scope, the basis for its calculation is the way in which the mass of our Galaxy affects the motions of the sun. An oversimplification of the problem might assert that the mass of the Galaxy acts as if concentrated in its central nucleus, hence the problem

Figure 13.15 *The structure of the Milky Way galaxy, in the 21-cm radio signal of neutral hydrogen. The darker regions represent higher concentrations of hydrogen. (Leiden Observatory, Netherlands, and Radio Physics Laboratory, Sydney, Australia)*

reduces essentially to a two-body problem like that of the sun and a planet. An approximation may be sought using Kepler's law:

$$m_{\text{gal}} + m_{\text{sun}} = \frac{(r_{\text{sun}})^3}{(p_{\text{sun}})^2}$$

where r is measured in astronomical units and p in years. The distance between the galactic center and the sun is approximately 30,000 light-years, which is equivalent to 2×10^9 A.U., and the period of the sun is 2×10^8 years. Since the mass of the sun is negligible in comparison to that of the Galaxy, the term m_{sun} may be dropped:

$$m_{\text{gal}} = \frac{(2 \times 10^9)^3}{(2 \times 10^8)^2} = \frac{8 \times 10^{27}}{4 \times 10^{16}} = 2 \times 10^{11} \text{ solar masses}$$

Thus, the total mass of the Galaxy would appear to be 200 billion times the mass of the sun. But since the luminosity of the sun is far more than that of the average star, we can conclude that its mass is likewise above average. Thus, we are led to believe that the number of stars in the Milky Way galaxy must exceed 200 billion.

Distribution of Stars
in the Milky Way Galaxy

We have characterized the Milky Way galaxy as being a flattened spiral, and this is its most apparent shape; however, observation of certain types of stars within the Galaxy reveal that it really has three shapes. As we trace these shapes, let us see if they give us a clue to the process whereby our Galaxy became a flattened spiral.

First, consider those star types thought to be oldest—stars typical of the globular clusters. We have already established the advanced age of such clusters upon the evidence that they have used up their gas and dust in star formation and, further, that their more massive stars have evolved off the main sequence. Astronomers refer to stars which inhabit the globulars as Population II objects. These include RR Lyrae and Type-II Cepheids, and long-period variables. The globular clusters form an almost spherical halo around the Galaxy, and they may still preserve the earlier shape of the cloud from which the Milky Way galaxy formed (Figure 13.16; see the halo labeled A).

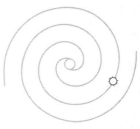

An intermediate system exists between the spherical halo of globulars and the flattened disk where new stars are forming. The intermediate system, as illustrated in Figure 13.16, is the semiflattened halo labeled B. It consists of semilong-period variables, subgiants, white dwarfs, G- to M-type dwarfs, and planetary nebulae. These types of objects are also

Figure 13.16   *The distribution of stars in the Milky Way (see text).*

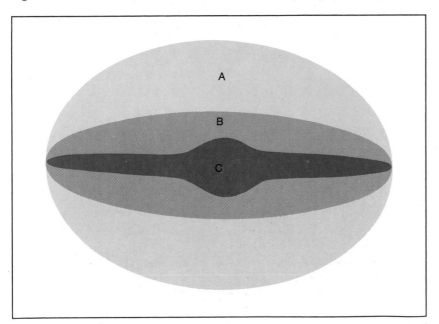

believed to be of moderately old age and may represent by their distribution the shape of the galactic cloud when it was still in the process of flattening due to rotation.

Finally, the flattened disk system (labeled C in Figure 13.16) is characterized by very hot, young (O-, B-, and A-type) stars, Type-I Cepheids, supergiants, open clusters, and interstellar gas and dust. Each of these types represents young stars or the material from which they are formed. Figure 13.17 shows Type I and Type II populations in the Andromeda galaxy.

Does this progression of shapes represent *evolution* of galaxies in the usual sense of the word? Does it mean that elliptical galaxies evolve into spirals, or vice versa, as some would suggest? Perhaps not, but it surely suggests that in its very early stages the Milky Way galaxy was

Figure 13.17   *Population I and II stars. (a) Andromeda Nebula, photographed in blue light, shows giant and supergiant stars of Population I in the spiral arms. The hazy patch at the upper left is composed of unresolved Population II stars. (b) NGC 205, companion of the Andromeda Nebula, photographed in yellow light, shows stars of Population II. The brightest stars are red and 100 times fainter than the blue giants of Population I. The very bright, uniformly distributed stars in both pictures are foreground stars belonging in our own Milky Way system. (Hale Observatories)*

(a)   (b)

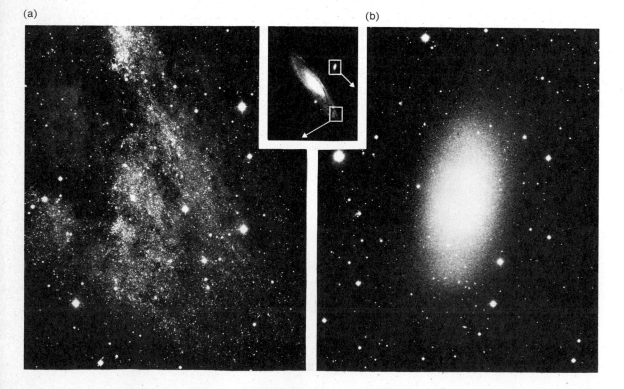

spherical, and that owing to rotation it flattened significantly in the first billion years of its existence as a separate entity in space.

The Galactic Nucleus

A number of phenomena within the central region of the Milky Way galaxy make astronomers suspect a very active nucleus. For instance, there is an armlike system of H I regions that are moving away from the nucleus at speeds of 100 to 200 km/sec. These regions contain gas with a total mass estimated at up to 2 million solar masses. Also, there is a "ring" of molecular clouds at a distance of approximately 200 pc from the center, expanding at the rate of 150 km/sec. There may be the equivalent of 100 million solar masses in these clouds. These are only two pieces of evidence that demand an explanation in terms of an expulsive force in the core of the Galaxy. Astronomers have found a very compact radio source which is no larger than the orbit of Saturn—about 20 A.U. in diameter. Mass estimates for this source run as high as 5 million solar masses. When the density of such a region is considered, it borders on being a massive black hole into which whole stars are falling, releasing tremendous amounts of energy.

As you continue the study of galaxies in the next chapter, you will see that compact and active nuclei may be very common in galaxies. Particularly note the discussion of BL Lacertae objects, which begins on page 451.

Questions

1. The term *nebula* includes what kinds of objects in addition to gas clouds?

2. What evidence does the casual (naked-eye) observer have that the Milky Way galaxy is not simply a sphere (ball) that surrounds us?

3. What role did the globular clusters have in helping us to find our true location in the Galaxy?

4. How much time would be required for light to travel from a point on the outer edge of our Galaxy to the opposite edge?

5. How do we know that our Galaxy is of a spiral nature? Specifically, how do we know it has "arms"?

6. The sun participates in the rotation of our Galaxy. What is its speed because of this rotation?

7. If the Milky Way galaxy contained 100 billion stars, and the average mass of these stars were twice the mass of the sun, find what the total mass of the Galaxy would be.

8. Indicate whether each of the following would be found in the halo (h) of the Galaxy or in its disk (d): (a) open clusters, (b) globular clusters, (c) Type-II Cepheids, (d) Type-I Cepheids, (e) interstellar gas, (f) interstellar dust, (g) very hot stars (O and B type).

9. Which do you think it is more reasonable to assume is the center of the Milky Way galaxy: the center of the distribution of globular clusters, or the sun? Why?

10. An optical picture of the spiral arms of the Milky Way galaxy has been made by a plot of what kinds of stars?

11. Why is the radio picture of the Galaxy more complete than the optical picture?

12. Why should a star in one of the arms of the Galaxy which is closer to the central nucleus appear to us to have larger Doppler shift in relation to the sun?

13. What evidence do we have that the Milky Way galaxy was once more like a sphere and then later flattened to become a spiral?

14. What factor probably influenced the Milky Way galaxy to become a flattened system?

Suggested Readings

BLAAUW. A., and SCHMIDT. M. (eds.), *Galactic structure*. Chicago: University of Chicago Press, 1965.

BOK, B. J., and BOK, P. F., *The Milky Way*, 4th ed. Cambridge, Mass.: Harvard University Press, 1973.

GINGERICH, OWEN (ed.), *Frontiers of astronomy* (Introduction to Chapter 5). San Francisco: W. H. Freeman & Company Publishers, 1970.

MAVRIDIS, L. N. (ed.), *Structure and evolution of the galaxy*. Dordrecht, Netherlands: Reidel, 1971.

SANDERS, R. H., and WRIXON, G. T., The center of the galaxy. *Scientific American* 230 (4), 66–77 (1974).

SCIENTIFIC AMERICAN, *New frontiers of astronomy* (readings). San Francisco: W. H. Freeman & Company Publishers, 1975.

WEAVER, HAROLD, Steps toward understanding the large scale structure of the Milky Way, *Mercury* 4 (5, pp. 18–24, and 6, pp. 18–29), (1975).

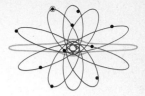

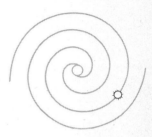

FOURTEEN

THE COSMOS

Suppose that we had the rare privilege of using the Palomar 5-m telescope to take a time-exposure photograph of the sky. Upon development of the film, numerous star images would be apparent. If we selected a small portion of the film and enlarged it sufficiently, we would likely see more galaxies than individual stars (Figure 14.1). For every star in our Galaxy (over 100 billion) there is one galaxy within the range of the 5-m telescope, and there are many more beyond its range, each containing billions of stars.

Classification of Galaxies

The Milky Way galaxy represents only one of the many forms in which galaxies are found. Hubble, in his early work with galaxies (during the 1920s), devised an extremely simple classification system for them. While some refinements are being suggested, his system remains in use to this day, essentially unchanged. The three main categories of galaxies include the *elliptical* type, the *spiral* type, and the *irregular* type (Figure 14.2).

The elliptical galaxies are subdivided in groups according to the

Figure 14.1 *Cluster of galaxies in the constellation of Hercules. (Hale Observatories)*

Figure 14.2 *Types of galaxies: (a) elliptical, NGC 4486; (b) spiral NGC 2841; (c) irregular, NGC 3034. (Hale Observatories)*

(a)   (b)   (c)

(a)   (b)

Figure 14.3   *Elliptical galaxies of (a) the EO-type and (b) the E5-type. (Hale Observatories)*

eccentricity of their shape. A spherical galaxy is designated E0, and galaxies which are more and more eccentric are designated successively E1, E2, . . . , E7. You will see examples of E0 and E5 ellipticals in Figure 14.3.

A further subdivision of elliptical galaxies is the dwarf systems, of which there are numerous examples within 1 million light-years of the Milky Way galaxy. They are intrinsically dim (luminosity about 0.001 that of the Milky Way), sparsely populated collections of stars, and yet they represent one of the most common types of galactic systems in the universe.

The spiral galaxies are flattened systems that by their appearance suggest rotation. Those with a very large, bright nucleus and tightly wound arms are called Sa-type spirals. The Milky Way galaxy is an example of the second classification, the Sb-type, in which the brightness is more evenly distributed between the nucleus and distinct arms. The third classification, type Sc, has a much smaller nucleus and bright, open arms. In fact most of the light of an Sc-type galaxy is concentrated in its arms. These three types are illustrated in Figure 14.4(a, b, c).

Within the general spiral class fall the barred spirals, a type which appear to have a barlike distribution of brightness running through their nuclei. The same general classification as for spirals applies here. Those with a high concentration of brightness in the bar and nucleus and tightly wound arms are called SBa, whereas the brightness of the SBb type is more evenly distributed between the arms and the nucleus. The SBc type may have a less distinct nucleus, but it has a bright barlike structure and open arms [Figure 14.4(d, e, f)]. Note that the SBab type shown in Figure 14.4(d) is midway between the SBa and SBb.

The S0-type galaxy shown in Figure 14.5 may at first glance resemble an elliptical type, yet if we could see this galaxy from an edge-on view, it would resemble the spiral type (Sa). Closer examination, however, reveals vast differences from either type. The S0 type shows no hint of arms, and

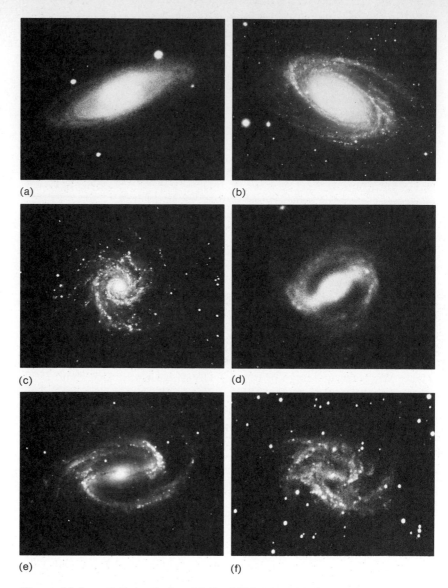

Figure 14.4 *Spiral galaxies: (a) Sa, NGC 2811; (b) Sb, NGC 3031; (c) Sc, NGC 628. Barred spiral galaxies: (d) SBab, NGC 175; (e) SBb, NGC 1300; (f) SBc, NGC 2525. (Hale Observatories)*

it shows no gas or dust which is characteristic of the spiral. The S0-type galaxy does seem to bridge the gap between the ellipticals and the spirals. However, this should not be taken as an indication that it is an intermediate step in the evolution of galaxies from elliptical to spiral types, or vice versa. Some observers have speculated that the S0 type may represent a galaxy that has passed through another galaxy. Had this happened, individual stars would have been so far apart compared to their size that

Figure 14.5   *An SO-type galaxy, NGC 1201. (Hale Observatories)*

they would not have collided, nor would their gravitational interaction have been apparent, but any gas and dust which they possessed would have interacted and been swept from both galaxies. Since S0 galaxies are found in regions of space where other galaxies are relatively close, this appears to be a plausible explanation for their lack of gas and dust. This absence of gas and dust clouds leaves the S0 type with predominantly type-II (old) stars.

The irregular type galaxies are so named because they have no symmetry or structure which can be specified. They fall into two subtypes: Irregular I, illustrated by the Large Magellanic Cloud (Figure 14.6); and Irregular II, like NGC 3034 (Figure 14.7). Irregular I galaxies are characterized by O- and B-type stars and regions of ionized hydrogen, and a few examples have a suggestion of spiral arms. Irregular II galaxies are characterized by the fact that their stars cannot readily be resolved in the telescope, hence they must be composed of dim stars; yet they show spectroscopic evidence of gas and dust and they eject material at high speed (as shown by the filaments that we see extending outward from the nucleus in Figure 14.7). M82 may be classified as an exploding galaxy, of which there are several additional types (see M87 on page 444, and others which follow in this chapter).

Hubble's Classification

Edwin Hubble suggested the orderly classification of galaxies shown in Figure 14.8, and some observers have taken this as a supposed evolu-

Figure 14.6 *The Large Magellanic Cloud. (Lick Observatory)*

Figure 14.7 *An Irregular II galaxy, NGC 3034, showing filaments extending 25,000 light-years outward from the nucleus. (Hale Observatories)*

tionary pattern as well. Some would suggest that galaxies evolve from right to left in Hubble's diagram, and others would suggest the reverse order; however, it is not clear whether galaxies suggest evolution at all. Note that the Hubble classification deals only with what may be called normal galaxies. We are concerned also about irregular, exploding, and compact galaxies, which will be described in subsequent sections of this chapter.

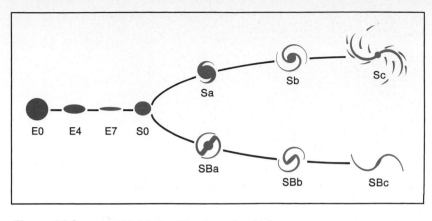

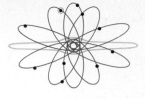

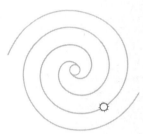

Figure 14.8   *Hubble's classification of galaxies.*

The Local Group

Our own local group of galaxies consists of 17 members that fall within a sphere approximately 2.5 million light-years in diameter. As can be seen in Figure 14.9, the Milky Way galaxy is not the center of the group, but we place it at the center of the diagram as an aid in visualizing distances to the various other members. Table 14.1 presents the variety of different types included in the local group in order of absolute magnitude.

Within the local group we see almost every type of galaxy represented except the S0 type, and the general range of mass, diameter, and luminosity represented in the local group is almost the same for the entire universe.

Clusters of Galaxies
and Intergalactic Media

Clustering is a phenomenon that seems to pervade the universe, and the largest subclass of the universe is a cluster of galaxies. The Local Group together with possibly a few additional galaxies represents the cluster to which the Milky Way belongs. In Figure 14.1 (page 432) we saw a galactic cluster in Hercules in which numerous spirals appear. The Coma cluster, shown in Figure 14.10, reveals virtually no spirals, but a predominance of ellipticals. Up to 1000 galaxies may be found in a cluster. Some observers have gone a step farther in defining a cluster of clusters—a supercluster; however, there is little agreement on the reality of this unit.

The space between the galaxies is not totally empty. Rather, a tenuous gas exists at a temperature of 100 million degrees Kelvin, and this gas may be responsible for a general X-ray background radiation emanating from clusters of galaxies.

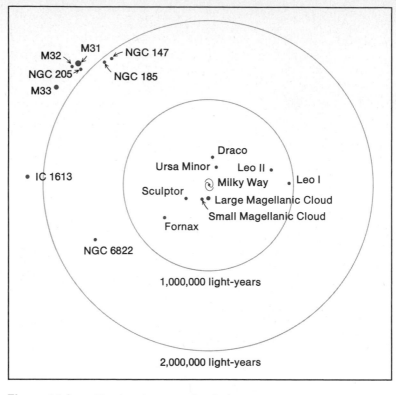

Figure 14.9 *The local group of galaxies.*

Table 14.1 *The Local Group of Galaxies*

GALAXY	TYPE	ABSOLUTE MAGNITUDE	DISTANCE (MILLION LIGHT-YEARS)	DIAMETER (LIGHT-YEARS)
M31 (Andromeda)	Sb	−21	2.2	100,000
Milky Way	Sb	−20	—	100,000
M33	Sc	−19	2.3	—
Large Magellanic Cloud	Irr	−18	0.16	30,000
Small Magellanic Cloud	Irr	−17	0.18	25,000
NGC 205	E5	−16	2.2	5,000
NGC 221 (M32)	E3	−16	2.2	8,000
NGC 6822	Irr	−15	1.5	9,000
IC 1613	Irr	−15	2.2	16,000
NGC 185	E_2	−15	1.9	8,000
NGC 147	E6	−15	1.9	10,000
Fornax System	E3 (Dwarf)	−13	0.6	22,000
Sculptor System	E3 (Dwarf)	−12	0.27	7,000
Leo I Systems	E4 (Dwarf)	−10	0.9	5,000
Draco System	E2 (Dwarf)	−10	0.33	4,500
Leo II System	E0 (Dwarf)	−10	0.75	5,200
Ursa Minor System	E4 (Dwarf)	−9	0.22	3,000

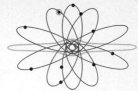

Figure 14.10   *The Coma Cluster of galaxies, showing a predominance of ellipticals. (Hale Observatories)*

Distances to Neighboring Galaxies

Distances to galaxies within our local group are relatively easy to determine. Recalling Henrietta Leavitt's work with Cepheid variables and the calibration of the period-luminosity relationship by Harlow Shapley and Walter Baade (Chapter 10), we understand that the distance to any galaxy may be determined if both the period and apparent magnitude of a single Cepheid in that galaxy can be determined. If a Cepheid cannot be isolated, then other methods must be used. If we assume that a supernova in another galaxy has approximately the same absolute magnitude as a supernova in our Galaxy, then we can estimate distance by observing its apparent magnitude and using the following equation:

$$\frac{L(10)}{L(r)} = \left(\frac{r}{10}\right)^2$$

where L is luminosity and r is distance. If a typical supernova has an absolute magnitude of -15 and an apparent magnitude of $+10$, then it would show an increase (brightening) of 25 mag if brought in to a distance of 10 parsecs, and this is equivalent to an increase in luminosity of 10^{10}. Substituting and solving for r, we get

$$\frac{10^{10}}{1} = \frac{r^2}{10^2}$$

$$r^2 = 10^{12}$$

$$r = 10^6$$

$$r = 1 \text{ million parsecs (3.26 million light-years)}$$

Furthermore, when we consider clusters of galaxies, we may consider the average absolute brightness of the 10 brightest galaxies in one cluster to be the same as the average for the 10 brightest galaxies in another, and we can thereby judge distance from their apparent magnitude—the dimmer the galaxies, the farther away the cluster.

Hubble's Red-Shift Law

In 1929 Edwin Hubble made a startling discovery. As he recorded the spectra of many galaxies, he noticed a relationship between the amount by which the spectral lines were red-shifted and the apparent magnitude of the galaxies. In general, the dimmer galaxies had larger Doppler shifts toward the red end of the spectrum. Looking only at the photographs and spectrograms in Figure 14.11, you will see this relationship. The red shift of each is shown by a horizontal arrow on the spectrograms—very short in the first, but lengthening with each successive spectrogram.

In an attempt to refine this relationship, Hubble made two assumptions. He assumed that the red shift of each galaxy is due to its motion away from the observer and that the larger red shifts represent proportionately greater recessional velocities. He also assumed that the dimmer galaxies are farther away. Converting red shift to velocity of recession and estimating the distance to nearby clusters of galaxies, he plotted several points on a graph similar to Figure 14.12. A relationship is evident from the diagonal distribution of these points; however, he had made an error in assessing the distance to the nearby clusters of galaxies, and as a consequence he obtained an incorrect value for the expansion rate, namely, 165 km/sec for each million light-years. This value turned out to be about 10 times too large, based upon a modern determination of distances. Figure 14.12 shows this modern determination. As you can see, it indicates a rate of recession of 15 km/sec for each million light-years. This is called *Hubble's constant*—a misnomer, because its value is subject to change with time. If the expansion of the universe is slowing down, then the value of Hubble's constant will be smaller in the future. Furthermore, its present value is subject to an accurate determination of the distance to certain clusters of galaxies. For instance, if the distance to the Virgo cluster is not 80 million light-years, as shown in Figure 14.11, then Hubble's constant is not 15 km/sec for each million light-years.

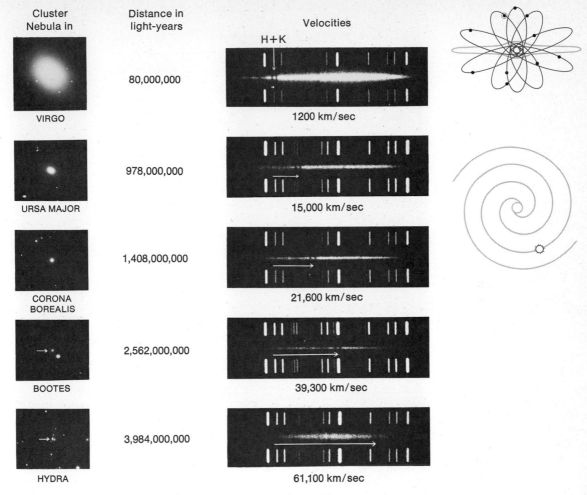

Cluster Nebula in	Distance in light-years	Velocities

H+K

VIRGO — 80,000,000 — 1200 km/sec

URSA MAJOR — 978,000,000 — 15,000 km/sec

CORONA BOREALIS — 1,408,000,000 — 21,600 km/sec

BOOTES — 2,562,000,000 — 39,300 km/sec

HYDRA — 3,984,000,000 — 61,100 km/sec

Figure 14.11 *Photographs of successively more distant galaxies; their distances are estimated by their apparent magnitude and thus are subject to reevaluation. The spectrogram of each galaxy is shown (on the right) with the Doppler shift, due to recessional velocity indicated by the horizontal arrows. Arrows indicate shift for calcium lines H and K. (Hale Observatories)*

Hubble's relationship can be expressed very simply as

$$H = v/r$$

where *v* is the velocity of recession and *r* is the distance to the cluster of galaxies. Hubble's red-shift law expressed the fact that the distance to any given cluster of galaxies appears to be directly proportional to its velocity of recession (based upon its red shift). While there is good confirmation for this relationship in the case of normal clusters of galaxies, controversy exists as to whether the red-shift law applies to quasars—objects whose

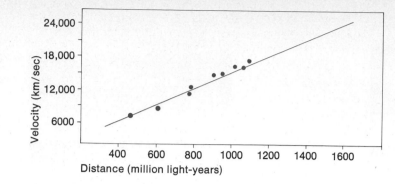

Figure 14.12 *The relationship between red shift and distance.*

red shifts indicate recessional velocities near the speed of light if Hubble's red-shift law applies. (We shall return to this topic when we discuss quasars.)

Radio Galaxies

In 1931 engineer Karl Jansky of the Bell Telephone Laboratories accidentally discovered radio signals coming from outer space. Did these signals merely represent a general background of radio noise, or were they associated with separate sources? This question was answered by Walter Baade in 1951 when, using the 5-m telescope at Mt. Palomar, he identified optically a definite source of radio energy at Cygnus A, a galaxy some 700 million light-years distant. The nature of this source was not realized until Baade's identification. This opened a new era for astronomy, the era of *radio observation.*

Figure 14.13 *(a) Cygnus A, a radio galaxy. (b) The radio map of Cygnus A shows that most of the radio energy is emitted from two regions widely separated from the optical image in the center of this view. (Hale Observatories)*

(a)   (b)

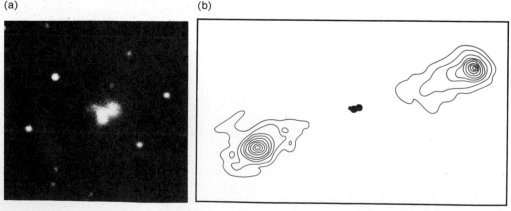

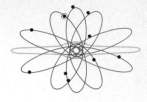

Cygnus A is a highly unusual galaxy, for its radiation in radio wavelength is about one million times greater than that of a normal galaxy. Yet to be discovered is the source of this prodigious outpouring of energy; perhaps a gravitational collapse or an enormous explosion is taking place (see Chapter 12). The emissions seem to arise from two regions about 50,000 light-years to either side of Cygnus A (Figure 14.13).

Radio sources can be classified as extended or compact according to their apparent diameter. Extended sources have an angular diameter of over 1 second of arc, and they probably range in diameter from 10 to 100 parsecs. Compact radio sources have angular diameters less than 1 second of arc and actual diameters less than 10 parsecs (Figure 14.14).

Thousands of discrete radio sources have been found, many of which have been identified with visible galaxies. A *normal radio galaxy* emits about one-millionth as much radio energy as it does optical energy. A *peculiar radio galaxy* emits about 100 times as much radio energy as the normal radio galaxy: M87 (Figure 14.15) is an example of an elliptical galaxy that emits approximately 100 times as much radio energy as the normal emitter NGC 1068. The radio emission seems to come from two regions at equal distances on either side of the galaxy. A short-exposure photograph shows a jetlike appendage, the light from which is highly polarized, indicating a synchrotron emission, a result of high-energy electrons moving in a magnetic field. The bright knots of the jet suggest several violent explosive events. The mass of M87 is over 100 times that of the Milky Way, and it is now known to have a central nucleus of about 5 billion solar masses, which is so dense that it may be likened to a

Figure 14.14 *Radio map of the sky showing intense sources near the galactic equator. (Ohio State University Radio Telescope)*

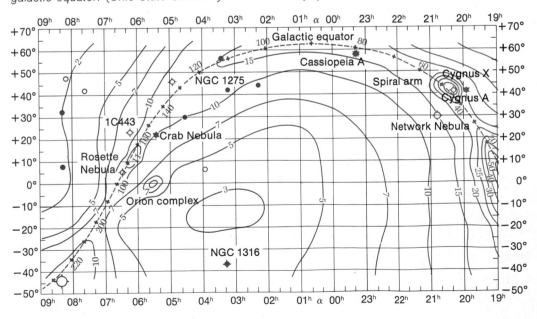

Figure 14.15 *An elliptical galaxy in Virgo, NGC 4486 (M87), with a jetlike appendage. (Lick Observatory)*

Figure 14.16 *A Seyfert galaxy, NGC 4151. (Hale Observatories)*

galactic "black hole." The infall of material onto that central core may be responsible for its tremendous outpouring of energy.

Seyfert Galaxies

In 1943 Carl K. Seyfert of Mt. Wilson Observatory described a class of galaxies that superficially resembled a normal spiral (Figure 14.16). The characteristics that seemed to separate these particular spirals included a very small, bright nucleus that showed bright emission lines within its spectrum. This is unusual, since the spectra of most galaxies show only broad absorption lines.

The light of several Seyfert galaxies has been observed to vary greatly in a period of only a few months. Several Seyferts are also strong radio emitters, again showing distinct variations in output. These properties, together with their general starlike appearance, suggest that they are

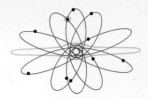

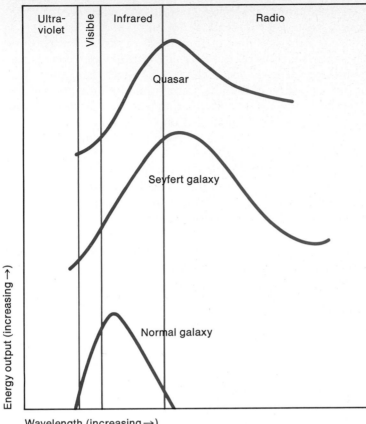

Figure 14.17 *The energy output of a quasar as compared to a Seyfert galaxy and a normal galaxy.*

very compact sources. The energy distribution of Seyfert galaxies over a wide spectrum of wavelengths is shown in Figure 14.17. Notice how similar this output curve is to that of a quasar.

Quasars

With the methods of radio astronomy firmly established, the search for new radio objects continued. Early in the 1960s observers discovered a type of radio source that appeared to be more starlike in dimension than galactic. These sources were termed *quasi-stellar radio sources* (contracted to *quasars*). The first such object was called 3C273 (Number 273 in the *Third Cambridge Catalogue of Radio Sources*). Attempts were made to find the optical image of 3C273; however, the ability of a radio telescope to pinpoint a source is rather limited. Because the radio wavelengths are long compared to light, the resolution of even the larger radio telescopes is rather poor (see Chapter 2). A unique approach to locating 3C273 precisely was used by Cyril Hazard of Parkes Radio Observatory in Australia. On a certain day when the moon was to pass directly in front of the radio source, blocking the radio signals, Hazard observed the exact time of eclipse and reappearance of the radio source. Using his knowledge of the exact position of the moon at all times, he then pinpointed the radio source. As soon as this information reached Maarten Schmidt at Palomar, he was able to locate 3C273 on earlier photographs. It was starlike in size but fuzzy, and it had a jet extending out a distance almost equal to its diameter (Figure 14.18).

Schmidt was puzzled by the spectrum of this object, which showed an unfamiliar placement of spectral lines. Finally, he hit upon the idea that he was seeing lines of hydrogen that should appear in the blue portion of the spectrum but had been red-shifted so far that they appeared in the red portion.

If the red shift of the spectral lines can be explained entirely as a phenomenon related to the speed with which an object is receding from the observer, then Schmidt had found an object receding at a rate of 45,000 km/sec. Furthermore, if this object was to fit Hubble's general pattern of red shift among the distant galaxies, then it had to be placed at a distance of 3 billion light-years.

Schmidt's analysis of the red shift was soon confirmed by his colleague Beverly Oke, who showed that the H_α (hydrogen-alpha) line, normally a red line, had been shifted into the infrared portion by an amount equal to Schmidt's estimate. But this was only the beginning, for today hundreds of quasi-stellar objects are known, including many which are radio-quiet, having been recognized by their unique spectra. In the spectra of quasar OQ172 such a large red shift has been found as to indicate a recessional velocity greater than 90 percent of the speed of light. If this object is to fit the Hubble red-shift pattern, then it must be placed at a distance of over 10 billion light-years.

These observations set the stage for a controversy that was to last many years. On the one hand, some astronomers believed that they were seeing the most distant objects ever seen by man, objects so far distant that they were looking backward in time to what was perhaps almost the

Figure 14.18   Quasar 3C273. (Hale Observatories)

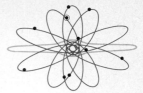

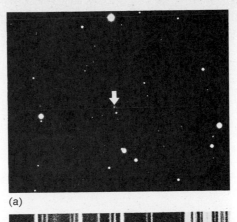

(a)

Figure 14.19 (a) *Source of radio noise, 3C295, in the constellation of Bootes. (b) Its very large red shift of spectral lines may indicate that it is one of the more distant objects of the known universe. (Hale Observatories)*

(b)

beginning of the present universe (Figure 14.19). And yet, if this was the case, then they were also looking at the brightest objects ever witnessed. In fact, they were hard pressed to explain a source of energy so great that it could be recorded 10 to 18 billion light-years away. No ordinary process in nearby galaxies could match the outpouring of energy necessary to be detectable at such distances—an energy output which must be 1000 times that of a normal galaxy. Other astronomers say that these objects are nearby and therefore need not be so bright.

Dr. Halton Arp, of Hale Observatories, may be characterized as the leader of the latter school of thought. He has observed that quasars are not distributed—either as to position in the sky or as to red shift—in the uniform way that one would expect if they were actually the more distant objects of an expanding universe. Dr. Arp finds alignments and grouping of quasars and other compact objects in apparent proximity with large elliptical and spiral galaxies. We say "apparent" because this is where the controversy lies. Various objects may appear close together in the sky yet be separated by vastly different distances from the observer. How can we tell if they are actually close to one another? Dr. Arp looks for bridges of material that physically connect the various objects and he looks for gravitational interaction between the members. He feels that it is too much of a coincidence for several objects to be so nearly aligned (see Figure 14.20), especially when this alignment is so closely correlated with the radio-emission lobes of the central galaxy (see Figure 14.21). Often the quasars or compact galaxies appear on both sides, as though they were ejected from the primary galaxy. Could this explain how NGC 7331 (an Sb spiral) could have a red shift indicating a recessional velocity

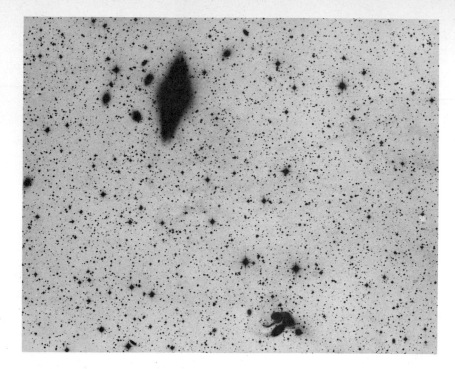

Figure 14.20 *A positive print showing a clustering of Arp objects in the region around NGC 7331 (a large spiral galaxy) and including Stephan's Quintet in the lower right. (H. Arp, Hale Observatories)*

of only 800 km/sec while the red shifts of the compact objects which appear in the same region indicate velocities between 6000 and 8000 km/sec greater? Or what of the five galaxies in Stephan's Quintet (see Figure 14.22), which fall within 0.5° of NGC 7331? Four of these objects have velocities in the range of 5700 to 6700 km/sec, whereas the fifth has an indicated velocity of 800 km/sec. If NGC 7331 and the compact galaxies which appear around it are actually close to each other, then we must look for a different explanation of their discordant red shifts—an explanation that does not depend simply upon an expanding universe in which larger red shifts correlate with greater distances. While his critics suggest that he is simply seeing what he wants to see, Dr. Arp would retort that there are far too many examples of the type cited to pass the matter off that lightly.

Let us return now to the possibility that quasars are distant objects, apparently very small. Some have suggested that they range from $\frac{1}{25}$ to $\frac{1}{100}$ the size of a normal galaxy, yet radiate up to 1 million times more energy than the average galaxy. Quasars have been seen to vary in energy output in as little as one week, and the conclusion is that they must be relatively small in order to show this variability. Theoretically an object which shows variations in as little as a week cannot be much over 1 light-week

Figure 14.21   *A radio map of the region surrounding the Sb Spiral galaxy NGC 7331, with Stephan's Quintet shown in the lower right. The shape of the radio emission lobes tends to suggest a connection between NGC 7331 and the other galaxies shown in the same region.*

in diameter, or such variations would be smoothed out. One light-week is equivalent to approximately 15 times the diameter of the solar system.

How can such small objects emit the tremendous amount of energy indicated, if they are in fact situated 10 to 18 billion light-years away? Thermonuclear reactions that take place in the stars of any normal galaxy simply do not release sufficient energy to be seen at these vast cosmic distances. We must look for more energetic reactions. One such source of energy that could be released in a relatively short period is gravitation. If the core of a supermassive galaxy collapsed, with the nuclear particles falling together, huge amounts of energy would be released in a very short time. The result of such a collapse could be the ejection of surrounding materials, such as is suggested by jets emanating from several quasars. Electrons ejected by such an event, if moving within a magnetic field, generate electromagnetic waves called *synchrotron radiation*. This process is capable of releasing large amounts of energy.

Figure 14.22 *The five galaxies in Stephan's Quintet lie within 0.5° of NGC 7331 (see Fig. 14.20). (Lick Observatory)*

An important factor to remember when talking of objects 10 to 18 billion light-years away is that we are also looking backward in time 10 to 18 billion years: we are seeing those objects as they existed at that time, and therefore we may be seeing galaxies in their initial stages of formation. The average density of the universe is thought to have been much greater at that time, a condition which may have permitted the formation of such dense galactic cores that they were in essence huge black holes (see Chapter 12). The gravitational collapse which would be a part of this picture could release energies in the amounts observed in quasars. Carrying this idea to the extreme, a quasar may be a galaxy which is collapsing into one gigantic black hole. Some observers have gone so far as to suggest that stars falling into such a black hole may appear in another space-time frame (see the subsequent section on cosmology). If such a phenomenon can take place, these objects may then be termed *white holes*, and they may explain the very energetic reactions which have been observed in the centers of relatively nearby galaxies.

One further theory focuses on the possibility that stars may collide in the very dense nucleus of the quasar, whereupon they coalesce to form such massive objects that a supernova-type reaction occurs. If this took place on a multiple basis and more or less continuously, the great outpouring of energy might be explained, together with its variability and occasional sudden brightenings of up to 100-fold. The ejection of material is also a natural consequence of such a supernova-type reaction.

These are rather imaginative theories, but they represent the type of thinking that may eventually lead to a solution to this fascinating problem.

BL Lacertae Objects

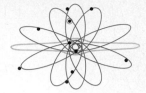

The very name of this object suggests its identification as a variable star in the constellation of Lacertae, and so it seemed in 1929 when this object was identified. It became the model for a whole collection of fuzzy variable "stars" called BL Lacs for short. In 1969 the discovery of the radio emission of these objects was the first clue that they had been incorrectly classified, for only galaxies and quasars have radio emissions of this type. Obviously, if a red shift could be found, it would settle the matter of whether BL Lacs belong to the Milky Way. No sharp spectral features were apparent, however, and the question was not settled. However, with improved techniques, weak emission and absorption lines are being found, and their red shifts indicate that the distances to the BL Lac Objects must be ranked with those of the quasars. This only compounds the questions about the quasars, raised in the last section, for if located at the great distances indicated, then these objects are even brighter than the quasars, and more variable. One might class them among the most violent quasars. Thus the BL Lac rose from its first identification as a lowly variable star in the Milky Way to become one of the most active objects in the entire universe. But the story doesn't end there.

A recent observation of BL Lacertae itself has revealed that this object is located at the center of a normal galaxy, and because of the very faint nature of that surrounding galaxy, the great "cosmological" distance of BL Lac is confirmed: it really is at the great distance indicated by its red shift. The question of energy source is only compounded by this determination. How can any object produce enough energy to be seen at such a distance? The variability both in light output and polarization indicates a very small nucleus of the order of a light-day. It appears that the emanations of the nucleus simply overpower those of the surrounding material, hence astronomers may be "seeing" closer to the heart of a violent quasar as they study the BL Lacs.

A Modern Classification of Galaxies

Hubble's classification of "normal" galaxies will continue to be useful; however, another scheme seems necessary when we consider the very active (violent) natures of many of the objects being observed today. It would appear that the BL Lacertae Objects are the most violent of all objects in the universe, and that their most intense activity is concentrated in a relatively small nucleus. Quasars follow as a close second, some with radio emissions and some without. The Seyfert would constitute the third classification, followed by the "normals."

Any classification seems to raise the question of a possible evolution among galaxies. Could a BL Lac Object be a quasar evolving into an elliptical or vice versa? There appears to be very little evidence to support any unified theory of evolution; however, perhaps our view of the universe may give a clue. When we look at very distant objects such as BL Lacs and quasars, we are seeing them as they existed in the past—in the early stages of the universe. In an expanding model of the universe, its density was much greater then, and this density may have influenced the nature of the galaxies as they formed. The very active small cores we see in these objects may have been a part of their early nature. As we look to nearer and nearer objects, we are seeing these objects in a later state of development, because light doesn't take so long to reach us. We see bright nuclei in certain middle-distance galaxies, such as Seyferts, but not so active as the quasars. We also see at near and middle distances the "normal" ellipticals with no apparent core emphasis. Some observers have suggested that a Seyfert is merely a nearby quasar—perhaps a quasar that has evolved with time to become a Seyfert. Likewise, a comparison can be made between the elliptical and the BL Lac, which has an elliptical around its violent core.

While this discussion does not solve the question of evolution, it does show the interrelationship of many different factors and the complexity of the astronomer's effort to read the past history of the universe.

Cosmology

Cosmology is the study of the universe as a whole: its large-scale structure, its organization, and its history. In a sense, some of the models of the universe which we considered in Chapter 1 represent early cosmologies; however, they were very limited views. What was the chain of events by which man expanded his view of the cosmos and gained hope that he might discover its history?

In 1915, Albert Einstein gave science one of its most profound theories: *the general theory of relativity*. The parts of this theory which are most pertinent to our discussion of cosmology are the following: (1) In order to describe an event, both the three dimensions of space and the fourth dimension of time must be used; hence, we live in a four-dimensional world of space-time. (2) Gravity is determined by the distribution of mass (and energy). (3) Space-time has a curvature which depends upon the distribution of mass (and energy). Einstein proposed a cosmology which embodied a static universe and, knowing that gravity tends to contract space, he introduced the idea of a cosmic force of repulsion which just balances the force of gravity. He later abandoned this concept of a repulsive force, calling it his greatest mistake, when Hubble demonstrated that the universe is expanding. While Einstein did not perfect a cosmological model, at the heart of every productive model which followed is his description of gravity.

In 1922, Alexander Friedmann found a solution to Einstein's equations for a time in the past history of the universe. His solution predicted a state of very high density and very high temperature (10 billion degrees

Kelvin). One might liken the high (perhaps infinite) density to that of the singularity of the black hole (see Chapter 12, page 407). Some would visualize this as the "primordial atom." It is thought that this superdense, superhot "atom" exploded, producing the "primordial fireball"—the Big Bang, which is the name given to the theory today. In its first few seconds, this explosion was dominated by a mixture of radiation and matter which interacted in a way that prevented the formation of atomic nuclei. Temperatures were measured in the billions of degrees Kelvin; however, rapid expansion produced rapid cooling as well. The density of the universe was so great that not even the neutrino could travel far without collision. Neutrons were continually being converted to protons by neutrino bombardment (neutrino + neutron $\longrightarrow$ proton + electron), hence the proton-to-neutron ratio increased throughout these early minutes. Positrons and electrons annihilated each other, producing additional energy to slow the cooling effect.

When the universe cooled to 3 billion atomic degrees Kelvin, atomic nuclei began to form. The hydrogen nuclei occurred in two isotopic forms: ^1_1H, hydrogen containing one proton, and ^2_1H, deuterium (heavy hydrogen) containing one proton and one neutron. The amount of deuterium formed is inversely proportional to the density of the matter at the time. In a very dense state, little deuterium is left, because most of it reacts with other nuclei (^1_1H or ^2_1H) to form helium. In a less dense situation, more of the deuterium will be left intact. Thus the deuterium abundance, as observed today is considered an important clue to the early state of the universe. While deuterium is very unstable at high temperatures, its formation was the steppingstone to the production of helium-3 and helium-4. The universe could now be characterized by the presence of ionized matter (protons, deuterium nuclei, helium nuclei, free electrons), neutrinos and electromagnetic radiation (photons), in strong interaction with each other. Within 35 minutes of the Big Bang, the temperature had cooled to 300 million degrees Kelvin and all nucleosynthesis had stopped: no more helium or heavier elements would be formed at this time. Only after about 100,000 years did the continued expansion result in cooling to about 4000°K, at which temperature free electrons could combine with the nuclei to form neutral atoms of hydrogen (73%), deuterium (trace), and helium (27%). With continued cooling (due to expansion) clusters of galaxies formed, and these clusters still demonstrate the expansion of the universe today. In the big bang theory, it is assumed that the galaxies formed in the first 100 million years or so.

In 1924 Edwin Hubble first demonstrated the existence of galaxies outside our own, when he recognized that the Andromeda Nebula was not a part of our own Galaxy. Only within the last half century, then, has man known that the universe is composed of clusters of galaxies, the Milky Way being only one member of a cluster called the Local Group. In 1929 Hubble discovered the red-shift law (discussed earlier), a relationship which plays an essential role in testing cosmological models. Hubble was joined by astronomers such as M. L. Humason and N. U. Mayall in his study of the red shifts of distant clusters of galaxies, and by 1936 two additional discoveries had been made, both of which were essential to the development of cosmological models:

1. These observers assumed that the brightest elliptical galaxies of one cluster had virtually the same absolute magnitude as the brightest ellipticals of any other cluster. This was significant because these brightest ellipticals could be used as an accurate distance indicator—the dimmer they appeared, the more distant they were. With knowledge of both their absolute magnitude and their apparent magnitude, solution of the equation

$$\frac{L(10)}{L(r)} = \left(\frac{r}{10}\right)^2$$

    was expected to provide the distance (as discussed earlier in this chapter). But certain corrections must be applied, as we shall see, and the process is not as simple as it may appear.

2. In all the astronomers' searching, clusters of galaxies appeared to be randomly distributed with regard to distance and with regard to direction. Clusters of galaxies do not favor one part of the universe over another part. This was the first observational evidence that the universe, on a large scale, appears to be *homogeneous* (similar in density throughout) and *isotropic* (the same in all directions). This was a very significant discovery, for it suggested that we do not live in some unusual part of the universe, but rather that what we see from our location is typical of the entire universe. This observation formed the basis of an assumption which is at the heart of most modern cosmological models. Called the *cosmological principle*, it is stated as follows:

The universe, on a large scale, appears the same from any location.

    This suggests that the different parts of the universe may have had a common origin and lends hope that its history can be discovered.

By about 1950 this cosmological principle had been expanded by a group at the University of Cambridge, headed first by Hermann Bondi and Thomas Gold and later by Fred Hoyle. They suggested that not only does the universe appear the same from any location, but it also appears the same at any time (past, present, or future). This concept, referred to as the *perfect cosmological principle*, formed the basis for the "steady state" model. This model does not deny the expansion of the universe,

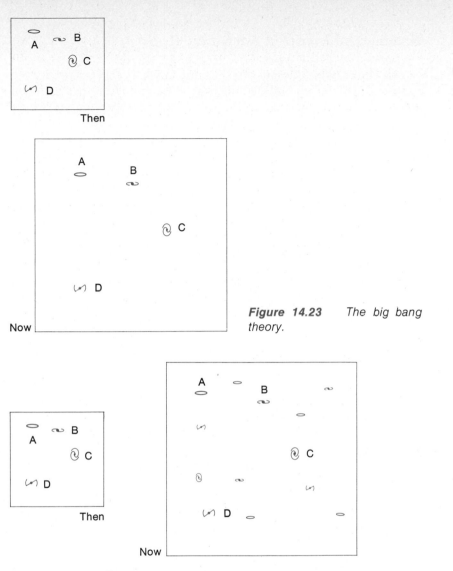

Figure 14.23 The big bang theory.

Figure 14.24 The steady state theory.

but rather it predicts its acceleration. In order to keep the distribution (density) of the universe constant, it suggests that as galaxies move apart, new matter comes into being to fill this space. Because it is assumed that the universe has always looked the same, there need be no creation event equivalent to the primordial fireball. The big bang theory and the steady state theory are illustrated in Figures 14.23 and 14.24. As we shall see, there are several lines of evidence against the steady state theory and in favor of the big bang model. One of the most convincing is the discovery of a form of radiation which may be a consequence of the primordial fireball.

Cosmic Background Radiation

If the universe began as a gigantic fireball, characterized by a temperature of the order of a billion degrees Kelvin, and if for a period of time a state of thermal equilibrium prevailed, its energy output would have a definite pattern, resembling that of the sun or stars in general. Because of the expansion of the universe, that radiation today should appear to have cooled and should still exhibit itself as a background radiation equivalent to that of a very cold blackbody, one at 3°K (−270°C). If this radiation could be found, then it would be direct evidence of the fireball (Figure 14.25).

About 20 years after it was initially predicted, blackbody radiation was found quite accidentally by two astronomers of the Bell Telephone Laboratory, Arno H. Penzias and Robert W. Wilson, who were testing certain very sensitive radio equipment. They experienced a form of radio

Figure 14.25 *The black-body radiation curves for the primordial fireball and for 3°K radiation, the latter falling almost entirely in the radio portion of the spectrum.*

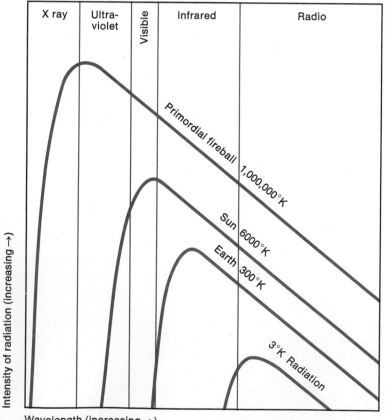

noise which showed a very remarkable similarity to that of a blackbody at approximately 3°K. Normally radiation is detected because of an increase in energy received when the radio telescope (antenna) is pointed toward the source, but the predicted 3°K radiation was thought to pervade space and thus would not be stronger in one direction than in another, unless we were moving through it. In 1964, Robert H. Dicke devised twin receiving antennae, of which one was attached to a reference source that was known to radiate at 4.2°K, and the other was pointed skyward. When the sky radiation was compared to the reference source, it became evident that 3°K radiation can be received from any direction. In the years that have followed, measurements have been taken in a variety of wavelengths, and this background radiation appears to fit the predicted curve quite closely. Sufficient evidence is now available to strongly suggest the existence of the fireball. The steady state theory provides no convincing explanation of this 3°K blackbody radiation.

This cosmic background radiation can be expected to form one of the best frames of reference by which the earth's motion in space may be specified. Because the radio antenna records an increase in temperature when pointed toward the constellation of Leo, observers compute for the earth a velocity of approximately 400 km/sec in that direction.

The Primordial Atom: Infinite or Finite?

To think of the matter of the entire universe being confined in the remote past to a very small volume—to a singularity—does not appear consistent with our observation of the cosmic background radiation. If this radiation had originated from only a limited region of space, it would have traveled outward at the speed of light and would not be detectable today (see Figure 14.26).

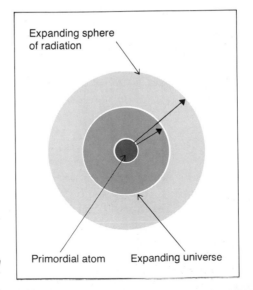

Expanding sphere of radiation

Primordial atom   **Expanding universe**

Figure 14.26   Radiation from a very small primordial atom.

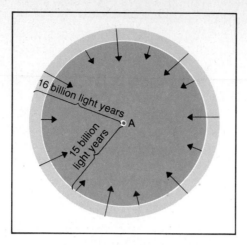

Figure 14.27 *Radiation from an infinite primordial atom.*

Rather, consider an infinite primordial atom which everywhere experienced the very high temperature of the Big Bang fireball, each point radiating energy in every direction. Subsequently a receiver located at point A would receive the signal which had originated at points just far enough away to be arriving now. For instance, we live about 15 billion years after the Big Bang; therefore we should be receiving a signal which was generated from places 15 billion light years away—that is, from all points on a sphere of radius 15 billion light years (see Figure 14.27).

If we could observe this background radiation a billion years hence, we would then sense it coming from points 16 billion light-years away. The radiation would appear cooler because it had been red-shifted more than we see it today. This model presents an infinite universe both before and after the Big Bang and does away with any need to visualize its point of origin.

Curvature of Space

Within the two models of cosmology we have presented, there are variations. For instance, the steady state model implies that the rate of expansion is speeding up (accelerating). On the other hand, all variations of the big bang model imply a slowing down (deceleration) of the rate of expansion. Different rates of deceleration are possible, however, and this is significant. In order to visualize these various possibilities, let us suppose that a ball is tossed upward with varying forces that impart different initial velocities in each case. Our normal experience with such a toss suggests that owing to the force of gravity, the ball will decelerate (slow down) eventually stop, and then return to the earth, only to bounce upward again and again [Figure 14.28(a)]. We might characterize its flight as an *oscillation*. On the other hand, if it were possible to give the ball a higher initial velocity, it might just overcome the force of gravity

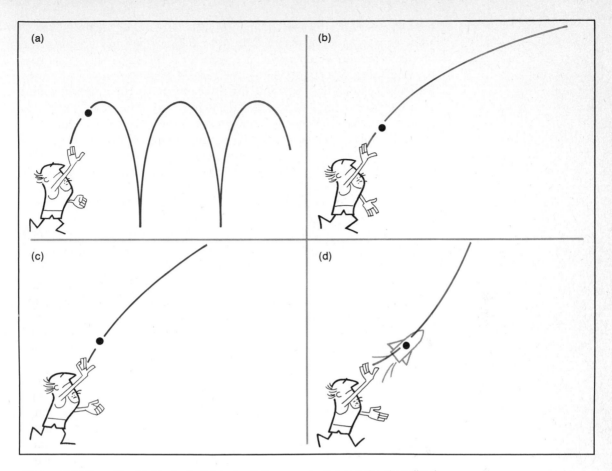

Figure 14.28 *The flight of a ball (shotput) thrown upward may be described in various ways depending upon the force applied: (a) oscillating (bouncing), if only a small initial force is applied; (b) parabolic—continually rising but also slowing down; (c) hyperbolic—continually rising at a greater rate than in (b); (d) accelerating—continually rising and speeding up, due to an added propulsive force (here applied by a rocket).*

and leave the earth forever [Figure 14.28(b)]. It would experience deceleration, but its flight would be arrested only at some great distance from the earth (infinity). The physicist would describe the flight of the ball under such conditions as *parabolic*. If still a greater force were used, the flight of the ball would appear as in Figure 14.28(c), continually decelerating, yet climbing higher without limit. Such a flight is termed *hyperbolic*. A fourth possibility involves the continual application of a force, say by means of a rocket attached to the ball [Figure 14.28(d)]. The force of the rocket produces an acceleration, and the ball is carried upward at increasing speed.

These four cases are analogous to various ways in which the universe might be expanding. The explosive force of the Big Bang event may have sent the material of the universe outward at such a rate as to allow gravity to bring it to a halt and collapse upon itself: this is the *oscillating model.* A plot of the scaling factor of the universe, as time goes by, would appear as in curve (a) in Figure 14.29. Had the outward force been just sufficient to overcome the force of gravity, the universe would continue to expand but would be slowing in its rate of expansion as shown in curve (b). The shape of such a curve is said to be *parabolic.* Curve (c) illustrates the expansion of the universe if still greater initial velocity had been imparted. This, too, represents a decelerating universe, but is said to be *hyperbolic.* Both the parabolic and hyperbolic models represent a universe which is infinite and in which radius has no meaning. The scaling factor in Figure 14.29 indicates only a change in the distance between galaxies when applied to these models.

The fourth model, that of an *accelerating expansion,* is illustrated by curve (d) in Figure 14.29. This represents a necessary condition of the steady state model; however, we have no indication that it is correct.

Figure 14.29 *The curves associated with the four models of the universe are shown with their present state made to coincide at the "now" point: (a) oscillating universe; (b) parabolic universe; (c) hyperbolic universe; and (d) accelerating universe. The straight reference line represents the best estimate of the Hubble "constant" at the present time. As is clear from this drawing, the "start-up" time of the universe is dependent upon the model that best describes its evolution.*

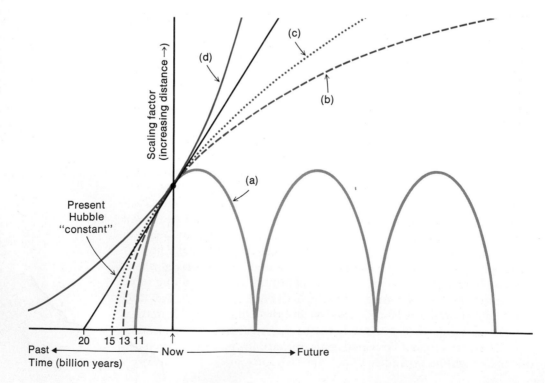

Astronomers' best determination of the Hubble "constant" at present is represented by the straight (solid) line in Figure 14.29, and points on each curve having the same slope as that reference line have been made to coincide (see the point labeled "now"). As you can see, each model predicts that Hubble's "constant" is not really constant but rather is changing. If it is changing as in curve (a) in Figure 14.29, then the universe will oscillate; furthermore, projecting the curve backward into the past, we see that it has only been 11 billion years since the last Big Bang. If the change in the rate of expansion is more accurately described by curve (b), then we must look back in time about 13 billion years for the Big Bang. Curve (c) indicates an even earlier beginning—about 15 billion years ago. If Hubble's "constant" were really constant, its straight-line graph would indicate an age for the universe of 20 billion years. As you can see in curve (d), the accelerating universe would require a still earlier beginning.

But how do these models relate to Einstein's concept of the curvature of space-time? To describe the evolving universe mathematically, he introduced a constant k. Depending upon the value of that constant, he termed the curvature of the universe either positive (if k is positive), zero (if $k = 0$), or negative (if k is negative). Positive curvature of space-time correlates with the oscillating model, zero curvature with the parabolic model, and negative curvature with the hyperbolic and the accelerating models.

We might try to visualize each of these curvatures of spacetime by considering models of one less dimension. A surface of zero curvature is a plane (a flat surface). If dots (which will represent galaxies) are distributed uniformly over such a surface, one may predict the ratio of number of dots in successive circles of radius 1, 2, 3, and so on. The area A of a circle is determined by the formula $A = r^2$, where r is the radius. Therefore, the area of successively larger circles will have the ratio of $1:4:9:$ etc. (the squares of the corresponding radii). Likewise, we would expect the ratio of counts of dots (galaxies) in successive circles to be $1:4:9:$ etc. On the other hand, if dots are distributed uniformly over a surface of positive curvature, that of a sphere, then successive radii produce less than the expected ratio, say $1:3.6:8:$ etc. This is graphically demonstrated by the gaps which develop in the flattening process shown in Figure 14.30(b), resembling a flattened orange peel. Negative curvature produces just the opposite effect: circles of successively larger radii produce more than the expected number of dots, for when a surface of negative curvature (hyperbolic or saddle shape) is flattened, wrinkles occur because of excess material. The ratio of dots may approach $1:4.5:10:$ etc.

One would expect the density of galaxies to be uniform in the case of zero curvature, to thin out with increasing distance in the case of positive curvature, or to increase with increasing distance in the case of negative curvature (Figure 14.30, bottom row). How can we possibly determine which model is correct?

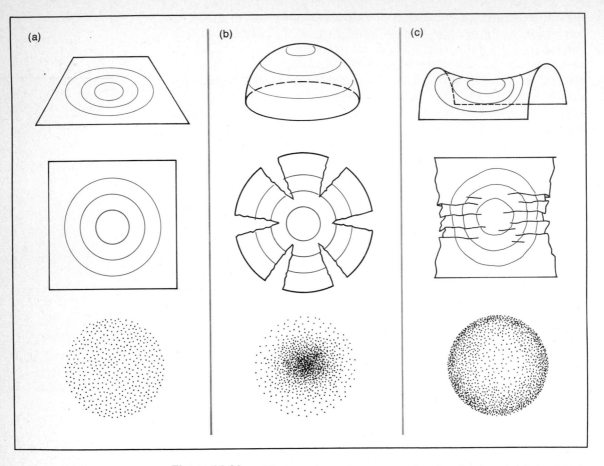

Figure 14.30   *The curvature of space may be visualized as: (a) flat, wherein we would find a uniform distribution of galaxies; (b) positive, wherein galaxies would appear to thin out with greater distances; or (c) negative, wherein the distribution of galaxies would appear more dense at greater distances.*

Tests for Cosmological Models

The effect of space-time curvature does not become apparent merely by our counting nearby galaxies. Therefore, the astronomer is concerned with galaxies which are beyond the reach of optical telescopes—that is, with radio galaxies. Recently a number of astronomers on the staff of Ohio State University completed a survey of over 8000 radio sources. They assumed that the radio sources located billions of light-years away are just as bright as those which are nearer. Based upon that assumption, their plot showed a thinning out with increased distance, somewhat resembling that of Figure 14.30(b), indicating positive curvature. We must be cautious at this point, however, for when we view very distant galaxies we are also looking backward to the time when the light left those sources (billions of years ago), and perhaps galaxies were intrinsically brighter (or dimmer) in the early stages of the universe. The very fact that such an evolutionary factor must be considered illustrates the uncertainty with which the astronomer must regard this survey of radio sources. Perhaps we may confirm or deny the results of the Ohio survey by taking still another approach, one typified by the work of Allan Sandage and William A. Baum.

Let us recall how Hubble first recognized the relationship of red shift to distance. He plotted the red shift of galaxies against their apparent magnitude (see Figure 14.11). Extending this procedure to the more distant clusters of galaxies, astronomers have produced the plot shown in Figure 14.31. The ways in which four models of the universe (oscillating, parabolic, hyperbolic, and accelerating) have been interpreted by cosmologists are shown superimposed upon this plot. You can see that the plot of nearby galaxies (lower left-hand corner) is not sufficient to differentiate among the models. As more and more distant galaxies are being added, however, there appears to be a fairly close correlation with

Figure 14.31   *Hubble diagram—a plot of distant galaxies shows a fair amount of scattering among the four theoretical models of the universe: (a) oscillating; (b) parabolic; (c) hyperbolic; or (d) accelerating (steady state). Although there is some tendency at present to favor the oscillating model, more data are necessary for a conclusive decision.*

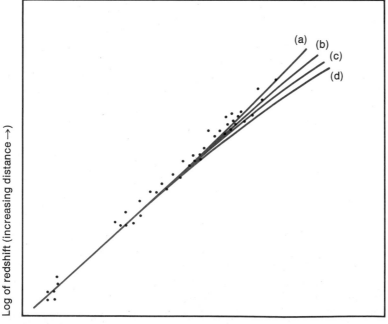

Log of redshift (increasing distance →)

Corrected apparent magnitude (decreasing brightness →)

the oscillating universe (positive curvature). Is this evidence conclusive? Perhaps not, for there is a fair amount of scattering in the points, and furthermore the location of the points on the graph is subject to correction with the advancement of techniques for measuring apparent magnitudes and for correcting these measurements for various effects. There is little chance, however, that the plot will change sufficiently to correspond to the accelerating universe. This observation has virtually sounded the death knell for the steady state theory.

We may summarize the primary task of the cosmologist as that of finding two numbers: the Hubble "constant" and the actual rate at which it is changing. The search is challenging the very limits of the art today. However, still one other test may be applied.

Mass-Energy Density of the Universe

Einstein predicted in his theory of relativity that the curvature of space-time is proportional to the mass-energy density of the universe. It is possible to compute the density associated with each model, and these figures are tabulated in Table 14.2.

At present, the best estimate of the density of the universe is 3×10^{-31} g/cm^3, about one-thirtieth of the mass-energy necessary to create an oscillating condition. This estimate is based upon the mass of all luminous material (stars, luminous nebulae, and the like), together with an estimate of nonluminous gas and dust in the galaxies and of all cosmic rays. Recent X-ray studies suggest a background of X-ray radiation which may be originating from nonluminous sources. Infrared astronomers are identifying large numbers of cool red dwarfs, unknown a few years ago, all of which may add to the known mass of the universe. One of the most significant sites of yet undetected mass may be the black hole (see Chapter 12). Astronomers are on the threshold of new techniques and new discoveries which may provide clues to enough mass-energy to close the universe—that is, to provide enough self-gravity to eventually stop the expansion and create the oscillating condition. More mass must be found if this test is to be reconciled to the indications set forth in Figure 14.31. The search goes on, and we cannot say for sure which model represents the universe. For that matter, who is to say that any of the proposed models is correct?

Table 14.2   *Mass-Energy Density Associated with Various Models of the Universe*

COSMOLOGICAL MODEL	MASS-ENERGY DENSITY
Oscillating (Closed)	More than 10^{-29} g/cm^3
Parabolic (Flat)	10^{-29} g/cm^3
Hyperbolic (Open)	Less than 10^{-29} g/cm^3

(a)

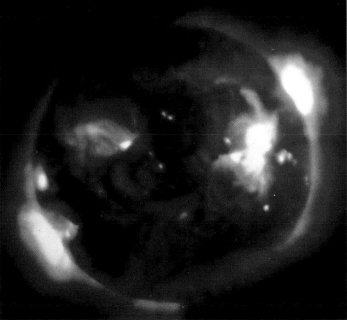

(b)

Plate 9 *(a) Total eclipse of the sun as seen on February 26, 1979, showing Baily's Beads. (Clifford Holmes, Riverside Astronomical Society) (b) An X-ray photograph of the sun taken aboard Skylab with the S-054 X-ray Spectrographic Telescope on May 28, 1973, revealing the corona with temperatures ranging upward to 1 million degrees Kelvin. The photo shows the whole range of coronal features in a broad spectral range. The active regions, bright points, interconnecting loops, filament cavities, coronal holes, and other features seen in the photograph are produced by the interaction of the sun's magnetic field and the ionized gas of the corona. (American Science and Engineering, Inc./NASA)*

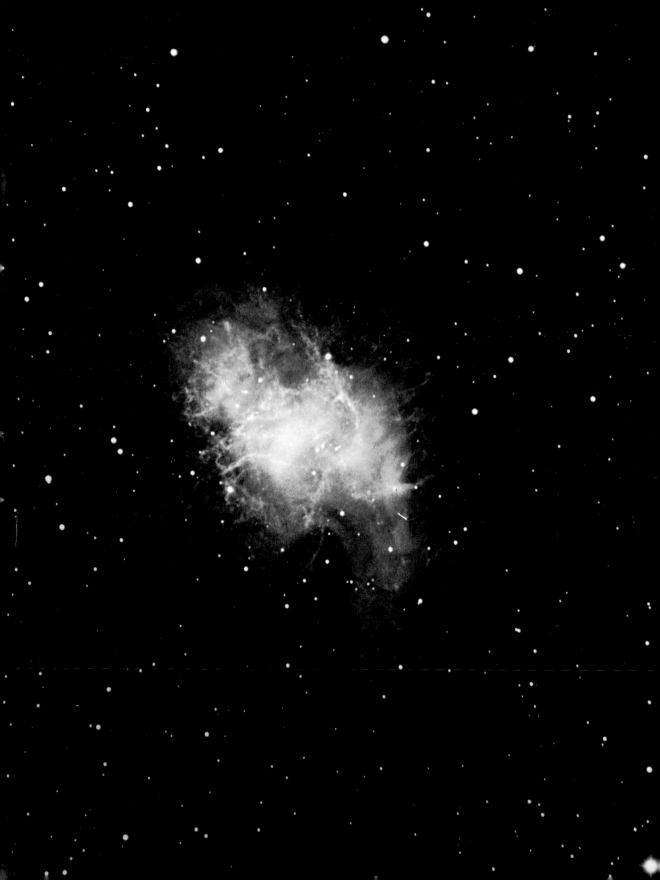

(a)

Plate 12 (a) Large Magellanic Cloud and (b) Small Magellanic Cloud—galaxies neighboring our own Milky Way but which are visible only from southern locations. (Cerro Tololo Inter-American Observatory)

(b)

Plate 13 Eta Carinae (NGC 3372), an emission nebula seen in the southern sky, as photographed in 1974 by the 1.5-m Schmidt camera. This emission nebula resulted from the explosion of a supernova recorded in 1843. Embedded in this nebula are numerous hot blue stars; the ultraviolet radiation of these stars excites the atoms of the nebula, causing it to emit light. (Cerro Tololo Inter-American Observatory)

Plate 14   The Trifid Nebula (M20; NGC 6514), a gaseous emission nebula in Sagittarius. A 4-m telescope photo taken in 1974. (Kitt Peak National Observatory)

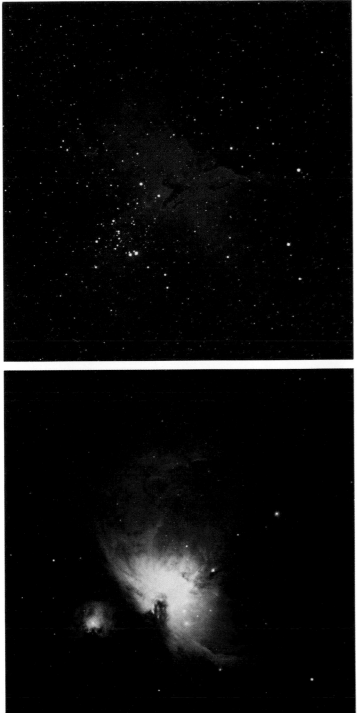

(a)

(b)

Plate 15   (a) Gaseous emission nebula and an associated galactic cluster in Serpens (M16;NGC 6611). Compact dust clouds, seen as dark globules, may be stars forming. Strong turbulence has been detected in the dark lanes. A 4-m telescope photo taken in 1974. (Kitt Peak National Observatory) (b) The Great Nebula (M42; NGC 1976), a gaseous emission nebula in Orion. Visible to the naked eye in Orion's sword, M42 has been known since the beginnings of recorded astronomy. Composed of mainly hydrogen gases, the Orion Nebula has a critical density about that required for star formation, and some star formation is believed to be occurring here. A 4-m telescope photo taken in 1974. (Kitt Peak National Observatory)

Plate 16   The Great Galaxy in Andromeda (M31; NGC 224), with satellite galaxies NGC 205 and 221. (Hale Observatories)

Deuterium Abundance

On page 453 we discussed the formation of deuterium as a part of the first few minutes following the Big Bang event. Because the percentage of material which remained in the form of deuterium (heavy hydrogen) depended upon the density of that early universe and because its present density is also related to that early density, the percentage of deuterium present in the universe today constitutes a test for cosmology. The deuterium abundance found indicates that the early density was too low to yield a closed universe, hence by this test it is "open" again. These calculations presume that no deuterium has been formed in a lasting way since the Big Bang.

The case for an open universe appears strong; however, most astronomers would agree that the tests are not conclusive. Certainly this question will not be settled for many years.

Each of these models corresponds to a particular state of space-time curvature as defined by Einstein; we shall discuss these states in the next chapter.

Questions

1. Astronomers estimate that the Milky Way galaxy contains at least 100 billion stars. How many galaxies do they estimate exist in the universe?

2. Sketch the following types of galaxies: (a) Sa, (b) Sc, (c) SBb, (d) E3.

3. What distinguishes an S0-type galaxy from the highly flattened elliptical galaxy or the Sa-type spiral galaxy?

4. What general type of galaxy is most prevalent in our local group?

5. In order to use a Cepheid variable to find the distance to a galaxy, what must the observer be able to determine in relation to that Cepheid?

6. Under what conditions is it safe to assume that because a galaxy looks dimmer, it must be farther away?

7. What relationship did Edwin Hubble find between the distances to galaxies and the red shift in their spectra?

8. Can we apply the red-shift law which Hubble developed to every distant object in the universe without any question as to its validity?

9. Normal galaxies emit much less radio energy than light energy; however, certain types of galaxies have a very high output of radio energy. Name several of these types.

10. Is it possible for a radio galaxy to emit as much as a million times as much radio energy as a normal galaxy?

11. What does the contracted word *quasar* stand for in full?

12. What characteristics do all quasars have in common?

13. What evidence exists for the idea that the quasars are the most distant and therefore the brightest objects ever sensed by astronomers? What evidence is there that quasars are relatively nearby?

14. If the quasars are very distant objects, then what may be their source of energy?

15. In what sense is time related to distance in all astronomical observations?

16. Is there more evidence that elliptical galaxies evolve into spirals or that spirals evolve into ellipticals? Or do galaxies not evolve except in their earliest stage of formation? Substantiate your answer.

17. What factor probably determines the ultimate shape of a galaxy?

18. If our universe is expanding, as indicated by the fact that all distant objects have red shifts, are we necessarily at the center of that universe? Would you still see the same expansion from any point of view in the universe? See if you can prove your answer by a drawing.

19. What must be happening in the universe in order to satisfy the steady state theory?

20. Within which theories of cosmology does the density of the universe change?

21. What recent findings tend to substantiate the big bang theory?

Suggested Readings

ALFVÉN, HANNES, *Worlds–antiworlds.* San Francisco: W. H. Freeman & Company Publishers, 1966.

BONDI, HERMANN, *Relativity and common sense.* New York: Doubleday, 1964.

DISNEY, M. J. AND VERON, PHILIPPE, BL lacertae objects. *Scientific American 237* (2), 32–39 (1977).

GORENSTEIN, P. AND TUCKER, W., Rich clusters of galaxies. *Scientific American* 239 (5), 110–128 (1978).

GOTT, J. R., III, GUNN, J., SCHRAMM, D. N., and TINSLEY, B. M., Will the universe expand forever? *Scientific American* 234 (3), 62 (1976).

GROTH, E. J., JAMES, P., PEEBLES, E., SELDNER, M., and SONEIRA, R. M., The clustering of galaxies. *Scientific American* 237 (5), 76 (1977).

KELLERMAN, KENNETH I., Extragalactic radio sources. *Physics Today* 26 (10), 38–47 (1973).

MARDER, L., *Time and the space-traveller.* Philadelphia: University of Pennsylvania Press, 1971.

MILLER, RICHARD A., The cosmic background radiation and the new aether drift. *Scientific American* 238 (5), 64 (1978).

SANDAGE, A., SANDAGE, M., AND KRISTIAN, J. (eds.), *Galaxies and the universe.* Chicago: University of Chicago Press, 1976.

SANDAGE, ALLEN R., Cosmology: A search for two numbers. *Physics Today* 23 (2), 34–41 (1970).

SCHATZMAN, E. L., The structure of the universe. New York: World University Library/McGraw-Hill, 1968.

SCHRAMM, DAVID N., The age of the elements. *Scientific American* 230 (1), 69–77 (1974).

SCIAMA, D. W., *Modern cosmology.* New York: Cambridge University Press, 1972.

STROM, S. E. and STROM, K. M., The evolution of disk galaxies. *Scientific American* 240 (4), 72 (1979).

WOLTJER, LODEWIJK (ed.), *Galaxies and the universe.* New York: Columbia University Press, 1968.

FIFTEEN

REL/ATIVITY

Man is continually refining his description of the universe, if not turning it upside down. The description that prevailed during the eighteenth and nineteenth centuries was due largely to the work of Sir Isaac Newton. The physical laws that he set forth in the late seventeenth century seemed to ring true for over 200 years, and they still serve to describe our everyday experiences. Newton had asserted, "I hold time and space to be absolute." By this he meant that two (or more) observers would always get the same answer when measuring the time and/or space interval between two events, even if the observers were moving with respect to each other. He thought that the time and the distance between two events could be treated separately, each being independent of the relative motion of the observers. Let us see why these assumptions are incorrect.

Relative Motion

Today we live with many aspects of relativity in our ordinary experiences. Consider a bullet fired from a hunter's gun. The bullet leaves the muzzle with a velocity of 200 m/sec. Now suppose that the same hunter were traveling in a car at 25 m/sec and fired his gun in the direction of the car's motion. What velocity would an observer riding in the car assign to the bullet? He would assign a velocity of 200 m/sec, because he would be moving with the gun. What velocity would an observer on the ground assign to the bullet? Within the limitations of his ability to measure, apparently he would assign simply the algebraic sum of the two velocities, 200 m/sec plus 25 m/sec, yielding 225 m/sec. This simple algebraic approach illustrates the Newtonian relationship mentioned above, and only as we approach very high velocities (representing a significant fraction of the speed of light) will it become obvious that this approach is incorrect.

Why did the two observers describe the velocity of the bullet differently? The reason is that they were in motion relative to each other; that is, they had different frames of reference. The observer in the car thought of the moving car as his frame of reference. In fact, he could just as well have thought of the car as "at rest," with the ground moving toward the rear of the car at 25 m/sec. The observer on the ground chose to think of the earth's crust as his frame of reference, and he described the car as moving 25 m/sec in a forward direction. Neither observer should be reprimanded for his view, for either one is acceptable. Indeed, these two frames of reference do not exhaust the possibilities.

Consider the fact that a point near the equator of the earth has a velocity of approximately 400 m/sec due to the earth's rotation and that the earth moves in its orbit around the sun at approximately 30,000 m/sec. An observer stationed at a point in space "above" the solar system could assume an entirely different frame of reference from those mentioned previously. The velocity of the bullet, as measured by this observer, within the limits of his ability to measure it, would still *seem* to be a simple algebraic sum of the four velocities mentioned thus far. What experiment would show that this answer is incorrect? Let the gun become a powerful searchlight in the nose of a rocket and let the car become that rocket ship (see Figure 15.1). Light travels 300,000 km/sec, and let us suppose the rocket ship is traveling toward earth at the rate of 100,000 km/sec (although this is not yet possible under modern technology). According to our ordinary experience, we would think that an observer on earth would measure the velocity of the light from the rocket as 400,000 km/sec, the algebraic sum of the two velocities. However, when the velocity of the light is measured by the observer on the ground it is found to be 300,000 km/sec, the same as that measured by an observer in the rocket. The velocity of the vehicle that carries the source of light adds nothing to the measured velocity of that light.

To see how this phenomenon was discovered, consider the late nineteenth-century experiment of Michelson and Morley. Observers of the eighteenth and nineteenth centuries, upon realizing that light is a wave phenomenon, were convinced that electromagnetic disturbances needed a medium through which their wavelike nature could be propagated. They called this medium the *ether* and thought of it as a frame of

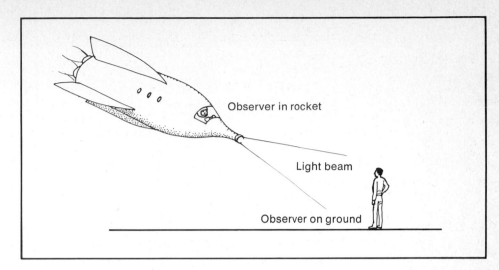

Figure 15.1 *Both the observer in the rocket and the observer on the ground measure the same speed of light.*

reference through which the earth (and other celestial bodies) moved. In 1881, Michelson tried to measure the velocity of the earth through the ether. He repeated his experiment with Morley in 1887.

Michelson-Morley Experiment

To illustrate the idea behind the experiment, consider a boat that makes different speeds, relative to the land, depending upon its direction of travel on a river (see Figure 15.2). Suppose the river flows steadily at 3 km/hr and that the boat can make 5 km/hr in still water. When going upstream, the boat will make only 2 km/hr (5 − 3), relative to the land. Going downstream, it will make 8 km/hr (5 + 3), relative to the land. Going across the stream, the boat has to head slightly upstream to compensate for the drift, and its velocity is computed by the Pythagorean relationship, $v^2 = (5)^2 - (3)^2 = 25 - 9 = 16$, $v = 4$ km/hr. These velocities are borne out by actual experiments, and the relationships that are expressed thereby seem to hold true for velocities encountered in such a situation (see Figure 15.2).

Note this interesting fact: If the boat made a trip of 8 km upstream and then returned, 5 hours would be required:

$$\text{time upstream} = \frac{8\text{ km}}{2\text{ km/hr}} = 4\text{ hr}$$

$$\text{time downstream} = \frac{8\text{ km}}{8\text{ km/hr}} = 1\text{ hr}$$

$$\text{round-trip time} = 5\text{ hr}$$

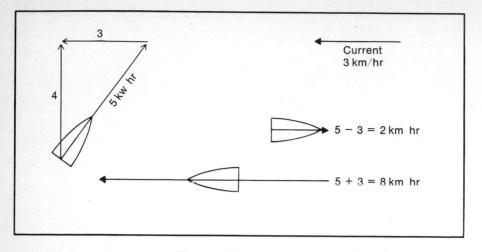

Figure 15.2 *When a boat moves upstream, its speed is slowed by the current; when it moves downstream, its speed is increased by the current, relative to the ground. In crossing the river, the boat must head slightly upstream in order to travel directly across the river.*

However, if the same distance were covered across the river and back, it would require only 4 hours:

$$\text{time across} = \frac{8 \text{ km}}{4 \text{ km/hr}} = 2 \text{ hr}$$

$$\text{time back} = \frac{8 \text{ km}}{4 \text{ km/hr}} = 2 \text{ hr}$$

round-trip time $= 4$ hr

Michelson and Morley, taking this same line of thought, devised an instrument whereby they treated the rate at which the earth "flows" through the ether as the rate of the river and the speed of light as the speed of the boat in still water. In Figure 15.3 you can see that they caused the light to reflect back and forth and in certain cases to be partially transmitted through half-silvered mirrors. If the ether really does exist and the earth moves through it, then the upstream-and-downstream round trip should take longer than the across-stream round trip. As the two beams are brought back to the eye, they interfere with one another, producing alternate light and dark bands. If the device is rotated 90°, interchanging the roles of the two beams, then one would expect a change in the interference pattern, because the time required for each beam to travel the set distance has been changed. The result of the Michelson and Morley experiment, to everyone's amazement, was that the interference pattern remained unchanged, even though their device

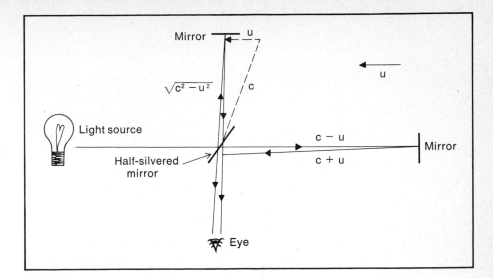

Figure 15.3 *The Michelson-Morley experiment.*

was capable of showing the expected change. The velocity of light was constant—it was not affected by the relative motion of the source. The results of this experiment produced a revolution in thought, eventually leading to the full-blown theory of relativity.

The Lorentz Contraction

Some thinkers remained bent on preserving the idea of the ether. Typical of this school were George F. Fitzgerald (1851–1901) of Ireland. and Hendrik A. Lorentz (1853–1928) of Holland, who, working independently, suggested that any device moving through the ether was shortened in the direction of its motion by just the amount necessary to allow the light to travel up and back in the same time as across and back. This *Lorentz contraction* is expressed mathematically as follows:

$$l = l_0 \sqrt{1 - \frac{v^2}{c^2}}$$

where l_0 is the length "at rest" relative to the earth, v is the velocity of the earth through the ether, and c is the speed of light. Lorentz and Fitzgerald held such a change in length to be a physical contraction caused by the pressure of passage through the ether. While recognizing the mathematical accuracy of the Lorentz contraction, Albert Einstein would offer still another explanation for its cause.

Figure 15.4   *Albert Einstein.*

Albert Einstein

Einstein's name is synonymous with the term *relativity* and rightly so, for he was truly interested in separating aspects of observation that are relative to the observer from those that are independent of the observer and therefore are "absolute." Einstein (Figure 15.4) felt no commitment to preserving the idea of the ether; rather, he postulated that the speed of light in a vacuum is constant, regardless of the motion of the source or of the observer. Thus it would be impossible to detect the motion of the earth in its orbit by means of an experiment using light. This posulate could explain the negative results of the Michelson-Morley experiment. However, Einstein extended his idea to a more fundamental assumption:

that no experiment can be performed to demonstrate the motion of an object if that object is moving uniformly (at a constant speed in a straight line).

Suppose, for example, you were riding in a train car that had no windows and that rode so smoothly that you felt no bumping or lurching. How would you know that you were moving with respect to the ground? You might try bowling a ball down the aisle to strike a set of pins at the end of the car; however, nothing about the car's motion would be revealed by such an experiment, for the ball would seem to move in precisely the same way as if the car were standing still, and the pins would be struck with the same force. One might conclude that the laws of nature, inside a moving car, are just the same as those experienced in a stationary car. This was Einstein's second postulate: The laws of nature are the same for observer A and for observer B, even though they are moving uniformly in relation to one another. If this assumption is to hold true and the speed of light is to remain constant, then the measure of certain other quantities must differ for observers who are moving relative to one another. Let us look now at the interrelationship between two such quantities—space and time.

Space-Time

In everyday life we think of space measure as a three-dimensional scheme, quite independent of the concept of time. We recognize that both the three dimensions of space and the single dimension of time are required to specify an event (to tell where and when an event happens), and in a sense we may have thought of time as a fourth dimension. Einstein went further. He recognized that the spacelike separation and the timelike separation of two events are totally dependent upon one another and cannot be treated independently. Thus Einstein conceived of a continuum of space-time. As we present the equations for time and space measurement, you will see this interdependence and the relationship of each to the relative motion of the observer.

A Thought Experiment

Consider the following thought experiment, even as Einstein must have done. Because light is the carrier of information, if you could travel away from a clock tower at the speed of light, no new information could reach you about that clock. The hands of the clock would seem frozen in place. Time would have stopped for you, as measured by that clock. If this is true, then an observer traveling at a speed a little less than the speed of light, relative to the clock tower, will see the hands of that clock move more slowly than does the person who stands at its base. On the other hand, if the observer who is moving rapidly away from the clock tower is wearing a wristwatch, he or she will see that watch running at its usual pace, whereas the person at the base of the tower will see that watch running slowly. Which observer is correct? Both are correct, for there is no absolute time by which their observations can be compared. Rather the rate at which time is measured, in any given frame of reference, is relative to the motion of the observer with respect to that frame. If an observer moves with a clock, the time it keeps can be labeled its "proper time"; however, if the observer moves in uniform motion, with velocity v, with respect to the clock, its rate will be slowed—its minutes and hours lengthened (dilated)—as follows:

$$t = \frac{t_0}{\sqrt{1 - \frac{v^2}{c^2}}}$$

where t_0 is proper time and c is the speed of light. Time dilation has been observed directly in the case of radioactive subatomic particles. When moving at high velocities, relative to the observer, they decay more slowly than does a like particle which is at rest relative to the observer. Since the radioactive decay rate of a given particle is as regular as the most accurate clock, we say that its clock runs slow from our point of view owing to its high relative velocity.

Based on his fundamental postulates, and working independently of Lorentz, Einstein also derived a relationship between the "proper" length of a rod and the length which would be measured by an observer who was moving uniformly at velocity v with respect to that rod. His results were in agreement with those of Lorentz:

$$l = l_0 \sqrt{1 - \frac{v^2}{c^2}}$$

where l is the measured length, l_0 is the "proper" length, and c is the speed of light. Einstein said simply that two observers, moving uniformly relative to each other, will *measure* the passage of time and distance between two points differently, hence time and distance are relative, whereas the speed of light and the laws of nature are not relative. How much shorter is the length of a given rod, as measured by an observer who is moving past it at half the speed of light, than its proper length as measured by an observer moving with the rod?

$$l = l_0 \sqrt{1 - \frac{v^2}{c^2}}$$

Let $v = \tfrac{1}{2}c$:

$$l = l_0 \sqrt{1 - \frac{(\tfrac{1}{2}c)^2}{c^2}}$$

$$l = l_0 \sqrt{1 - \tfrac{1}{4}}$$

$$l = l_0 \sqrt{\tfrac{3}{4}} = 0.866 l_0$$

The rod measures approximately 13 percent shorter for the moving observer. If you could move down a street at half the speed of light, all the buildings would appear skinny and tall, for in the direction of your travel the widths of the buildings would diminish by 13 percent while their heights would appear normal to you. The contraction effect applies only to the direction of relative motion.

One further quantity must be considered: that of mass. It has been effectively demonstrated that the mass of an object increases with higher and higher velocities relative to the observer. Electrons have been accelerated to speeds close to that of light, at which speeds they become harder and harder to accelerate, indicating that their mass is significantly greater than when they were "at rest" relative to the observer. At velocity v with respect to the observer, mass (m) is given by

$$m = \frac{m_0}{\sqrt{1 - \dfrac{v^2}{c^2}}}$$

where m_0 is the "rest mass" of the object and c is the speed of light; hence mass is also a relative quantity. What is the source of this increase in mass? To continuously apply a force, energy must be continuously expended. As the object gains velocity, relative to the observer, it also gains kinetic energy. Einstein recognized an equivalency between energy and mass when he stated: $E = mc^2$. We see that equivalency here, in that an object moving relative to the observer not only possesses its rest mass but gains mass due to its energy of motion. As its velocity approaches the speed of light, the term $\sqrt{1 - (v^2/c^2)}$ tends toward zero $[1 - c^2/c^2 = 1 - 1 = 0]$, hence m tends toward infinity (a very large value). As the mass approaches infinity, more and more energy is required to produce any additional velocity. Thus the speed of light seems to be a limit beyond which no material object can travel.

The equivalency of mass and energy is also demonstrated in the sun's conversion of a portion of its mass to energy every second (see Chapter 7) and in the process whereby photons and neutrinos (massless entities) are transformed into particles which possess rest mass (see Chapter 14, page 453). In a less obvious way, we add mass-energy to our watch every time we wind it, and it loses mass-energy as it runs down.

The concepts we have presented thus far are expressions of Einstein's 1905 theory of special relativity—"special" in the sense of the restrictions he placed upon its application. It describes only those situations in which two observers are traveling at constant velocities with respect to each other, and it does not cover the highly probable situation in which one observer is accelerating with respect to the other. Remembering that since acceleration can take several forms—speeding up, slowing down, or changing direction—we need only think of the influence of gravity to envision situations where Einstein's special theory would not apply.

In 1915, Albert Einstein made one of the most significant contributions ever made to science. His general theory of relativity is a fundamental attempt to explain gravity by means of equations which describe how space-time is affected by a gravitational field and how objects will move in such a field. We stated some of the basic assumptions of this theory in the last chapter (page 452), and we saw how solutions to Einstein's equations predicted events we now characterize as the Big Bang birth of the universe (page 453). To present the equations or to attempt to solve them would be beyond the level of this text; in fact their solution has challenged the finest minds of the world. Let us be content to review some of the predictions which flow out of certain solutions and see how they have been confirmed.

The general theory predicts that light will appear to be bent as it passes a massive object—or, more accurately, space-time will be warped near a massive object in such a way that a light ray which follows the most efficient path will appear to us to be bent. Its most efficient path is not a straight line as in unwarped space. This predicted effect of gravity was proven true in 1919 when, during a total solar eclipse, stars near the sun were photographed. When the apparent positions of these stars were compared to their positions photographed months earlier, small deviations in position were observed (see Figure 15.5). The light from stars that appeared near the edge of the sun had been bent by an angle which was within 15 percent of the predicted value (1.75 seconds of arc). Einstein was elated by this observation.

A byproduct of the Viking mission to Mars was a confirmation of this same phenomenon. When Mars passed behind the sun (at conjunction), its radio communication signal passed near the sun, and a very

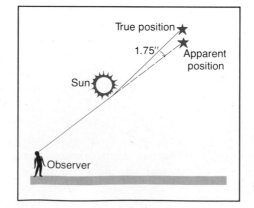

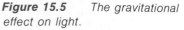

Figure 15.5   *The gravitational effect on light.*

careful observation which utilized Viking landers, Viking orbiters, and two radio antennas on earth confirmed Einstein's prediction to an accuracy of 1 percent.

Moving beyond Kepler's description of the elliptical orbits of the planets, Einstein's equations predicted a slight deviation from the perfectly closed orbit. They predicted that the perihelion (closest point to the sun) of each planet's orbit would gradually occur in new places. This effect, referred to as the precession of the perihelion, is illustrated in Figure 15.6 for Mercury, which shows the effect most obviously. Mercury's perihelion point moves through an angle of 5600 seconds of arc per century—a phenomenon that had long been known but remained unaccounted for. This precession is in exact agreement with Einstein's equations.

Also coming out of the general theory is the prediction that massive stars will collapse in their old age and become so dense as to produce a surface gravity that will prevent any form of radiation from leaving the star, since photons of electromagnetic radiation possess mass and experience a gravitational attraction for the body which generates them. When radiation can no longer escape such an object, then it becomes a black hole. Astronomers are quite certain that such objects exist; an example is Cygnus X-1.

Lest we conclude that the general theory is proven beyond any shadow of doubt (something that never happens in science), we should note that there may be alternate explanations for any or all of these experimental results. In fact, rival theories concerning gravitation do exist, and their predictions are quite similar to Einstein's. Experimental evidence is not yet sufficiently accurate to allow us to test one against the other and decide among them. We may, however, have a great deal of confidence in both the special and general theories.

The one expressed goal of Einstein's career was to develop a unified field theory which would unite all the forces of nature with the particles of nature and provide a unified explanation for both the microworld of the atom (a quantum approach) and the macroworld of the universe (and the "in between" where we live). This goal he failed to achieve, but it has motivated a great amount of human effort in that direction, and many workers feel that we are moving close to such an achievement.

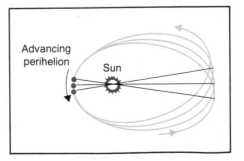

Figure 15.6 *The advancing perihelion of Mercury.*

Questions

1. If a train travels eastward at 80 km/hr and a passenger runs down the aisle toward the rear of the train (westward) at 6 km/hr, how will someone on the ground describe the passenger's motion as he passes?

2. How will someone seated in the train describe the runner's motion in Question 1?

3. Why are the answers to Questions 1 and 2 different?

4. If modern rockets (and/or space probes) can achieve speeds no greater than 30 km/sec, what fraction of the speed of light does this represent?

5. When the Michelson-Morley apparatus was rotated 90°, the roles of the beams were interchanged. Explain what this sentence means.

6. What changes did Lorentz predict in moving objects?

7. Einstein said that it is impossible to determine the motion of an object, say a smoothly riding railroad car with no windows, by performing an experiment in that car. What would happen if one threw a ball straight upward in a moving car? Wouldn't it fall toward the rear of the car? If not, why not?

8. Einstein's correction for the observed length of a rod, moving with respect to the observer, appears to be the same (mathematically) as that of Lorentz. What is different, however, about their interpretation of the meaning of that correction?

9. Determine the measured length of a rod, which has a "proper" length of one meter, if the observer is moving at three-fourths the speed of light ($\frac{3}{4}c$) with respect to the rod.

$$[Answer: \frac{\sqrt{7}}{4} \text{ m.}]$$

10. How does the equation $E = mc^2$ relate to the sun's source of energy?

11. What limitation is placed on the special theory of relativity?

12. If scientists placed a very accurate (radioactive) clock on board the fastest jet and flew it around the earth, leaving a twin clock at home, how would they compare upon return to earth?

13. List several specific predictions of Einstein's general theory that have apparently been demonstrated.

Suggested Readings

BRONOWSKI, J., *The ascent of man*. Boston: Little, Brown, 1973.

CALLAHAN, J. J., The curvature of space in a finite universe. *Scientific American* 235 (2), 90 (1976).

CASPER, BARRY M., and NOER, RICHARD J., *Revolution in physics*. New York: W. W. Norton & Co., Inc., 1972.

EINSTEIN, ALBERT, *The meaning of relativity*. New York: Princeton University Press, 1956. (A classic.)

FRENCH, A. P., *Special relativity*. New York: W. W. Norton & Co., Inc. (paper), 1968.

GARDNER, MARTIN, *The relativity explosion*. New York: Vantage Books, Random House, 1976.

KAUFMANN, WILLIAM J., *The cosmic frontiers of general relativity*. Boston: Little, Brown, 1977.

MOLLER, C., *The theory of relativity*. Oxford: Clarendon Press, 1972.

RUSSELL, BERTRAND, *ABC of relativity*. London: George Allen and Unwin, Ltd., Ruskin House, 1969.

TAYLOR, E. F., and WHEELER, J. A., *Spacetime physics*. San Francisco: W. H. Freeman & Company Publishers, 1966.

TREFIL, J. S., It's all relative when you travel faster than light: Tachyons. *Smithsonian* 7, 132–139 (1976).

VAN FLANDERN, T. C., Is gravity getting weaker? *Scientific American* 234 (2), 44–56 (1976).

WILLIAMS, L. PEARCE (ed.), *Relativity theory: Its origins and impact on modern thought*. New York: John Wiley, 1968.

SIXTEEN

EXTRATERRESTRIAL LIFE

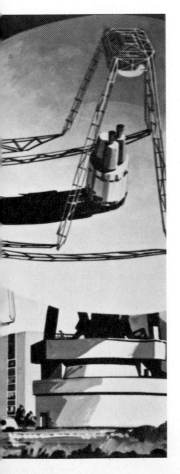

It would never have occurred to an ancient observer to ask, "Am I alone in the universe?"—for his or her universe consisted of only a flat earth with a canopy of stars overhead. Not until humans realized that the sun is a star, and a rather ordinary star at that, did they begin to see the possibility of other planetary systems. But what is the probability that other systems exist? To answer this question we must look to theories concerning the origin of stars in general.

Astronomers believe that stars form as a result of condensations within vast clouds of gas and dust, owing largely to gravity but perhaps triggered by shock waves and influenced also by magnetic fields acting on plasma. As a huge gas and dust cloud condenses, it spins faster and faster (assuming that at least a small rotational component exists in the original cloud) and flattens to form a disk of material surrounding the protostar (a star in the making). Then, depending upon the protostar's mass, which ultimately determines its temperature, planets may form in the disk as we believe they did surrounding the sun. In the case of very massive stars, which are also very hot, materials may be blown away by the solar wind before planets can form. In the case of low-mass stars, the star may remain so cool that, even if planets formed, life could not exist on them. So it seems reasonable to assume that only stars having about the mass and temperature of the sun may have planets circling them, some of which are suitable for life. Out of the 100 billion stars in the Milky Way galaxy, it seems reasonable to assume that at least 100,000 are similar to the sun in mass and surface temperature. Current observations seem to confirm this speculation in that numerous examples of stars with a temperature similar to that of the sun have been found. Some of these stars appear to have dark companions—perhaps planets in orbit.

As in our own solar system, not all planets are habitable. Those closer to the sun than the earth are far too hot, and those beyond Mars are far too cold, so any system of planets must have a limited habitable zone. This conclusion is based on the very narrow view that life must be like our own. If we were to take a broader view of life, then much larger probabilities open up. Even within our present knowledge we should broaden our estimate a bit, for microorganisms exist in highly unlikely places. One species is able to live within the fuel tanks of jetliners, subsisting on kerosene. Others are known to survive a wide range of temperatures.

In order to fully consider the question of life in the universe, we must look more closely and ask, "What is life?" What makes the living essentially different from the nonliving? We think of the ability to reproduce like kind as being unique among living organisms, yet certain nonliving compounds are able, in a sense, to reproduce themselves. Likewise certain living organisms, such as viruses, may lie dormant (appearing to have no reproductive process) for many years, then begin replication again. Perhaps the division between the living and the non-living is not as sharp as one might think. We do know that in all earth-forms that are clearly alive, a code within each cell directs the replication process. The encoding molecule is called deoxyribonucleic acid (DNA) (Figure 16.1). Should we impose this restriction on life elsewhere in the universe? Must life be as we know it—essentially based on organic molecules known as carbon rings? Perhaps living organisms in other solar systems sustain life on silicon molecules or others.

Besides self-replication, one further process seems to be character-istic of living organisms: a process whereby certain materials and/or energies are utilized from the environment to sustain life. Examples are the process of photosynthesis in plants and various metabolic processes in animals. In our search for life on planets of our solar system, we must use our knowledge of such processes. For instance, the primary search for life on Mars (via the Viking probes) attempted the detection of gases typi-cally released as a product of metabolism. On the other hand, if we are to search for life on planets orbiting other stars, we must look to another characteristic of at least some living creatures, namely intelligence. Only if we can communicate with other intelligent beings, using some form of electromagnetic energy, can we traverse the great distances involved in our quest. In fact, for communication to be possible we must assume that intelligent beings on another planet have a radio technology equating or

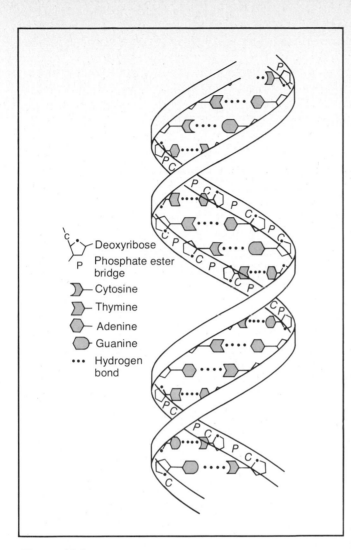

Legend:
- C — Deoxyribose
- P — Phosphate ester bridge
- Cytosine
- Thymine
- Adenine
- Guanine
- ••• Hydrogen bond

Figure 16.1 *The DNA double helix molecule.*

surpassing our own. That is, they must possess radio telescopes, transmitters, and receivers at least as adequate as our own. This does not preclude the possibility that other intelligent beings may have advanced far beyond us in this and in many other respects. When we think of the history of radio technology in relation to the history of the earth, we realize that this technology has "peaked up" in only the last few years. Figure 16.2 suggests the technological explosion that is occurring in our own lifetime.

If we hope to communicate with another civilization, is it essential that its technology peak up at the same time as ours? Perhaps not; in fact it may be more advantageous if it is otherwise. To understand such a statement, let us review the relationship of time and distance. Radio

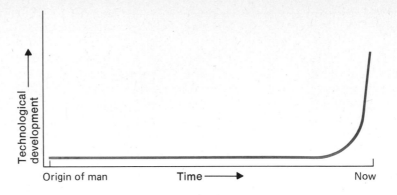

Figure 16.2   *The peaking of technological development.*

signals travel at the speed of light, hence if an intelligent community with radio capabilities were located at a distance of 10,000 light-years, their radio signal would require 10,000 years to reach us. If they had only recently developed radio capabilities and decided to beam a message in our direction, we would not receive it for another 10,000 years. Likewise our message, if sent today, would not reach them for 10,000 years. More advantageous would be the coincidence that the distant civilization perfected the radio 10,000 years ago and beamed a message then. We would just be receiving the information now. You can see how unlikely two-way communication seems to be, unless civilizations are very close. If you had only the possibility of one-way communication, within the span of several generations, what would you want to convey to an extraterrestrial civilization—or what would you hope they communicated to you? What code or language would you use? On what wavelength would you transmit and/or listen?

Perhaps one could overcome the language barrier by transmitting pictures, as we have done from space probes or as we have been doing for a quarter-century via television. Television and radio signals have been traveling outward from the earth for many years without any concerted effort to communicate with alien beings. Wouldn't it be a strange twist if those signals were the first to reach them? If we intentionally designed a visual message, it might include information as to our location in the universe, in the Galaxy, and in the solar system, the nature of our being (man and woman), and the nature of our technology, but perhaps more vital than any of these would be how we have solved our social problems so as to have survived so long. Perhaps a more advanced civilization might give us a clue as to how to solve some of the problems that most perplex us. Our civilization could never be the same after such a contact.

The question that remains, assuming we believe it worthwhile to attempt communication, is a choice of one or more wavelengths on which to broadcast. Certain wavelengths may be especially appropriate. The most prevalent element in the universe is hydrogen, and this atom emits a characteristic radio signal of 21-cm wavelength. Certainly any technical civilization would be aware of this fact and would most likely be listening at that wavelength, simply because so much can be learned about the universe in this way. The hydroxyl radical (OH) also produces a natural radio signal at 18.3 cm wavelength. It is interesting to note that many astronomers refer to this range of wavelengths between 21 cm and 18.3 cm as the "water hole" of space—the "place" where many different beings may gather to share their tales. Note that the combination of hydrogen (H) and the hydroxyl radical (OH) yields water (H_2O). Attempts are currently being made to reserve this band of wavelengths for a search for life among the stars, free from earthly communication interference.

The question, during the past few years, centers about the commitment we are willing to make to the search for extraterrestrial life through communication. One proposal included a complex of radio antennae like that shown in Figure 16.3. The estimated cost of such a project is between 6 and 10 billion dollars. A group of Soviet astronomers have proposed a huge radio antennae placed in solar orbit, capable of mapping the entire cosmos in three dimensions, to be used incidentally to detect the signals of alien beings. More reasonable would be the

Figure 16.3 *The Cyclops proposal for a radio telescope array. (NASA–Ames)*

commitment of a certain percentage of existing radio telescope time to this search. This we can do with only a reassessment of priorities.

We now have the technological skill to receive many channels of information simultaneously and, with the aid of properly programmed computers, to recognize signals from intelligent beings. If such a commitment is made, perhaps we will find that we are not alone in the universe.

Questions

1. Why does it seem reasonable to assume that many other solar systems may exist in the Milky Way galaxy?

2. If the probability of stars' having planets is one in a million, then what do you think may be the probability of life existing elsewhere in the universe? Substantiate your view.

3. What is the function of the DNA molecule in living cells?

4. Suppose intelligent beings existed on a planet 1000 light-years away and they possessed a radio technology compatible with ours. What time delays would affect the effort to carry on a dialogue? How would you solve the problem of such a delay?

5. Why would the 21-cm radio wavelength be a natural one for communication with extraterrestrial beings?

6. Devise a code or language with which one might communicate with extraterrestrial beings.

Suggested Readings

BRACEWELL, RONALD N., *Intelligent life in outer space.* San Francisco: W. H. Freeman & Company Publishers, 1974.

CANNON, W. H. and JENSEN, O. G., Terrestrial timekeeping and general relativity—A discovery. *Science 191*, 489–491 (1976).

KUIPER, T. B. H., and MORRIS, M., Searching for extraterrestrial civilizations. *Science 196*, 616–621 (1977).

MACVEY, JOHN W., *Whispers from space.* New York: Macmillan, 1973.

PONNAMPERUMA, C. and CAMERON, A. G. W., *Interstellar communication: Scientific perspectives.* Boston: Houghton Mifflin Co., 1974.

SAGAN, CARL, *The cosmic connection: An extraterrestrial perspective.* New York: Doubleday, 1973.

SAGAN, CARL, Seeking other worlds. *Newsweek 90*, 46–47 (1977).

STEENO, D., Extraterrestrials—Who are they? Where are they? *Space World* N-2-158, 41–42 (1977).

YOUNG, R. S., *Extraterrestrial biology.* New York: Holt, Rinehart & Winston, 1966.

Glossary

aberration of starlight The apparent displacement in the location of a star due to the orbital motion of the earth.

absolute magnitude The apparent magnitude of a star if viewed from a distance of ten parsecs.

absolute zero The temperature at which the linear motions of all molecules stop. A temperature of $0°K$ (Kelvin), equivalent to $-273°C$ (Celsius) and $-460°F$ (Fahrenheit).

absorption spectrum Dark lines on the background of a continuous spectrum.

acceleration The change in velocity. It may be an increase or decrease in velocity, or a change in direction.

achromatic lens A lens system composed of two or more elements that are designed to correct for chromatic aberration.

active sun The sun during times of unusually large numbers of sunspots, flares, and other events.

albedo The percentage of light that a planet or moon reflects.

alpha particle A positively charged particle that consists of two protons and two neutrons; hence, a helium nucleus.

altitude The angle at which an object appears above the horizon as measured along its vertical circle.

angle of incidence The angle between the incoming ray and the normal (perpendicular) to the reflecting or refracting surface.

angstrom (Å) A unit of length equal to 10^{-10} m, used to measure very small wavelengths.

angular diameter The angle that the diameter of an object makes as measured at the observer's eye.

angular distance The angle between two objects as viewed on the celestial sphere.

annular eclipse An eclipse of the sun that occurs when the apparent diameter of the moon is not as great as the apparent diameter of the sun, thus leaving a ring of sunlight showing around the moon.

antimatter Particles that appear to possess properties opposite to those of matter.

aphelion A point in the orbit of a planet at which it is farthest from the sun. (*Helios* is a Greek word for sun.)

apogee A point in the orbit of an earth satellite at which it is farthest from the earth. (*Ge, gee,* and *geo* are combining forms—suffixes or prefixes—meaning earth.)

apparent magnitude A measure of the brightness of a star or other celestial object as seen from earth.

apparent solar day The interval between two successive transits of the sun's center across the observer's meridian.

Arctic Circle The parallel of latitude 66.5° N. Within this circle the sun is not seen on the day of the winter solstice.

ascending node The point on the orbit of a body at which it crosses the celestial equator from south to north.

association A very loose cluster of stars that are thought to have a common origin.

asteroid A small body in orbit around the sun; a minor planet or planetoid.

astrology A study of the supposed influence of the positions of the sun, moon, planets, and stars upon human affairs.

astrometric binary A binary-star system in which only one component is visible but in which the presence of the second component is deduced from the perturbations (disturbances) that it produces upon the orbital motion of the first.

astrometry The branch of astronomy that is primarily concerned with the accurate measurement of positions and motions of stars.

astronomical unit (A.U.) The average distance between the earth and the sun. By international agreement, 1 A.U. = 92,870,000 miles, or approximately 149,790,000 km.

astronomy The science whereby celestial objects are described according to their location, motion, size, composition, and appearance.

astrophysics The branch of astronomy that applies the methods and tools of physics to the study of celestial objects.

atmospheric refraction The bending of light rays from celestial objects due to refraction by the earth's atmosphere. This phenomenon is most noticeable near the observer's horizon.

atom The smallest particle of an element that retains the properties which characterize that element.

atomic mass unit One-twelfth the mass of an atom of the most common form of carbon (^{12}C); approximately the mass of the hydrogen atom.

atomic number The number of protons in the nucleus of a given atom.

atomic weight The average mass of an atom of a given element as measured in atomic mass units.

aurora The light display that is produced by ionized atoms, usually in the polar regions; the northern and southern lights—*aurora borealis* and *aurora australis*.

autumn equinox The point on the celestial equator at which the sun crosses from north to south.

azimuth The angle measured eastward along the horizon from the north point to the vertical circle which passes through a given object.

Balmer lines The series of spectral lines, either bright or dark, that are produced by electron transitions up from energy level 2 or down to energy level 2, in the hydrogen atom. These lines lie in the visible portion of the spectrum.

barred spiral A galaxy characterized as having a "bar" (armlike extensions) through its nucleus. Spiral arms extend from the ends of the bar.

barycenter A point around which two objects that lie in each other's gravitational field seem to orbit. It constitutes a center of mass of the system.

beta particle A negatively charged particle; an electron.

"big bang" theory A theory concerning the evolution of the universe, which states that the expansion of the universe is the result of a primeval explosion.

binary star A double-star system in which the components revolve around a common point situated between them, their barycenter.

black body Theoretically, a body that is a perfect radiator, for instance, one that absorbs and reemits all radiation which falls on it.

black dwarf Thought to be the final stage in the evolution of some stars, a state in which all energy of the star has been exhausted and in which it no longer radiates.

black hole A star which has collapsed under the influence of gravity to such an extent that its surface gravity prevents further radiation of energy.

blink microscope An instrument in which two different photographs may be viewed alternately. If the two photographs represent the same region of the sky but were taken at differing times, then stars that have moved or changed in their apparent brightness may be easily recognized in the instrument. Stars that have moved will seem to jump back and forth. The image of stars that have varied in brightness will appear to change in size on the photograph.

Bode's law (More correctly called the *Bode-Titius relationship*.) A sequence of numbers that approximate the distances from the sun to the planets, measured in astronomical units.

Bohr atom A model of the atom, devised by Niels Bohr, that depicts the electrons in orbit around the nucleus.

bolide A very bright meteor or fireball, sometimes accompanied by a sound.

bolometric magnitude A measure of the total radiation of a star as received above the earth's atmosphere, measured in the full spectrum of electromagnetic radiation.

bright-line spectrum An array of colorful lines against a dark background, produced by an excited, low-pressure gas.

brightness A measure of the actual luminosity of an object.

calorie A unit of heat energy, the amount needed to raise the temperature of 1 g (1 cm³) of water 1°C.

candlepower A unit of light intensity.

carbon cycle The series of nuclear reactions, involving carbon, that transforms hydrogen into helium.

cardinal points The four main points of the compass: north, east, south, and west.

Cassegrain reflector A telescope that utilizes a convex secondary mirror to bring the light rays to a focus near the primary lens. In order that the image formed at this point may be viewed, a hole is made in the primary mirror and an eyepiece inserted there.

Cassini's division A wide gap in the ring system of Saturn, between the outer and middle ring.

celestial equator The projection of the earth's equator onto the sky; hence a great circle that is 90° from each pole in the sky.

celestial mechanics A branch of astronomy that deals primarily with the motions and mutual gravitation of objects in space.

celestial meridian The great circle through the celestial poles and the observer's zenith.

celestial navigation The art of finding one's way from one point to another on the earth's surface by means of observation of the position of the sun, moon, planets, and stars.

celestial poles Extensions of the earth's axis of rotation; points on the celestial sphere about which the sky appears to rotate daily.

celestial sphere An apparent sphere of very large radius, centered on the observer, upon which the location of the stars may be specified.

center of gravity The point through which gravity seems to act no matter how the object is turned. The center of mass of the object.

centripetal force A force, directed toward the center of curvature, that diverts a body from a straight path into a curved one.

Cepheid variables A class of pulsating stars that vary in light output.

Ceres The first asteroid (minor planet) to be discovered; also the largest known.

chromatic aberration A defect in a simple lens whereby light of different colors is brought to different focal points.

chromosphere That portion of the sun's atmosphere which lies directly above the photosphere.

circumpolar stars Those stars near the celestial poles which always appear above the observer's horizon.

cluster of galaxies A grouping of galaxies composed of hundreds or thousands of member galaxies.

cluster of stars A grouping of stars that is held together by mutual gravitation; they thus possess a common motion.

cluster variables A large class of pulsating variable stars that have a period of less than one day and are usually found in globular star clusters.

color index (C.I.) The difference between the photographic and the visual magnitudes of a star: C.I. = $m_p - m_v$.

coma A defect in a telescope by which the rays of light that enter the telescope at an angle to its axis are not brought to the same focal point but form *comma*-like (,) images.

coma (of a comet) The fuzzy, gaseous component of the comet's head.

comet A swarm of bodies composed of frozen gases with solid particles at their centers. Comets usually have very elongated orbits around the sun.

compound A substance composed of two or more elements.

concave lens or mirror A lens or reflecting surface curved somewhat like the inside of a sphere.

conduction The transfer of energy (usually heat energy) by the direct passing of energy from atom to atom.

configuration A particle form or arrangement of the sun, moon, planets, and/or stars.

conic section A curve formed by cutting a circular cone with a flat surface. Such a cut may produce a circle, ellipse, parabola, or hyperbola.

conjunction A lining up of celestial objects so that they appear to have the same right ascension.

constellation A group of stars that by its shape suggests an object, person, or animal. Today, a constellation includes a definite region of the sky around such a configuration.

continuous spectrum The uninterrupted band of color produced by a heated solid, liquid, or gas under pressure.

convection The transfer of energy (heat) by the motion of the medium that "carries" the energy.

convex lens A lens in which one or both surfaces are curved like the outside of a sphere and thus are thicker at the center than near the outside.

Copernican system A system of planets that revolve around the sun (heliocentric).

core The central portion of a planet or of any celestial body.

corona The outer atmosphere of the sun, seen only when the central disk of the sun is covered.

corona (of the galaxy) The halo or sphere of objects that surround the central nucleus of the Milky Way galaxy.

coronagraph An instrument for photographing the outer atmosphere of the sun by artificially covering the disk

of the sun's image at the focal plane of the telescope.

cosmic rays High-energy particles, largely protons, that strike the earth.

cosmogony The study of the origin and evolution of the material in the universe.

cosmological principle The assumption that, in general, the universe would appear the same from any location within itself.

cosmology The study of the organization and structure of the universe and its evolution.

cosmos The entire universe, seen as an orderly, self-inclusive system.

Coudé focus A system by which an arrangement of secondary mirrors directs the light gathered by the primary mirror along the polar axis of the telescope to the Coudé focal point. This point remains fixed as the telescope moves to follow a star.

crater A depression in the surface of the earth or moon.

crepe ring The innermost visible ring of Saturn; a dark ring.

crescent That phase of the moon which shows it less than half full; also applicable to Venus, or any planet when viewed from space.

crust The outer layer of the earth, moon, or other such body.

dark nebula A gas or dust cloud that obscures the light of stars and galaxies behind it.

declination The smallest angle between a given object and the equator.

deferent The larger orbit in the Ptolemaic system, along which the center of the epicycle of a planet supposedly moves.

deflection of starlight The bending of light as it passes close to a massive object; a bending due to gravity.

density The mass of an object divided by its volume and usually measured in grams per cubic centimeter.

descending node The point along the orbit of an object at which it passes from north to south of the celestial equator or of some other reference plane.

deuterium (Also called "heavy hydrogen.") A form of hydrogen in which the nucleus of each atom contains one proton and one neutron.

diffraction The absorption and reemission of light as it passes an object or passes through a small opening.

diffraction grating A system of finely ruled lines that diffract the light and by interference produce a spectrum.

diffraction pattern A series of bright and dark lines produced by the interference of light.

diffuse nebula A bright or dark nebula of irregular shape; not of the planetary form.

diffusion (of light) The scattering of light from an irregular surface.

direct motion The typical motion of

a planet—west to east—as seen against the background of stars.

disk (of our Galaxy) The flattened portion of the Milky Way galaxy; hence, the spiral arms.

disk (of a planet) The round shape of a planet as seen in a telescope; of measureable size.

dispersion The separation of white light into its component colors by means of refraction or diffraction.

diurnal circle The apparent path of an object in the sky during one day, due to the earth's rotation.

diurnal motion The daily apparent motion of all objects, due to the earth's rotation.

Doppler shift The change in observed wavelength of sound, radio, or light, due to the motion of the source, the observer, or both.

double star A star system that is composed of two stars, each influenced by the other's gravitational field. Their dual nature is revealed by telescopic or spectroscopic observation.

dwarf (star) A star of less than average mass and luminosity; a small star.

dyne A unit of force in the metric system—the force necessary to give a 1-g mass an acceleration of 1 cm/sec/sec.

earthquake A release of stresses that have been stored within the earth's crust, producing a movement of the crust.

earthshine (earthlight) That light which is reflected by the earth's atmosphere and which illuminates the dark portion of the moon.

eccentric A point within a circle but off-center.

eccentricity A measure of the degree to which an ellipse is elongated. It may be found by dividing the distance between the foci by the length of the major axis.

eclipse The partial or total darkening of an object that passes within the shadow of another object.

eclipse path The path on the earth's surface swept out by the shadow of the moon during a total eclipse of the sun.

eclipse season A period of time during which eclipses of the moon and sun may take place, approximately one month in duration and occurring approximately 6 months apart.

eclipsing binary (stars) A binary-star system in which the plane of revolution of the two stars is seen almost edge-on. The light of each star is periodically diminished by the passage of the other star in front of it.

ecliptic The plane of the earth's orbit projected onto the sky. The apparent path of the sun on the celestial sphere.

electromagnetic radiation A disturbance that is transmitted from its source to the observer by means of changing electrical and magnetic fields. Includes radio, infrared, visible light, ultraviolet, X rays, and gamma rays.

electromagnetic spectrum The full array of electromagnetic disturbances (radio, infrared, visible light, ultraviolet, X rays, gamma rays).

electron A subatomic particle that carries a negative charge and is thought to move about the nucleus of the atom.

element Any one of more than 100 fundamental substances that can not be broken down into simpler forms by chemical processes.

elements of an orbit Those particular quantities that describe the size, shape, and orientation of the orbit of a body in space. Such quantities are used to determine the location of the body at any given time.

ellipse A closed curve that may be obtained by passing a plane completely through a circular cone; a curve that describes the orbits of bodies in space.

elliptical galaxy A galaxy whose visible shape is that of an ellipse.

elongation The apparent angle between the sun and the specified object, measured at the observer's eye. (The *elongation* of Mercury is 14° when it appears in the sky 14° from the sun.)

emission line A bright spectral line produced by a downward transition of an electron.

emission nebula A bright cloud of gas that has been excited and is producing its own light.

emission spectrum A system of bright lines produced by an excited, low-pressure gas.

energy The capacity to do work.

energy level (in the atom) The possible energies that an electron may possess if excited by some external source such as light or an electrical current.

ephemeris A table that states the position of celestial bodies at various times.

epicycle In the Ptolemaic system, a small circle, the center of which moves along the deferent. The planet moves on the epicycle, thus the motion of the planet might be described as a circle moving on a circle.

equation of time The difference between the apparent solar time minus the average (mean) solar time; the amount of time by which the noonday sun appears on the observer's meridian, either ahead or behind the average noon time.

equator A great circle on the earth lying halfway between the poles of the earth.

equatorial mount A telescope mounting designed so that one axis is parallel to the earth's axis and hence only one motion is necessary to drive the telescope to compensate for the earth's rotation.

equinox Either of the two intersection points of the celestial equator and the ecliptic.

erg A unit of energy in the metric system; the amount of energy expended when a force of 1 dyne moves an object through a distance of 1 cm.

escape velocity The velocity at which a body overcomes the pull of gravity of another body and moves off into space.

evolution (of the cosmos) Progressive changes in the universe or in objects found in it.

excitation The process whereby the electrons of an atom are given greater energy than they normally possess.

extragalactic Beyond our own Milky Way galaxy.

eyepiece A lens system that magnifies the image formed by the objective lens at the prime focus.

filar micrometer A device attached to a telescope, at the eyepiece, to measure small angles between stars.

fireball A spectacular meteor usually visible for several seconds.

fission The breaking up of heavier atomic nuclei into two or more lighter atomic nuclei.

flare A sudden, temporary brightening of a given region on the sun's surface. It represents a tremendous outpouring of energy.

flare star A star that increases suddenly and unexpectedly in brightness.

flash spectrum A bright-line spectrum produced by the lower atmospheric "layers" of the sun and seen only for an instant before and after the total phase of a solar eclipse.

flocculi Bright regions usually just above sunspots and visible only in a spectroheliogram. (Also called *plages.*)

fluorescence Occurs when light of one wavelength is absorbed and then reemitted at another wavelength. Ultraviolet light may excite certain atoms to produce visible light.

focal length The distance from the center of a lens or mirror to the prime focus of the telescope.

focal ratio (*f*-number) The focal length of a lens or mirror divided by its diameter.

focus The point at which converging rays of light meet.

forbidden lines Spectral lines that are not usually produced under laboratory conditions but may be produced under certain conditions of space, for instance, very low pressure.

force The influence which can change the speed and/or direction of an object.

Foucault pendulum A device employing a swinging pendulum, used to prove the rotation of the earth.

Fraunhofer lines Absorption (dark) lines in the spectrum of the sun (or star).

frequency The number of waves that pass a given point per second.

fringes The alternate bright and dark regions produced by the interference of light.

full moon The phase of the moon that occurs when it is directly opposite the sun in the sky, revealing its fully lighted face.

fusion The process whereby heavier elements are created by means of nuclear reactions within lighter ones.

galactic center The point around which a given galaxy rotates. In the Milky Way galaxy, this point is located toward the constellation of Sagittarius.

galactic cluster An open cluster of stars situated in the spiral arms of the Galaxy.

galactic equator A great circle on the celestial sphere that indicates the plane of the Galaxy.

galactic poles Points on the celestial sphere 90° from the galactic equator.

galaxy (Capitalized when our own Galaxy, the Milky Way, is meant.) A fundamental collection of stellar material, usually containing millions to hundreds of billions of stars.

gamma rays Electromagnetic disturbances of wavelength shorter than X rays and carrying the most energy of all such disturbances.

gauss A unit of magnetic flux density; a measure of the strength of a magnetic field at a given point.

gegenschein (counterglow) A region directly opposite the sun that appears to glow, perhaps due to the reflection of sunlight from interplanetary particles.

geomagnetic field The earth's magnetic field.

geomagnetic poles The points on the earth's surface that seem to possess the same magnetic properties as the poles of a bar magnet.

giant (star) A star of very large radius.

gibbous The phase of the moon or a planet during which it appears more than half full but less than full.

globular cluster A large globe-shaped system of stars, usually found in the halo that surrounds the nucleus of the Galaxy.

globule A small, dense, dark nebula; possibly a cloud of gas that is in the process of forming a star.

granulation (of the photosphere) The mottled appearance of the visible "surface" of the sun, due to rising columns of hot gases.

gravitation The mutual force of attraction that masses exert on each other.

gravitational constant (G) A number that allows us to express the force of gravitational attraction in terms of the masses of the objects involved and the distance separating these objects.

When the mass is measured in grams and the distance in centimeters, then the force may be expressed in dynes; specifically $G = 6.668 \times 10^{-8}$ dyne-cm^2/gm^2.

gravitational energy The energy that may be released by the partial or total collapse of a system.

gravity (of the earth) The force of attraction that the earth exerts on a given object. The weight of an object.

great circle The largest circle that can be drawn on a sphere; a circle that divides a sphere into two equal parts.

greatest elongation The largest angle of separation between the sun and either Mercury or Venus.

Greenwich meridian A portion of a great circle on the earth that passes through the poles and through a given point at the Royal Greenwich Observatory in England. This meridian is also referred to as the "prime meridian," from which the longitude of any point on the earth is measured.

Gregorian calendar The modern calendar, which was introduced by Pope Gregory XIII in the sixteenth century.

H I (hydrogen-one) region A region in space in which neutral hydrogen gas is found.

H II (hydrogen-two) region A region in space in which ionized hydrogen gas is found.

half-life The time required for one-half of the atoms in a sample of a radioactive element to change spontaneously into another element.

halo (of the Galaxy) System of globular clusters, stars, and gas that surrounds the nucleus of the Galaxy.

harmonic law Kepler's third law of planetary motion: *The cubes of the semimajor axes of the planetary orbits compare in the same way as the squares of the periods of the planets.*

harvest moon The full moon nearest the time of the autumn equinox.

head (of a comet) That portion of the comet, exclusive of any tail, which includes the nucleus and coma.

heavy elements Elements that have many protons and neutrons in their nuclei. (The term sometimes refers to any element other than hydrogen or helium.)

heliacal rising The simultaneous rising of any object with the sun.

heliocentric system A sun-centered system.

Hertzsprung-Russell (H-R) diagram A plot of a group of stars according to their absolute magnitude and spectral class (temperature).

horizon (celestial) A great circle 90° from the observer's zenith.

horizontal system A system by which the location of an object is specified in terms of its angle above the horizon (altitude) and the angle it makes relative to due north (azimuth).

Azimuth is measured in an easterly direction from due north along the horizon.

hour angle The angle between the zero hour circle and any given hour circle.

hour circle A great circle on the celestial sphere that runs through the celestial poles, hence is perpendicular to the celestial equator.

Hubble constant A number that expresses the apparent relationship between the distance to the galaxy and its speed of recession.

hydrostatic equilibrium A condition in which the inward gravitational attraction exactly balances the outward pressure force at every point within a star.

hyperbola A curve that may be produced by cutting a circular cone with a plane parallel to the axis of the cone. It may represent the path of certain comets.

hypothesis An idea that is assumed to be true and upon which a given theory is built.

ideal gas A gas that obeys certain laws that describe the relationships between pressure, volume, and temperature.

image A representation of an object that can be seen by the eye, usually because light rays are brought to a focus.

inclination (of an orbit) The angle between the plane of the orbit of a given body and another specified reference plane, such as that to the celestial equator or the galactic equator.

Index Catalogue (IC) A listing of star clusters, nebulae, and galaxies that supplements the New General Catalogue (NGC).

index of refraction A number found by dividing the speed of light in a vacuum by its speed in a given transparent substance. This number would then be the index of refraction of the transparent substance.

inertia The property of matter which resists any change in its velocity: an object at rest tends to remain at rest; an object in motion tends to remain in motion in a straight line.

inferior conjunction The configuration of Mercury or Venus when that planet is directly between the earth and the sun.

inferior planet A planet whose orbit lies between that of the earth and the sun: thus, Mercury and Venus.

infrared An electromagnetic radiation of wavelength just longer than red light.

intensity The brightness of a source of light.

interference The reinforcing or canceling of waves, as in light or water waves.

interferometer An optical device by

which the diameters of the largest nearby stars may be determined.

International Date Line A line opposite the prime (Greenwich) meridian, at approximately 180° longitude. When a traveller crosses this line, the date is changed by one day.

interstellar dust Microscopic dust-like grains that exist in the space between stars.

interstellar gas Very diffuse gas that exists in the space between stars.

interstellar lines Dark spectral lines that are due to absorption by interstellar gases. These lines are seen against the spectrum of the star itself.

interstellar matter Gas and dust in the space between stars.

ion An atom that has gained or lost one or more electrons; hence, a charged atom.

ionization The process whereby an atom gains or loses electrons.

ionosphere A region of the earth's upper atmosphere in which many atoms are ionized.

irregular galaxy A galaxy of irregular shape, neither spiral nor elliptical.

irregular variable A star whose variations in brightness do not occur at regular intervals.

island universe The name once given to galaxies.

isotope One of several forms that atoms of a given element may take, each varying in the number of neutrons contained in the atomic nucleus.

Jolly balance An apparatus used to measure the mass of the earth. (May also refer to an instrument used to determine the density of a given object.)

Jovian planet A Jupiter-like planet; hence, one of relatively low density—Jupiter, Saturn, Uranus, and Neptune.

Julian calendar A calendar based on the apparent motions of the sun, introduced by Julius Caesar.

Julian day The number assigned to a given day (in the Julian calendar) based on a starting time of January 1, 4713 B.C.

Jupiter The fifth planet from the sun and the largest in the solar system.

Kepler's laws Three basic statements that describe the motions of the planets.

kinetic energy The energy that a body possesses due to its motion.

Kirchhoff's laws Three statements that describe the formation of the continuous, the emission, and the absorption spectra.

Kirkwood's gaps Voids in the spacing of particles in the rings of Saturn or in the asteroid belt.

latitude An angular measure of a point on the surface of the earth that indicates its distance north or south of the equator.

law of areas Kepler's second law of planetary motion: A line joining any given planet with the sun sweeps out equal areas in equal times.

law of the red shift For distant galaxies, their recessional velocities are proportional to their distance. Their recessional velocities are measured by their red shift, hence the red shift becomes an indicator of distance.

law of reflection The angle of reflection is equal to the angle of incidence.

law of refraction A ray of light is bent toward the normal when passing from one medium to another in which it travels more slowly.

law of universal gravitation Any two objects in space experience a mutual and equal force of attraction. This force is proportional to the product of their masses and inversely proportional to the square of the distance separating the objects.

leap year A year containing 366 days; occurs once every four years.

librations (of the moon) Motions that over a period of time allow the observer to see more than one-half the moon's surface.

light An electromagnetic disturbance that is visible to the eye.

light curve A graph that indicates the variations in light output of a variable star or an eclipsing binary system as they occur over a period of time.

light-gathering power (of a telescope) A measure of the amount of light a telescope can collect.

light-year The distance light travels in one year—approximately 9.5 trillion kilometers (6 trillion miles).

limb The visible edge of a planet, the moon, or the sun.

limb darkening The phenomenon of decreased brightness of the sun near its edge.

limiting magnitude The dimmest magnitude observable in a given telescope under specified conditions.

line of apsides Major axis of an ellipse.

line broadening Any phenomenon that results in the broadening of the spectral lines. (Pressure, for example, may cause line broadening.)

line of nodes The line connecting two nodes of an orbit; hence, a line that lies both in the plane of the object's orbit and also in the plane of another's orbit.

local apparent noon The time at which the center of the sun is on the observer's meridian.

Local Group The cluster of galaxies to which the Milky Way galaxy belongs.

local mean noon The time at which the average sun is on the observer's meridian.

local mean time The hour angle of the average sun.

local meridian An imaginary line running north and south through the observer's zenith (in the sky) or through the observer's position (on earth).

longitude The angle between the prime meridian and a meridian through the point on the surface of the earth. This angle is measured along the equator.

long-period variable A variable star whose period of variation exceeds 100 days.

luminosity A measure of actual brightness; the sun may be used as a reference star and other stars compared to it.

lunar eclipse An eclipse of the moon, occurring when the moon passes within the earth's shadow.

lunar month A period of time ($29\frac{1}{2}$ days) based upon the cycle of phases of the moon, i.e. full moon to full moon.

Lyman series Those lines in the spectrum which occur in the ultraviolet portion, created by upward or downward electron transitions that originate or terminate with the first (lowest) energy level of the hydrogen atom.

magnetic field A region of space where magnetic forces may be detected.

magnetic pole One of two points on a magnet (or in a magnetic field) where the greatest density of lines of force exist.

magnifying power The number of times that the apparent size of an object is increased by viewing it through a telescope, as compared to naked-eye observation.

magnitude A number that is used to indicate the brightness of an object, either apparent or absolute, as specified.

main sequence A line on the H–R diagram that represents the majority of stars.

major axis (of an ellipse) A line drawn through the foci of an ellipse; also, the longest line that may be drawn joining two points of an ellipse. It may refer specifically to the length of such a line.

major planet One of the four largest planets (Jupiter, Saturn, Uranus, and Neptune).

mantle (of the earth) The layer of the earth that lies between the crust and the core.

maria Latin for "seas" (singular: *mare*); name of certain regions, once thought to be sealike, on the moon or Mars.

Mars The fourth planet from the sun.

maser Acronym for **m**icrowave **a**mplification by **s**timulated **e**mission, a device utilizing natural oscillations of atoms or molecules for amplifying electromagnetic waves in the microwave region of the spectrum.

mass A measure of the amount of atomic material in an object.

mass-luminosity relation The observed relationship that, in general, the more massive stars are, the brighter they are.

mean solar day The period of time between successive passages of the mean (average) sun across the observer's meridian; the average length of an apparent solar day.

mean sun An imaginary sun that moves at a constant speed along the celestial equator, making one circuit of the sky in one year.

Mercury The planet nearest the sun.

meridian A great circle on the earth that passes through a given point and both poles; a great circle on the celestial sphere that passes through the observer's zenith and both celestial poles.

Messier Catalogue (M) A listing of nebulae, star clusters, and galaxies compiled by Charles Messier in 1787.

meteor The bright streak of light that occurs when a particle from outside the earth's atmosphere enters that atmosphere and is heated due to friction.

meteor shower Numerous meteors that seem to radiate from a given point in the sky. Showers usually occur when the earth passes through collections of material left in the path of a comet.

meteorite That portion of a meteoroid which survives its flight through the earth's atmosphere and strikes the ground.

meteoroid A meteor particle while it is still in space, without relation to the phenomenon it causes once it has entered the earth's atmosphere.

micrometeorite A very small meteoroid which does not create friction sufficient to burn up in the atmosphere, but which falls to the earth.

micrometer (μm) A unit of length equal to 10^{-6} m.

Milky Way A diffuse band of light composed of millions of stars and nebulae. It encircles the sky and represents the flattened disklike portion of the Milky Way galaxy (our Galaxy).

minor axis (of an ellipse) The smallest diameter of an ellipse.

minor planet An asteroid, ranging in size from several hundred kilometers to less than 1 km in diameter.

Mohorovičić discontinuity The boundary between the crust and the mantle of the earth, named for the Yugoslav geologist Andrija Mohorovičić.

molecule A combination of two or more atoms; the smallest particle of a substance that still retains all the properties of that substance.

momentum Mass multiplied by the velocity of an object; a measure of the state of motion of an object.

monochromatic Limited to one color or wavelength.

moving cluster (of stars) A group of stars that are moving in the same direction and at the same speed.

n-body problem A problem that involves the gravitational effects of several bodies on each other.

nadir A point opposite the observer's zenith.

nautical mile The average length of an arc on the earth's surface that makes an angle of one minute (1/60 degree) at the center of the earth.

navigation The art of finding one's position on the earth by means of celestial observations.

neap tides The less extreme tides that occur when the moon is near first or third quarter phase.

nebula A cloud of gas or dust in space; the term may sometimes refer to galaxies or star clusters (because of their fuzzy appearance).

nebular hypothesis The theory that the solar system was formed from a nebula.

Neptune The eighth planet from the sun.

neutrino A neutral particle of little or no mass that carries away energy from a nuclear reaction.

neutron A particle, within the nucleus of an atom, that has no charge but has mass approximately equal to that of a proton.

New General Catalogue (NGC) A listing of nebulae, star clusters, and galaxies that succeeded the one compiled by Charles Messier.

new moon That phase of the moon which occurs when it is between the earth and the sun.

Newtonian reflector A reflecting telescope that, by use of a flat diagonal mirror, brings the rays of light from a distant object to a focal point near the side of the tube.

Newton's laws The three statements set forth by Sir Isaac Newton regarding the motions of objects.

node The intersection points of the plane of the moon's or planet's orbit with another plane, such as that of the earth's orbit or the celestial equator.

north celestial pole The point on the celestial sphere that is determined by the extension of the earth's axis in a northerly direction.

north point A point on the observer's celestial horizon directly under the north celestial pole.

nova A star that suddenly brightens and then fades again, hence is seen as a "new" star.

nuclear Refers to the nucleus of the atom.

nuclear fission A process of nuclear change that results in lighter elements being formed from heavier ones.

nuclear fusion A process of nuclear change that results in heavier elements being formed from lighter ones.

nucleon A constituent of the nucleus of an atom: a neutron or proton.

nucleus (of an atom) The central portion of the atom, containing almost the entire mass of the atom.

nucleus (of a comet) The collection of solid (frozen) particles that compose the head of the comet.

nucleus (of a galaxy) The central, more dense portion of the galaxy.

nutation The small variations in the movements of the earth's poles. The principal movement is precession.

objective The main lens or mirror in a telescope, used to bring rays of light to a focal point.

oblate spheroid A sphere, such as the earth, that is flattened by rotation.

oblateness A measure of the degree to which a sphere is flattened; a number obtained by dividing the difference between the major and minor axes by the major axis itself.

obliquity of the ecliptic The angle between the ecliptic plane and that of the celestial equator.

obscuration The absorption of starlight by interstellar dust.

occular An eyepiece.

occultation The passing of one object behind a larger one, say, the passing of a moon behind Jupiter.

open cluster A loosely formed cluster of stars, usually found in the disk of the galaxy.

opposition The configuration of a planet when it is directly opposite the sun as seen from earth. The planet is seen 180° from the sun in the sky.

optical binary Two stars that merely appear close together because they happen to line up from the observer's point of view. They may actually be separated by a great distance and therefore they do not influence each other gravitationally.

optics A branch of physics that deals with light and its properties.

orbit A closed path along which a body moves as it revolves around a point in space.

orbital plane The plane in which a given body moves in its orbit; an imaginary flat surface that is determined by the motion of the body.

outer planet One of the planets beyond the asteroid belt (Jupiter, Saturn, Uranus, Neptune, and Pluto).

ozone layer A "layer" of the earth's atmosphere, composed of a special form of oxygen (O_3), which filters much of the ultraviolet radiation of the sun.

parabola A curve that is made by cutting a circular cone with a plane parallel to one of its elements (edges).

paraboloid A concave surface, the cross section of which is a parabola. The shape used for the primary mirror of most reflecting telescopes.

parallax The apparent shift in position of an object due to the motion of the observer.

parallax (stellar) The apparent shift of a star against the background of more distant stars due to the motion of the earth around the sun; the angle subtended (cut off) by the radius of the earth's orbit (1 A.U.) at the distance of the star.

parsec The distance to a star that exhibits 1 second of heliocentric parallax (1 parsec = 3.26 light-years).

partial eclipse An eclipse of the moon or sun in which the object is not completely obscured.

penumbra That portion of the shadow of an object in which the light for an extended source is not completely obscured.

penumbral eclipse An eclipse of the moon in which the moon merely passes through the penumbra of the earth's shadow.

perfect radiator A black body—a body that absorbs all radiation falling upon it and reemits all the radiation.

periastron That point in the orbit of a member of a binary star system at which it is nearest its companion star.

perigee The point in the orbit of an earth satellite at which it is nearest the earth (*-gee*).

perihelion The point in the orbit of an object that revolves around the sun (*helios*) at which it is nearest the sun.

period The interval of time necessary to complete one rotation, one revolution, or one cycle.

period-luminosity relation The relationship between the period and the absolute magnitude of certain variable stars.

periodic comet A comet whose orbit is elliptical and hence one that returns to perihelion at regular intervals. A comet whose return is predictable.

perturbation Any gravitational disturbance that causes a body to deviate from its primary orbital path. Such disturbances may be caused by the presence of a third object.

phases (of the moon or planet) Changes in the portion of the illuminated "face" of the moon or planet that is visible from the earth.

photoelectric effect The emission of electrons from the surface of a substance caused by light striking it.

photoelectric magnitude A measure of the brightness of an object as indicated by a photomultiplier.

photographic magnitude A measure of the brightness of an object as indicated on a blue-sensitive photographic plate.

photometry The science of measuring the apparent brightness of celestial objects.

photomultiplier A light-sensitive cell in which the electric current generated by light is amplified so that it can be more easily and accurately measured.

photon A unit of electromagnetic energy; a certain quantity of light energy.

photosphere The apparent (visible) "surface" of the sun; the layer of the sun from which light seems to radiate.

photovisual magnitude A measure of the apparent brightness of an object, using film that is sensitive to the same region of the spectrum as the human eye.

plage A bright region just above the sun's surface as seen in a spectroheliogram.

Planck's constant A number that relates the energy "carried" by a photon of light to its wavelength.

planet One of the nine main bodies that revolve around the sun and reflect its light; any similar body revolving around a star in another (possible) solar system.

planetarium A projection device that is capable of creating an artificial sky on a domed ceiling and showing the motions of celestial objects greatly speeded up in time.

planetary nebula A spherical shell of gas that surrounds a very hot star and is expanding relatively slowly.

planetoid A minor planet or asteroid.

Pluto The ninth planet from the sun.

polar axis (of a telescope) That axis, set parallel to the earth's axis, about which the telescope turns to compensate for the earth's rotation.

polarization A filtering process in which only those light rays whose disturbances lie in a given plane are allowed to pass.

Population I and II (stars) Two classes of stars that appear to be quite different in evolutionary state and in location within the Galaxy: type-I stars are found primarily in the spiral arms of the Galaxy, whereas type-II stars are found elsewhere in it.

positron A particle that has approximately the same mass as the electron but that carries a positive charge.

potential energy That type of energy which an object possesses because of its position; the capacity of an object to do work by reason of its position.

pound A unit of force (not mass) in the English system.

precession The slow gyration of the earth's axis, which sweeps out a circle in the sky over a 26,000-year period. This motion of the earth causes a continuous change of the polar positions in the sky.

precession of the equinoxes The slow westward shift of the equinoxes along the ecliptic due to the precession of the earth.

prime focus The point at which the objective of a telescope brings the light rays to a focus without the use of any secondary mirrors or lens.

prime meridian The meridian that runs through the Royal Observatory of Greenwich, England. The longitudes of points on the earth are measured from this meridian.

primeval atom A single mass composed of all the matter of the universe.

primeval fireball The expanding ball of matter that resulted from the explosion of the primeval atom as depicted in the "big bang" theory of the origin of the universe.

Principia Newton's great work, in which he described the motions of objects under the influence of gravity: *Philosophiae Naturalis Principia Mathematica*.

prism A triangular shape (of glass or other transparent material) that is utilized to disperse light into its spectrum.

prominence A protrusion from the limb of the sun which appears as a flame or loop, best seen in the light of hydrogen.

proper motion The rate at which a star's position in the sky changes, measured in seconds of arc per year.

proton One of the basic subatomic particles that compose the nucleus, carrying a positive charge.

proton-proton cycle An atomic reaction that occurs in the core of a star whereby four hydrogen nuclei combine to form a helium nucleus. This reaction is the source of energy of the star (or sun).

protostar The mass of material that is in the process of forming a star.

pulsar An object that emits brief pulses of radio energy and that has also been observed optically by use of special photographic equipment; possibly a very dense (neutron) star that is spinning very rapidly.

pulsating variable A variable star that changes size at regular intervals, its variation in light output being directly related to its variation in size.

quadrature The configuration of a planet or moon as seen 90° from the sun.

quantum mechanics The study of the structure of atoms and how they interact with one another.

quarter moon A half-full moon as seen when it appears 90° from the sun, one-quarter or three-quarters of the way around its orbit.

quasar Contraction of the term *quasi-stellar radio source;* a starlike object that has a very large red shift in its spectrum, hence is presumed to be very distant—perhaps the most distant object yet known—and probably a galaxy that is emitting much more energy than is normal.

quiet sun The sun at a time of very low activity.

R Corona Borealis variables The class of variable stars that exhibit irregular and sudden decreases in brightness.

RR Lyrae variables The class of variable stars that have periods of less than one day.

radar telescope A radio telescope that is also capable of sending a radio signal into space and then listening for its echo (reflection).

radial velocity That part of the velocity of an object which is measured along the observer's line of sight.

radiant (of a meteor shower) A point in the sky from which a number of meteors seem to originate during a meteor shower.

radiation The process whereby energy is transferred from one point to another through empty space.

radiation pressure The small force that electromagnetic radiation exerts on matter which it intercepts.

radio astronomy That branch of astronomy which is primarily concerned with receiving and analyzing the radio energy received from celestial objects.

radio telescopes A large parabolic reflector that collects the radio energy from one region of the sky and concentrates (focuses) that energy at a focal point. This energy is then amplified and recorded by electronic equipment.

radioactive element An element whose nucleus spontaneously disintegrates to produce a lighter element. Energy is also released in this process.

rays (lunar) A system of bright streaks that seem to radiate from certain craters on the moon.

reaction force The equal but opposite force that accompanies every force.

real image An image, formed at the focus of a telescope, that can be photographed; an image formed by light rays that converge after passing through a lens or after reflecting from a mirror.

red giant A very large, cool star.

red shift The shifting of spectral lines toward the red end of the spectrum due to the relative motion of the source away from the observer.

reddening (interstellar) The reddening of starlight as the result of the scattering of blue light when the light of a star passes through clouds of gas and dust in space.

reflecting telescope A telescope in which the primary objective is a concave mirror; a telescope that depends on the principle of reflection for its operation.

reflection The process whereby the direction of travel of light rays is changed by an optical surface.

reflection nebula A cloud of interstellar dust that is visible because it reflects starlight.

refracting telescope A telescope that depends on the principle of refraction for its operation, hence a telescope that has a lens or lens system as its objective.

refraction The bending of light as it passes from one transparent medium to another of different density.

relativity A theory formulated by Albert Einstein that deals with the measurement of various events as observed by two different observers, themselves in motion.

resolution The ability of a telescope to separate objects that appear close together; its ability to show detail.

retrograde motion (of a planet) The apparent westward motion of a planet as seen against the background of stars.

reversing layer (of the sun) A thin layer of solar atmosphere, just above the photosphere, that produces the dark-line spectrum of the sun.

revolution The motion of a body around a given point in space (for example, the earth's *revolution* around the sun).

right ascension The smallest angle between the zero hour circle and a given celestial object.

rill (or rille) A crevasse in the surface of the moon.

rotation The spinning of a body on its own axis.

saros An 18-year cycle during which the circumstances that produce similar eclipses recur.

satellite Any body that revolves around a larger body (the moon, for instance, is a satellite of the earth).

Saturn The sixth planet from the sun.

scale (of an image) The size of an extended image compared to its apparent size in the sky measured in centimeters (or inches) per degree.

Schmidt camera A telescope that utilizes both a spherical mirror and a weak refracting lens to produce a camera that is capable of photographing a wide field of view.

science The branch of knowledge which seeks systematically to describe phenomena of nature.

scientific method An approach in which the researcher first observes certain pertinent phenomena, then formulates a theory that seems to be consistent with those observations, and finally tests his theory by determining whether it will accurately predict future events.

Sculptor-type system A very small elliptical galaxy similar to the galaxy in Sculptor.

secondary mirror A mirror that reflects the light gathered by the primary mirror; a mirror second in line.

"seeing" conditions Those conditions within the earth's atmosphere which affect the quality of image formed in the telescope.

seismic waves Vibrations that travel through the interior of the earth due to earthquakes.

seismograph An instrument that records the time, type, and strength of seismic (earthquake) waves.

seismology The science that deals with the origin and transmission of seismic waves in the earth.

semimajor axis One-half the major axis of an ellipse. It represents the average distance from the sun to a given planet or comet.

separation The angular distance between two stars in a visual binary system.

shell star A star that is surrounded by a sphere or shell of gas.

shower (of meteors) Numerous meteors that seem to radiate from a given point in the sky. Such showers usually occur when the earth passes through collections of meteoric material left in the path of a comet.

sidereal day The length of a day as measured by the successive passages of any given star across the observer's meridian. A day as measured by the stars.

sidereal month The length of time required for the moon's revolution around the earth as measured by the stars.

sidereal revolution The period of revolution of one body around another with respect to the stars.

sidereal time Star time; the hour angle of the vernal equinox; the right ascension of the observer's meridian at the given time.

sidereal year The time required for the earth's revolution around the sun with respect to the stars.

siderite An iron-nickel meteorite.

singularity The central core of a black hole, characterized by infinite density and infinite tidal forces.

small circle Any circle on the surface of a sphere that is smaller than a great circle.

solar activity Prominences, sunspots, plages, flares, etc.; activities that occur on or above the photosphere of the sun.

solar apex The direction in which the sun is moving with respect to the average motion of the nearest stars.

solar constant The amount of solar radiation received at the distance of the earth, measured in ergs per square centimeters per second.

solar day The average time required for two successive passages of the sun across the observer's meridian.

solar eclipse An eclipse of the sun.

solar flare A sudden outburst of energy from the sun, causing a brightening of a given region, usually near a sunspot.

solar parallax The angle subtended (cut off) by the equatorial radius of the earth as seen from a distance of 1 A.U.

solar system the system of all objects that revolve around the sun: the planets, moons, comets, meteoroids, etc.

solar time Time as based on the sun; the hour angle of the sun plus 12 hr.

solar wind The outflow of particles from the sun.

solstices Either of the two points on the ecliptic where the sun reaches its maximum declination north or south of the equator; the longest and the shortest days of the year.

south celestial pole A point on the celestial sphere determined by extending the earth's axis southward until it intersects that sphere.

south point The point of intersection of the observer's meridian with his southern horizon.

space motion The velocity of a star with respect to the sun.

specific gravity The density of a given body or substance compared to that of water; numerically equal to the density of the body measured in grams per cubic centimeter.

spectral class The classification of a star with respect to characteristics of its spectrum.

spectrogram The photograph of a spectrum.

spectrograph The instrument used to photograph a spectrum.

spectroheliogram The photograph of the sun, taken in the light of a single spectral line of an element such as hydrogen.

spectroheliograph The instrument used to photograph the sun in the light of a single spectral line of an element such as hydrogen or calcium.

spectroscope An instrument in which an observer may view the spectrum of an element, a star, or the sun.

spectroscopic binary star A star system in which the true binary nature of the system is revealed by the periodic shifting of spectral lines. Such a system can not be separated optically.

spectroscopic parallax A method whereby the distance to a star is determined by observing its spectral characteristics, converting this to absolute magnitude by use of an H–R diagram, and comparing that to the star's apparent magnitude.

spectrum The rainbow of colors produced when light is dispersed by refraction or diffraction.

spectrum analysis The determination of such characteristics of a light source as velocity, temperature, and pressure by studying the spectrum of the source.

spectrum binary A system of stars whose true binary nature is revealed by the presence of spectral lines associated with two stars of different temperatures.

speed The rate at which the distance to an object changes without regard to its direction of travel.

spherical aberration A defect in a lens or mirror that is due to its incorrect shape. Light rays that pass near the center of the lens (or mirror) are brought to a different focus as compared to those which pass near the outer edge.

spicule A jet of hot material rising in the atmosphere of the sun.

spiral arms (of a galaxy) The curved, armlike structures that surround the nucleus of certain galaxies.

spiral galaxy A flattened galaxy composed of a central nucleus and a system of arms which spiral out from that nucleus.

sporadic meteor A meteor that does not appear to be associated with a known shower of meteors.

spring tides The most extreme tides produced when the moon, sun, and earth are aligned, that is, when the moon is new or full.

standard time The time used within a given time zone, computed as the average solar time for that zone.

star A spherical mass of gas that radiates various forms of energy owing to nuclear reactions within its core.

star cloud A region of the sky in which the stars are so close together that they appear as a luminous cloud.

star cluster A grouping of stars that is held together by mutual gravitation; the stars thus possess a common motion through space.

star map A map showing the positions and magnitudes of stars, designed to be held over the observer's head.

steady state theory The theory that the universe has always been as it is today, that matter is being continually created to replace matter which is converted into energy. The density of the universe would thus remain at the same level.

Stefan's law The assertion that the total amount of energy radiated from a body in a given time depends upon the absolute temperature of the body, raised to the fourth power.

stellar evolution The life cycle of a star. Stars change in size, pressure, luminosity, and structure.

stellar parallax The angle subtended (cut off) by the radius of the earth's orbit (1 A.U.) at the distance of the star.

stratosphere The layer of the earth's atmosphere between the troposphere and the ionosphere.

subdwarf star A star that, owing to its smaller size, is less luminous than a main-sequence star of the same spectral class.

subgiant star A star of luminosity between that of a normal giant and a main-sequence star of the same spectral class.

sublimate A process whereby a solid turns directly into a gaseous state without passing through a liquid state, for instance, Dry Ice.

summer solstice The point on the ecliptic where the sun appears farthest north of the equator; the longest day of the year.

sun The star about which the earth revolves.

sunspots Regions of the sun that appear dark because they are temporarily cooler than the surrounding region.

sunspot cycle The period over which both the number and location (latitude) of the sunspots vary (approximately 11 years).

supergiant A large star of very high luminosity.

superior conjunction A configuration that occurs when a planet appears to line up with the sun on its (the sun's) far side. This term is used only in reference to Mercury or Venus.

superior planet A planet whose orbit is beyond that of the earth.

supernova An exploding star that temporarily increases in luminosity, perhaps a million times brighter than it was before the eruption.

synchrotron radiation A type of radiation that results from charged particles being accelerated by a magnetic field.

synodic month The time required for the moon to complete its cycle of phases (29.5 days).

synodic period The time required for a planet (or moon) to move from a given configuration back to that same configuration again, as seen from earth; for example, the period between successive oppositions of a superior planet.

syzygy A lining up of any three celestial objects, as in a conjunction or opposition.

T Tauri stars A class of variable stars that show very rapid and irregular pulsations.

tail (of a comet) The gases that are forced away from the head of the comet by the solar wind.

tangential velocity That part of a star's space velocity which is perpendicular to its radial velocity; a measure of the speed with which a star crosses the observer's line of sight.

tektites Glasslike objects that have traveled in the earth's atmosphere. These objects are thought to have been formed by meteorite impact on either the earth or the moon.

telescope An optical instrument that makes possible the observation and photographing of objects too dim and/or too distant to be seen with the naked eye; hence, a light-gathering device.

telluric (spectral) lines Spectral lines that are produced by elements in the earth's atmosphere.

temperature A measure of the average speed with which the molecules of a substance (or atoms of a gas) are moving.

terminator The line between the sunlit and dark portion of the moon (or planet).

terrestrial planet An earthlike planet, similar to the earth in density (Mercury, Venus, Mars).

theory A set of ideas that are consistent with observed phenomena.

thermocouple A device for measuring the intensity of infrared radiation.

thermodynamics A branch of physical science that deals with the way in which heat moves from one body to another.

thermonuclear reactions Nuclear changes that result from high-temperature and high-pressure conditions.

tide The deformation of land and/or water masses by the differential gravitational attraction of another body. The moon and sun create tides on the earth; the earth and sun create tides on the moon.

time zone A zone on the earth's surface, approximately 15° wide, within which the hour used is uniform.

ton (English) A unit of force (weight) equivalent to 2000 pounds.

ton (metric) A unit of force (weight) equivalent to 1 million grams.

total eclipse (1) An eclipse of the sun during which the disk of the moon completely covers the photosphere of the sun. (2) An eclipse of the moon during which the moon lies completely within the umbra of the earth's shadow.

trail (of a meteor) The temporary luminous streak produced by the passage of a meteoroid through the earth's atmosphere.

transit (1) The passage of a body across the face of a larger body. (2) The passage of a body across a given meridian. (Mercury transits the sun. The sun transits the prime meridian.)

transverse wave A wave in which particles are disturbed in a direction perpendicular to the direction in which the wave is traveling.

triangulation A process whereby an inaccessible side of a triangle may be determined from the measurement of accessible sides and angles.

Trojan minor planet An asteroid that orbits the sun in approximately the same orbit as Jupiter but is located 60° ahead of or behind the planet as viewed from the sun.

Tropic of Cancer The parallel of latitude that lies 23.5° north of the equator, the limit of the sun's apparent travel in a northerly direction.

Tropic of Capricorn The parallel of latitude that lies 23.5° south of the equator, the limit of the sun's apparent travel in a southerly direction.

tropical year The time required for the earth to make one revolution around the sun as measured by the vernal equinox (approximately 365.25 days).

troposphere The layer of the earth's atmosphere just above the earth and extending to an elevation of about 15 km.

twinkle The apparent changes in the brightness and color of a star due to the motion of the earth's atmosphere.

ultraviolet radiation That part of the electromagnetic spectrum with wavelengths just shorter than visible light, approximately in the range of 100 to 4000 Å.

umbra (1) The completely dark central portion of a shadow. (2) The darkest portion of a sunspot.

Universal Time Average Greenwich time.

universe All of space that is occupied by matter and/or radiation.

Uranus The seventh planet from the sun.

Van Allen belts Regions that surround the earth consisting of high-energy charged particles whose motions are directed by the earth's magnetic field.

variable star A star that exhibits changes in luminosity and/or color; a pulsating star.

vector A quantity that has both magnitude and direction.

velocity A vector quantity that denotes both the speed and direction of motion.

velocity of escape The velocity at which a body overcomes the pull of gravity of another body and moves off into space.

Venus The second planet from the sun.

vernal equinox The point on the celestial equator at which the sun crosses on its way northward; an intersection of the celestial equator and the ecliptic; the position of the sun on March 21.

vertical circle A great circle that passes through the zenith of the observer and is perpendicular to the horizon.

visual binary A binary system in which the two components are visible as separate stars in a telescope.

volume A measure of the amount of space occupied by an object.

Vulcan An imaginary planet once thought to orbit between Mercury and the sun.

walled plain (of the moon) A very large crater on the moon.

wandering of the poles A shifting of the body of the earth in relation to its axis of rotation.

watt A unit of power, equivalent to 10 million ergs of energy used up in 1 sec.

wavelength The distance from any point on a wave to the next similar point on the succeeding wave, as crest to crest.

weight A measure of the force exerted on one object by another due to gravity; specifically, the gravitational force exerted by the earth on a given mass.

west point A point on the celestial horizon that is located 270° from the north point, measured in a clockwise direction.

white dwarf An old star that has collapsed due to its exhausted fuel supply and yet has a large portion of its original mass; hence, a very dense, hot star.

Widmanstätten figures A definite pattern of crystal formations often seen in the interior of a polished meteorite.

Wien's law A statement relating T, the temperature of a body (in degrees Kelvin), to λ, the wavelength of its maximum radiation: $\lambda = 3000 \ \mu m/T$.

winter solstice A point on the ecliptic at which the sun reaches its maximum distance south of the celestial equator; the shortest day of the year.

Wolf-Rayet stars A class of very hot stars that eject shells of gas at high velocity.

X rays Electromagnetic radiation of short wavelength between that of ultraviolet rays and of gamma rays.

year The time required for one revolution of the earth around the sun.

Zeeman effect The splitting or broadening of spectral lines, which indicates the presence and strength of magnetic fields at the source.

zenith A point on the celestial sphere directly over the head of the observer.

zodiac A band on the celestial sphere that is centered on the ecliptic and contains the twelve constellations usually associated with astrology.

zodiacal light A faint light seen along the ecliptic, possibly due to sunlight being scattered by interplanetary dust.

zone of avoidance A region toward the center of the Milky Way galaxy where few—if any—other galaxies are seen, since they are obscured by clouds of gas and dust.

Appendixes

Appendix 1 Temperature Conversion Charts

FAHRENHEIT	CELSIUS	KELVIN	EXAMPLES
27 million °F	Approximately 15 million °C	15 million °K	Core of the sun
10,337°F	5727°C	6000°K	Surface temperature of the sun
700°F	371°C	644°K	Probable maximum temperature of Venus
212°F	100°C	373°K	Boiling point of water
100°F	38°C	311°K	Normal body temperature
68°F	20°C	293°K	Normal room temperature
32°F	0°C	273°K	Freezing point of water
0°F	−18°C	255°K	
−100°F	−73°C	200°K	Minimum temperature on the earth's surface
−198°F	−128°C	145°K	Polar cap of Mars
−230°F	−146°C	127°K	Average temperature of Saturn
−297°F	−183°C	90°K	Dark side of Mercury
−459°F	−273°C	0°K	Absolute zero; all translational molecular motion stops

Appendix 2 International System of Units (SI) with English Equivalents

INTERNATIONAL SYSTEM	ENGLISH SYSTEM
Length (units most commonly used):	
1 micrometer (μm) = 0.000001 m	
1 millimeter (mm) = 0.001 m	= 0.03937 in.
1 centimeter (cm) = 0.01 m	= 0.3937 in.
1 meter (m) = 1.00 m	= 39.37 in.
1 kilometer (km) = 1000 m	= 0.6214 miles
1 megameter (Mm) = 1,000,000 m	
1.6093 km	= 1 mile
2.5400 cm	= 1 in.

Note: SI units are based on powers of ten (see Appendix 3), and the prefix indicates the power to be taken. The following list of prefixes may be used with any unit [for instance, *kilo*gram (kg), a unit of mass; or *nano*second (nsec), a unit of time]:

pico (p) = 10^{-12}	centi (c) = 10^{-2}	kilo (k) = 10^{3}
nano (n) = 10^{-9}	deci (d) = 10^{-1}	mega (m) = 10^{6}
micro (μ) = 10^{-6}	deka (da) = 10^{1}	giga (g) = 10^{9}
milli (m) = 10^{-3}	hecto (h) = 10^{2}	tera (t) = 10^{12}

INTERNATIONAL SYSTEM	ENGLISH SYSTEM
Mass:	
1 milligram (mg) = 0.001 g	
1 gram (g) = 1.000 g	= 0.0022046 lb
1 kilogram (kg) = 1000 g	= 2.2046 lb
453.6 g	= 1 lb = 16 oz
28.3495 g	= 1 oz
Time:	
1 nanosecond (nsec) = 0.000000001 sec	
1 microsecond (μsec) = 0.000001 sec	
1 millisecond (msec) = 0.001 sec	
1 second (sec) = 1.0 sec	

In writing very large or very small numbers, it is convenient to use the following system of notation:

$$10^1 = 10$$
$$10^2 = 10 \times 10 = 100$$
$$10^3 = 10 \times 10 \times 10 = 1000$$
$$10^4 = 10 \times 10 \times 10 \times 10 = 10,000$$

Following this pattern:
$$10^{12} = 1,000,000,000,000$$

A light-year is approximately equivalent to 6,000,000,000,000 miles, which could be written $6 \times 1,000,000,000,000$ miles, or 6×10^{12} miles—a much simpler notation.

In a very similar way:
$$10^{-1} = 0.1 = 1/10$$
$$10^{-2} = 0.01 = 1/100$$
$$10^{-3} = 0.001 = 1/1000$$

Following this pattern:
$$10^{-7} = 0.0000001$$

The wavelength of blue light is approximately 0.0000005 m, but this number is equal to 5×0.0000001, therefore it may be written 5×10^{-7} m.

Summary:

If given 7×10^9, move the decimal nine places to the right, which produces 7,000,000,000.

If given 7×10^{-9}, move the decimal nine places to the left, which produces 0.000000007.

Pi $(\pi) = 3.14159$
$\cong 22/7$

Velocity of light, $c = 2.99793 \times 10^{10}$ cm/sec
$\cong 300,000$ km/sec
$\cong 186,000$ mi/sec

Constant of gravitation,
$G = 6.67 \times 10^{-8}$ dyne-cm^2/g^2

Angstrom unit, $\text{Å} = 10^{-10}$ m

Astronomical unit, A.U. $= 1.49598 \times 10^{11}$ m
$\cong 150,000,000$ km
$\cong 93,000,000$ mi

Parsec $= 206,265$ A.U.
$= 3.262$ light-years

Light-year $= 9.4605 \times 10^{15}$ m
$\cong 9.5 \times 10^{12}$ km
$\cong 6 \times 10^{12}$ mi

Mass of the sun, $m_s = 1.991 \times 10^{33}$ g

Mass of the earth, $m_e = 5.98 \times 10^{27}$ g

Mass of the proton, $m_p = 1.672 \times 10^{-24}$ g

Mass of the neutron $= 1.674 \times 10^{-24}$ g

Mass of the electron $= 9.108 \times 10^{-28}$ g

Charge of the electron
$= 1.60 \times 10^{-19}$ coulomb

Avogadro's number
$= 6.025 \times 10^{23}$ molecules/mole

Planck's constant $= 6.625 \times 10^{-34}$ joule-sec

Acceleration due to gravity $= 980$ cm/sec^2
$= 32$ ft/sec^2

[a] Round-number approximations are indicated by $\cong$.

Appendix 5 Orbital Data of the Planets

PLANET	SYMBOL	SEMIMAJOR AXIS (A.U.)	SIDEREAL PERIOD	SYNODIC PERIOD	ECCENTRICITY OF ORBIT	INCLINATION OF ORBIT	AVERAGE ORBITAL SPEED (km/sec)
Mercury	☿	0.387	87.97 days	116 days	0.2056	7.0°	47.8
Venus	♀	0.723	224.7 days	584 days	0.0068	3.4°	35.0
Earth	⊕	1.000	365.26 days	—	0.0167	0.0°	29.8
Mars	♂	1.524	687.0 days or 1.88 years	780 days	0.0934	1.8°	24.2
(Ceres[a])	①	2.77	4.60 years	467 days	0.0765	10.6°	17.9
Jupiter	♃	5.20	11.86 years	399 days	0.0484	1.3°	13.1
Saturn	♄	9.54	29.46 years	378 days	0.0557	2.5°	9.7
Uranus	♂ or ♅	19.18	84.01 years	370 days	0.0472	0.8°	6.8
Neptune	♆	30.06	164.79 years	367.5 days	0.0086	1.8°	5.4
Pluto	♇	39.44	248 years	366.5 days	0.2502	17.2°	4.7

[a] An asteroid.

Appendix 6 Physical and Rotational Data for the Planets

PLANET	DIAMETER km	DIAMETER E[a] = 1	MASS (E[a] = 1)	DENSITY (WATER = 1)	PERIOD OF ROTATION	INCLINATION OF EQUATOR TO ECLIPTIC[b]	ALBEDO	SURFACE GRAVITY (E[a] = 1)	VELOCITY OF ESCAPE (km/sec)
Mercury	4,880	0.38	0.05	5.2	58^d 15^h	7°	0.07	0.39	4.3
Venus	12,108	0.95	0.82	5.3	243^d 4^h	176°	0.76	0.90	10.3
Earth	12,750	1.00	1.00	5.5	23^h 56^m	23°27'	0.39	1.00	11.2
Mars	6,800	0.53	0.11	3,8	24^h 37^m	25°	0.18	0.38	5.1
Jupiter	142,800	11.23	317.9	1.3	9^h 50^m	3°	0.50	2.58	59.5
Saturn	120,000	9.41	95.2	0.7	10^h 14^m	26°45'	0.61	1.11	35.6
Uranus	50,800	3.98	14.6	1.3	10^h 49^m	98°	0.61	1.07	21.4
Neptune	49,500	3.88	17.2	1.7	15^h 40^m	29°	0.62	1.40	23.6
Pluto	3,000	0.23	0.002	1.1	6^d 9^h 17^m	?	0.50	?	?

[a] E is the earth.
[b] An inclination greater than 90° indicates retrograde rotation.

PLANET	SATELLITE	DISCOVERER	MEAN DISTANCE FROM PLANET (km)	SIDEREAL PERIOD (DAYS)[a]	INCLINATION OF ORBIT TO PLANET'S EQUATOR	DIAMETER OF SATELLITE (km)	APPROXIMATE MAGNITUDE AT OPPOSITION
Earth	Moon	—	384,405	27.322	23.5°	3476	−12.5
Mars	Phobos	A. Hall (1877)	9,380	0.319	1°	27	11.5
	Deimos	A. Hall (1877)	23,500	1.262	2°	12	12.0
Jupiter	V	Barnard (1892)	180,500	0.498	∼0°	160	13.0
	I Io	Galileo (1610)	421,800	1.769	∼0°	3658	5.5
	II Europa	Galileo (1610)	671,400	3.551	∼0°	3100	5.7
	III Ganymede	Galileo (1610)	1,070,000	7.155	∼0°	5300	5.0
	IV Callisto	Galileo (1610)	1,884,000	16.689	∼0°	5000	6.3
	XIII	Kowal (1974)	11,100,000	239	26.7°	10	20
	VI	Perrine (1904)	11,470,000	250.57	27.6°	120	14.0
	VII	Perrine (1905)	11,800,000	260.10	24.8°	40	17.5
	X	Nicholson (1938)	11,850,000	263.55	29.0°	20	19.0
	XII	Nicholson (1951)	21,200,000	617.0r[a]	147.0°	20	18.5
	XI	Nicholson (1938)	22,600,000	692.5r	164.0°	24	19.0
	VIII	Melotte (1908)	23,500,000	735.0r	145.0°	20	17.5
	IX	Nicholson (1914)	23,700,000	758.0r	153.0°	22	19.0
	XIV & XV[b]						
Saturn	Janus	A. Dollfus (1966)	168,700	0.749	∼0°	300	14.0
	Mimas	W. Herschel (1789)	185,800	0.942	∼0°	500	12.0
	Enceladus	W. Herschel (1789)	238,300	1.370	∼0°	600	12.0
	Tethys	Cassini (1684)	294,900	1.888	∼0°	1200	10.5
	Dione	Cassini (1684)	377,900	2.737	∼0°	1000	11.0
	Rhea	Cassini (1672)	527,600	4.518	∼0°	1500	10.0
	Titan	Huygens (1655)	1,222,600	15.945	∼0°	5800	8.3
	Hyperion	Bond (1848)	1,484,100	21.277	∼0°	400	13.0
	Iapetus	Cassini (1671)	3,562,900	79.331	14.7°	1300	11.0
	Phoebe	W. Pickering (1898)	12,960,000	550.45r	150.1°	300	14.0
Uranus	Miranda	Kuiper (1948)	130,000	1.414r	0°	400	19.0
	Ariel	Lassell (1851)	191,000	2.520r	0°	1200	15.0
	Umbriel	Lassell (1851)	266,000	4.144r	0°	1000	16.0
	Titania	W. Herschel (1787)	436,000	8.706r	0°	1800	14.0
	Oberon	W. Herschel (1787)	583,400	13.463r	0°	1600	14.0
Neptune	Triton	Lassell (1846)	355,500	5.877r	160°	4000	14.0
	Nereid	Kuiper (1949)	5,567,000	359.881	27.7°	500	19.0
Pluto	Charon	Christy (1978)	20,000.	6.39	—	1000	20.0

[a]The notation "r," when shown with the sidereal period of satellite, indicates that the satellite orbits the planet in retrograde motion.
[b]A fourteenth and a fifteenth moon of Jupiter has been tentatively identified.

Appendix 8 The Twenty Brightest Stars

STAR	RIGHT ASCENSION (1950) (h) (m)	DECLI-NATION (1950) (°) (')	DIS-TANCE (PARSECS) (pc)	PROPER MOTION (")	SPECTRA OF COMPONENTS[a,c] A	B	C	VISUAL MAGNITUDES OF COMPONENTS[b,c] A	B	C	ABSOLUTE VISUAL MAGNITUDES OF COMPONENTS[c] A	B	C
Sirius	6 42.9	−16 39	2.7	1.32	A1V	wd	—	−1.47	+7.1	—	+1.4	+10.5	—
Canopus	6 22.8	−52 40	30	0.03	F0lb	—	—	−0.72	—	—	−3.1	—	—
α Centauri	14 36.2	−60 38	1.3	3.68	G2V	K5V	M5V	−0.01	+1.5	+10.7	+4.4	+5.8	+15
Arcturus	14 13.4	+19 27	11	2.28	K2III	—	—	−0.06	—	—	−0.3	—	—
Vega	18 35.2	+38 44	8.0	0.34	A0V	—	—	+0.04	—	—	+0.5	—	—
Capella	5 13.0	+45 57	14	0.44	G0II	M1V	M5V	+0.09	+10.2	+13.7	−0.7	+9.5	+13
Rigel	5 12.1	−8 15	250	0.00	B8la	B9	—	+0.10	+6.6	—	−6.8	−0.4	—
Procyon	7 36.7	+5 21	3.5	1.25	F5IV-V	wd	—	+0.38	+10.7	—	+2.7	+13.1	—
Betelgeuse	5 52.5	+7 24	200	0.03	M2lab	—	—	+0.41v	—	—	−5.5	—	—
Achernar	1 35.9	−57 29	20	0.10	B5V	—	—	+0.47	—	—	−1.6	—	—
β Centauri	14 00.3	−60 08	90	0.04	B1III	—	—	+0.63	—	—	−4.1	—	—
Altair	19 48.3	+8 44	5.1	0.66	A7IV,V	—	—	+0.77	—	—	+2.2	—	—
α Crucis	12 23.8	−62 49	120	0.04	B1IV	B3	—	+1.39	+1.9	—	−4.0	−3.5	—
Aldebaran	4 33.0	+16 25	16	0.20	K5III	M2V	—	+0.86v	+13	—	−0.2	+12	—
Spica	13 22.6	−10 54	70	0.05	B1V	—	—	+0.91	—	—	−3.6	—	—
Antares	16 26.3	−26 19	120	0.03	M1lb	B4V	—	+0.92v	+5.1	—	−4.5	−0.3	—
Pollux	7 42.3	+28 09	12	0.62	K0III	—	—	+1.16	—	—	+0.8	—	—
Fomalhaut	22 54.9	−29 53	7.0	0.37	A3V	K4V	—	+1.19	+6.5	—	+2.0	+7.3	—
Deneb	20 39.7	+45 06	430	0.00	A2la	—	—	+1.26	—	—	−6.9	—	—
β Crucis	12 44.8	−59 24	150	0.05	B0.5IV	—	—	+1.28	—	—	−4.6	—	—

[a] The Roman numerals after the spectral classifications have the following meanings: Ia or Ib, *supergiant;* II or III, *giant;* IV, *subgiant;* V, *main-sequence star* (see pages 229–230). The notation "wd" indicates *white dwarf.*

[b] The notation "v" following the magnitude indicates a *variable star.*

[c] When entries are shown in both A and B columns, the star is known to be a binary system. When an entry is also shown in the C column, the system is known to have three components.

M	NGC	RIGHT ASCENSION (1950) (h) (m)		DECLI-NATION (1950) (°) (')		APPARENT VISUAL MAGNITUDE	DESCRIPTION
1	1952	5	31.5	+21	59	8.4	Crab Nebula in Taurus; remains of supernova
2	7089	21	30.9	−1	02	6.4	Globular cluster in Aquarius
3	5272	13	39.8	+28	38	6.3	Globular cluster in Canes Venatici
4	6121	16	20.6	−26	24	6.5	Globular cluster in Scorpius
5	5904	15	16.0	+2	16	6.1	Globular cluster in Serpens
6	6405	17	36.8	−32	10	5.3	Open cluster in Scorpius
7	6475	17	50.7	−34	48	4.1	Open cluster in Scorpius
8	6523	18	00.1	−24	23	6.0	Lagoon Nebula in Sagittarius
9	6333	17	16.3	−18	28	7.3	Globular cluster in Ophiuchus
10	6254	16	54.5	−4	02	6.7	Globular cluster in Ophiuchus
11	6705	18	48.4	−6	20	6.3	Open cluster in Scutum
12	6218	16	44.7	−1	52	6.6	Globular cluster in Ophiuchus
13	6205	16	39.9	+36	33	5.9	Globular cluster in Hercules
14	6402	17	35.0	−3	13	7.7	Globular cluster in Ophiuchus
15	7078	21	27.5	+11	57	6.4	Globular cluster in Pegasus
16	6611	18	16.1	−13	48	6.4	Open cluster with nebulosity in Serpens
17	6618	18	17.9	−16	12	7.0	Swan or Omega Nebula in Sagittarius
18	6613	18	17.0	−17	09	7.5	Open cluster in Sagittarius
19	6273	16	59.5	−26	11	6.6	Globular cluster in Ophiuchus
20	6514	17	59.4	−23	02	9.0	Trifid Nebula in Sagittarius
21	6531	18	01.6	−22	30	6.5	Open cluster in Sagittarius
22	6656	18	33.4	−23	57	5.6	Globular cluster in Sagittarius
23	6494	17	54.0	−19	00	6.9	Open cluster in Sagittarius
24	6603	18	15.5	−18	27	11.4	Open cluster in Sagittarius
25	(4725)ᵃ	18	28.7	−19	17	6.5	Open cluster in Sagittarius
26	6694	18	42.5	−9	27	9.3	Open cluster in Scutum
27	6853	19	57.5	+22	35	7.6	Dumbbell Planetary Nebula in Vulpecula
28	6626	18	21.4	−24	53	7.6	Globular cluster in Sagittarius
29	6913	20	22.2	+38	21	7.1	Open cluster in Cygnus
30	7099	21	37.5	−23	24	8.4	Globular cluster in Capricornus
31	224	0	40.0	+41	00	4.8	Andromeda galaxy
32	221	0	40.0	+40	36	8.7	Elliptical galaxy; companion to M31
33	598	1	31.0	+30	24	6.7	Spiral galaxy in Triangulum
34	1039	2	38.8	+42	35	5.5	Open cluster in Perseus
35	2168	6	05.7	+24	21	5.3	Open cluster in Gemini
36	1960	5	33.0	+34	04	6.3	Open cluster in Auriga
37	2099	5	49.1	+32	33	6.2	Open cluster in Auriga
38	1912	5	25.3	+35	47	7.4	Open cluster in Auriga
39	7092	21	30.4	+48	13	5.2	Open cluster in Cygnus
40	—	12	20	+58	20	—	Close double star in Ursa Major
41	2287	6	44.9	−20	41	4.6	Loose open cluster in Canis Major
42	1976	5	32.9	−5	25	4.0	Orion Nebula

ᵃIndex Catalogue (IC) number.

M	NGC	RIGHT ASCENSION (1950) (h) (m)	DECLI- NATION (1950) (°) (')	APPARENT VISUAL MAGNITUDE	DESCRIPTION
43	1982	5 33.1	−5 19	9.0	Northeast portion of Orion Nebula
44	2632	8 37	+20 10	3.7	Praesepe; open cluster in Cancer
45	—	3 44.5	+23 57	1.6	The Pleiades; open cluster in Taurus
46	2437	7 39.5	−14 42	6.0	Open cluster in Puppis
47	2422	7 34.3	−14 22	5.2	Loose group of stars in Puppis
48	2458	8 11	−5 38	5.5	Open cluster in Hydra
49	4472	12 27.3	+8 16	8.5	Elliptical galaxy in Virgo
50	2323	7 00.6	−8 16	6.3	Loose open cluster in Monoceros
51	5194	13 27.8	+47 27	8.4	Whirlpool spiral galaxy in Canes Venatici
52	7654	23 22.0	+61 20	7.3	Loose open cluster in Cassiopeia
53	5024	13 10.5	+18 26	7.8	Globular cluster in Coma Berenices
54	6715	18 51.9	−30 32	7.3	Globular cluster in Sagittarius
55	6809	19 36.8	−31 03	7.6	Globular cluster in Sagittarius
56	6779	19 14.6	+30 05	8.2	Globular cluster in Lyra
57	6720	18 51.7	+32 58	9.0	Ring Nebula; planetary nebula in Lyra
58	4579	12 35.2	+12 05	8.2	Barred spiral galaxy in Virgo
59	4621	12 39.5	+11 56	9.3	Elliptical spiral galaxy in Virgo
60	4649	12 41.1	+11 50	9.0	Elliptical galaxy in Virgo
61	4303	12 19.3	+4 45	9.6	Spiral galaxy in Virgo
62	6266	16 58.0	−30 02	6.6	Globular cluster in Ophiuchus
63	5055	13 13.5	+42 17	10.1	Spiral galaxy in Canes Venatici
64	4826	12 54.2	+21 57	6.6	Spiral galaxy in Coma Berenices
65	3623	11 16.3	+13 22	9.4	Spiral galaxy in Leo
66	3627	11 17.6	+13 16	9.0	Spiral galaxy in Leo; companion to M65
67	2682	8 48.4	+12 00	6.1	Open cluster in Cancer
68	4590	12 36.8	−26 29	8.2	Globular cluster in Hydra
69	6637	18 28.1	−32 24	8.9	Globular cluster in Sagittarius
70	6681	18 40.0	−32 20	9.6	Globular cluster in Sagittarius
71	6838	19 51.5	+18 39	9.0	Globular cluster in Sagitta
72	6981	20 50.7	−12 45	9.8	Globular cluster in Aquarius
73	6994	20 56.2	−12 50	9.0	Open cluster in Aquarius
74	628	1 34.0	+15 32	10.2	Spiral galaxy in Pisces
75	6864	20 03.1	−22 04	8.0	Globular cluster in Sagittarius
76	650	1 38.8	+51 19	11.4	Planetary nebula in Perseus
77	1068	2 40.1	−0 12	8.9	Spiral galaxy in Cetus
78	2068	5 44.2	+0 02	8.3	Small reflection nebula in Orion
79	1904	5 22.1	−24 34	7.5	Globular cluster in Lepus
80	6093	16 14.0	−22 52	7.5	Globular cluster in Scorpius
81	3031	9 51.7	+69 18	7.9	Spiral galaxy in Ursa Major
82	3034	9 51.9	+69 56	8.4	Irregular galaxy in Ursa Major
83	5236	13 34.2	−29 37	10.1	Spiral galaxy in Hydra
84	4374	12 22.6	+13 10	9.4	S0 type galaxy in Virgo
85	4382	12 22.8	+18 28	9.3	S0 type galaxy in Coma Berenices
86	4406	12 23.6	+13 13	9.2	Elliptical galaxy in Virgo

M	NGC	RIGHT ASCENSION (1950)		DECLINATION (1950)		APPARENT VISUAL MAGNITUDE	DESCRIPTION
		(h)	(m)	(°)	(′)		
87	4486	12	28.2	+12	40	8.7	Elliptical galaxy in Virgo
88	4501	12	29.4	+14	42	10.2	Spiral galaxy in Coma Berenices
89	4552	12	33.1	+12	50	9.5	Elliptical galaxy in Virgo
90	4569	12	34.3	+13	26	9.6	Spiral galaxy in Virgo
91[b]	4571(?)	—	—	—	—	—	
92	6341	17	15.6	+43	12	6.4	Globular cluster in Hercules
93	2447	7	42.4	−23	45	6.0	Open cluster in Puppis
94	4736	12	48.6	+41	24	8.3	Spiral galaxy in Canes Venatici
95	3351	10	41.3	+11	58	9.8	Barred spiral galaxy in Leo
96	3368	10	44.1	+12	05	9.3	Spiral galaxy in Leo
97	3587	11	12.0	+55	17	12.0	Owl Nebula; planetary nebula in Ursa Major
98	4192	12	11.2	+15	11	10.2	Spiral galaxy in Coma Berenices
99	4254	12	16.3	+14	42	9.9	Spiral galaxy in Coma Berenices
100	4321	12	20.4	+16	06	10.6	Spiral galaxy in Coma Berenices
101	5457	14	01.4	+54	36	9.6	Spiral galaxy in Ursa Major
102[b]	5866(?)	—	—	—	—	—	
103	581	1	29.9	+60	26	7.4	Open cluster in Cassiopeia
104	4594	12	37.4	−11	21	8.3	Spiral galaxy in Virgo
105	3379	10	45.2	+13	01	9.7	Elliptical galaxy in Leo
106	4258	12	16.5	+47	35	8.4	Spiral galaxy in Canes Venatici
107	6171	16	29.7	−12	57	9.2	Globular cluster in Ophiuchus

[b] Items of doubtful identification.

Appendix 10 The Greek Alphabet

A	α	alpha	H	η	eta	N	ν	nu	T	τ	tau
B	β	beta	Θ	θ	theta	Ξ	ξ	xi	Υ	υ	upsilon
Γ	γ	gamma	I	ι	iota	O	o	omicron	Φ	ϕ	phi
Δ	δ	delta	K	κ	kappa	Π	π	pi	X	χ	chi
E	ϵ	epsilon	Θ	λ	lambda	P	ρ	rho	Ψ	ψ	psi
Z	ζ	zeta	M	μ	mu	Σ	σ	sigma	Ω	ω	omega

Appendix 11   The Constellations

CONSTELLATION NAME[a]	DESCRIPTION	POSITION IN SKY R.A.[b]	DEC.[b]	CONSTELLATION NAME[a]	DESCRIPTION	POSITION IN SKY R.A.[b]	DEC.[b]
Andromeda	Princess of Ethiopia	1^h	+40°	Leo Minor	The Lion Cub	10^h	+35°
Antlia	The Air Pump	10^h	−35°	Lepus	The Hare	6^h	−20°
Apus	The Bird of Paradise	16^h	−75°	Libra	The Beam Balance	15^h	−15°
Aquarius	The Water Bearer	23^h	−15°	Lupus	The Wolf	15^h	−45°
Aquila	The Eagle	20^h	+5°	Lynx	The Lynx	8^h	+45°
Ara	The Altar	17^h	−55°	Lyra	The Lyre	19^h	+40°
Aries	The Ram	3^h	+20°	Mensa	The Table Mountain	5^h	−80°
Auriga	The Charioteer	6^h	+40°	Microscopium	The Microscope	21^h	−35°
Boötes	The Bear Driver	15^h	+30°	Monoceros	The Unicorn	7^h	−5°
Caelum	The Sculptor's Chisel	5^h	−40°	Musca	The Fly	12^h	−70°
Camelopardus	The Giraffe	6^h	−70°	Norma	The Carpenter's Square	16^h	−50°
Cancer	The Crab	9^h	+20°	Octans	The Octant	22^h	−85°
Canes Venatici	The Hunting Dogs	13^h	+40°	Ophiuchus	The Serpent Holder	17^h	0°
Canis Major	The Greater Dog	7^h	−20°	Orion	The Great Hunter	5^h	+5°
Canis Minor	The Lesser Dog	8^h	+5°	Pavo	The Peacock	20^h	−65°
Capricornus	The Sea Goat	21^h	−20°	Pegasus	The Winged Horse	22^h	+20°
Carina	The Keel (of Argo Navis)	9^h	−60°	Perseus	The Hero	3^h	+45°
Cassiopeia	Queen of Ethiopia	1^h	+60°	Phoenix	The Phoenix	1^h	−50°
Centaurus	The Centaur	13^h	−50°	Pictor	The Painter's Easel	6^h	−55°
Cepheus	King of Ethiopia	22^h	+70°	Pisces	The Fishes	1^h	+15°
Cetus	The Sea Monster	2^h	−10°	Piscis Austrinus	The Southern Fish	22^h	−30°
Chamaeleon	The Chameleon	11^h	−80°	Puppis	The Stern (of Argo Navis)	8^h	−40°
Circinus	The Compasses	15^h	−60°	Pyxis	The Compass Box (of Argo)	9^h	−30°
Columba	The Dove (of Noah)	6^h	−35°	Reticulum	The Net	4^h	−60°
Coma Berenices	Berenice's Hair	13^h	+20°	Sagitta	The Arrow	10^h	+10°
Corona Austrina	The Southern Crown	19^h	−40°	Sagittarius	The Archer	19^h	−25°
Corona Borealis	The Northern Crown	16^h	+30°	Scorpius	The Scorpion	17^h	−40°
Corvus	The Crow (or Raven)	12^h	−20°	Sculptor	The Sculptor's Workshop	0^h	−30°
Crater	The Cup	11^h	−15°	Scutum (Sobieski)	The Shield (of John Sobieski[c])	19^h	−10°
Crux	The Southern Cross	12^h	−60°	Serpens	The Serpent	17^h	0°
Cygnus	The Swan	21^h	+40°	Sextans	The Sextant	10^h	0°
Delphinus	The Dolphin	21^h	+10°	Taurus	The Bull	4^h	+15°
Dorado	The Swordfish	5^h	−65°	Telescopium	The Telescope	19^h	−50°
Draco	The Dragon	17^h	+65°	Triangulum	The Triangle	2^h	+30°
Equuleus	The Foal	21^h	+10°	Triangulum Australe	The Southern Triangle	16^h	−65°
Eridanus	The River	3^h	−20°	Tucana	The Toucan	0^h	−65°
Fornax	The Laboratory Furnace	3^h	−30°	Ursa Major	The Greater Bear	11^h	+50°
Gemini	The Twins	7^h	+20°	Ursa Minor	The Lesser Bear	15^h	+70°
Grus	The Crane	22^h	−45°	Vela	The Sail (of Argo Navis)	9^h	−50°
Hercules	Hercules	17^h	+30°	Virgo	The Maiden	13^h	0°
Horologium	The Clock	3^h	−60°	Volans	The Flying Fish	8^h	−70°
Hydra	The Water Serpent	10^h	−20°	Vulpecula	The Fox	20^h	+25°
Hydrus	The Water Snake	2^h	−75°				
Indus	The American Indian	21^h	−55°				
Lacerta	The Lizard	22^h	+45°				
Leo	The Lion	11^h	+15°				

[a] Constellations with declinations between −50° and −90° are difficult or impossible to see from the United States.
[b] R.A., right ascension; Dec., declination.
[c] King John III of Poland (1624–1697).

Appendixes   **509**

Appendix 12   The Periodic Table

Atomic weight ──▶ 1.00797

Atomic number ──▶ 1 H ◄── Symbol

Hydrogen ◄── Name

Metals

Nonmetals

IA								
1.00797 1 H Hydrogen	IIA							
6.939 3 Li Lithium	9.012 4 Be Beryllium							
22.99 11 Na Sodium	24.31 12 Mg Magnesium	IIIB	IVB	VB	VIB	VIIB	VIII	VIII
39.10 19 K Potassium	40.08 20 Ca Calcium	44.96 21 Sc Scandium	47.90 22 Ti Titanium	50.94 23 V Vanadium	52.00 24 Cr Chromium	54.94 25 Mn Manganese	55.85 26 Fe Iron	58.93 27 Co Cobalt
85.47 37 Rb Rubidium	87.62 38 Sr Strontium	88.91 39 Y Yttrium	91.22 40 Zr Zirconium	92.91 41 Nb Niobium	95.94 42 Mo Molybdenum	(98)* 43 Tc Technetium	101.07 44 Ru Ruthenium	102.9 45 Rh Rhodium
132.91 55 Cs Cesium	137.34 56 Ba Barium	Lanthanides 57–71	178.49 72 Hf Hafnium	180.95 73 Ta Tantalum	183.85 74 W Tungsten	186.2 75 Re Rhenium	190.2 76 Os Osmium	192.2 77 Ir Iridium
(223) 87 Fr Francium	(226) 88 Ra Radium	Actinides 89–103						

Lanthanides ──▶

138.91 57 La Lanthanum	140.12 58 Ce Cerium	140.91 59 Pr Praseodymium	144.24 60 Nd Neodymium	(145) 61 Pm Promethium	150.4 62 Sm Samarium	151.96 63 Eu Europium

Actinides ──▶

(227) 89 Ac Actinium	232.04 90 Th Thorium	(231) 91 Pa Protactinium	238.03 92 U Uranium	(237) 93 Np Neptunium	(242) 94 Pu Plutonium	(243) 95 Am Americium

*Atomic weights in parentheses indicate radioactive isotopes of longest half-life.

			IIIA	IVA	VA	VIA	VIIA	O
							1.00797 $_1$H Hydrogen	4.0026 $_2$He Helium
			10.811 $_5$B Boron	12.01 $_6$C Carbon	14.01 $_7$N Nitrogen	15.999 $_8$O Oxygen	18.998 $_9$F Fluorine	20.182 $_{10}$Ne Neon

VIII	IB	IIB	26.98 $_{13}$Al Aluminum	28.09 $_{14}$Si Silicon	30.97 $_{15}$P Phosphorus	32.06 $_{16}$S Sulfur	35.45 $_{17}$Cl Chlorine	39.95 $_{18}$Ar Argon
58.71 $_{28}$Ni Nickel	63.54 $_{29}$Cu Copper	65.37 $_{30}$Zn Zinc	69.72 $_{31}$Ga Gallium	72.59 $_{32}$Ge Germanium	74.92 $_{33}$As Arsenic	78.96 $_{34}$Se Selenium	79.91 $_{35}$Br Bromine	83.80 $_{36}$Kr Krypton
106.4 $_{46}$Pd Palladium	107.9 $_{47}$Ag Silver	112.4 $_{48}$Cd Cadmium	114.82 $_{49}$In Indium	118.69 $_{50}$Sn Tin	121.75 $_{51}$Sb Antimony	127.6 $_{52}$Te Tellurium	126.90 $_{53}$I Iodine	131.30 $_{54}$Xe Xenon
195.09 $_{78}$Pt Platinum	196.97 $_{79}$Au Gold	200.59 $_{80}$Hg Mercury	204.37 $_{81}$Tl Thallium	207.19 $_{82}$Pb Lead	208.98 $_{83}$Bi Bismuth	(210) $_{84}$Po Polonium	(210) $_{85}$At Astatine	(222) $_{86}$Rn Radon

157.25 $_{64}$Gd Gadolinium	158.9 $_{65}$Tb Terbium	162.5 $_{66}$Dy Dysprosium	164.9 $_{67}$Ho Holmium	167.3 $_{68}$Er Erbium	168.9 $_{69}$Tm Thulium	173.0 $_{70}$Yb Ytterbium	175.0 $_{71}$Lu Lutetium
(247) $_{96}$Cm Curium	(247) $_{97}$Bk Berkelium	(251) $_{98}$Cf Californium	(254) $_{99}$Es Einsteinium	(253) $_{100}$Fm Fermium	(256) $_{101}$Md Mendelevium	(253) $_{102}$No Nobelium	(257) $_{103}$Lr Lawrencium

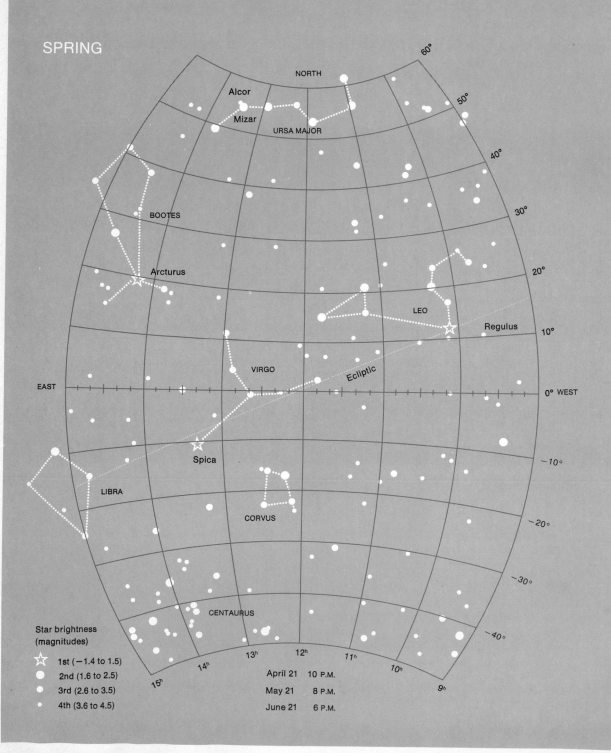

SPRING

Star brightness
(magnitudes)

☆ 1st (−1.4 to 1.5)

● 2nd (1.6 to 2.5)

● 3rd (2.6 to 3.5)

· 4th (3.6 to 4.5)

April 21   10 P.M.

May 21   8 P.M.

June 21   6 P.M.

These star maps show the brighter stars and the prominent constellations as they appear on the dates and at the times indicated. To use these maps, face the south and hold the book overhead with top of the map toward the north and the right-hand edge toward the west. The brightest stars are indicated by the star symbol (☆) and the names are indicated. (Star maps were designed by the author.)

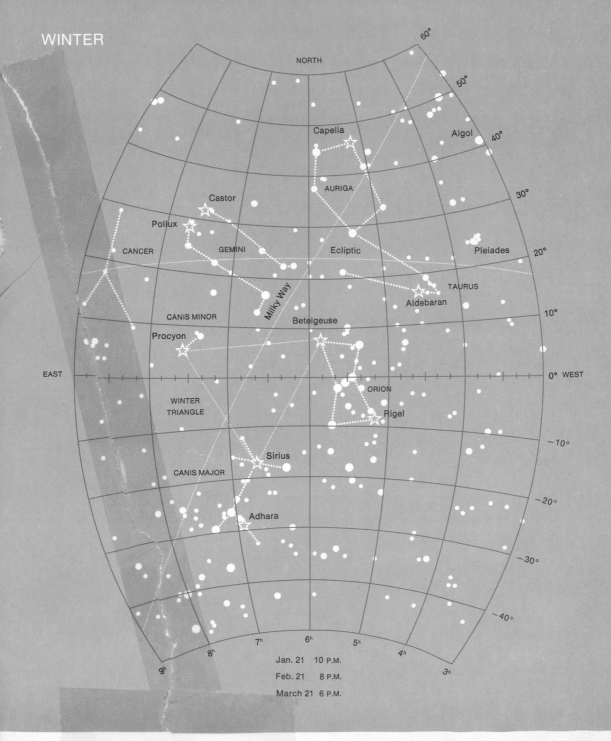

WINTER

NORTH

60°

50°

Capella

Algol

40°

AURIGA

30°

Castor

Pollux

Ecliptic

Pleiades

20°

CANCER

GEMINI

TAURUS

Milky Way

Aldebaran

10°

CANIS MINOR

Betelgeuse

Procyon

EAST

0° WEST

WINTER
TRIANGLE

ORION

Rigel

−10°

Sirius

CANIS MAJOR

−20°

Adhara

−30°

−40°

9ʰ

8ʰ

7ʰ

6ʰ

5ʰ

4ʰ

3ʰ

Jan. 21 10 P.M.

Feb. 21 8 P.M.

March 21 6 P.M.

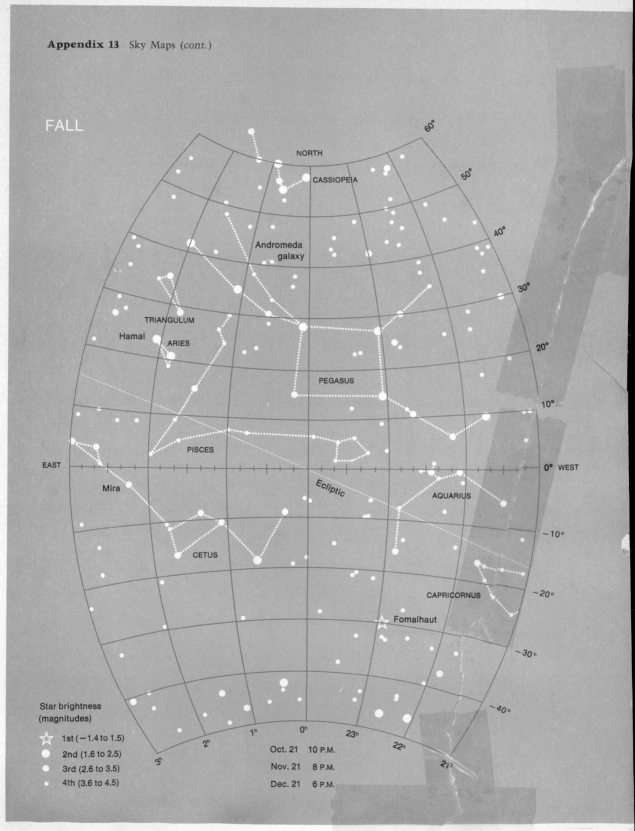

FALL

NORTH

60°

50°

CASSIOPEIA

40°

Andromeda galaxy

30°

TRIANGULUM

20°

Hamal

ARIES

PEGASUS

10°

PISCES

EAST

0° WEST

Mira

Ecliptic

AQUARIUS

−10°

CETUS

CAPRICORNUS

−20°

Fomalhaut

−30°

−40°

3ʰ   2ʰ   1ʰ   0ʰ   23ʰ   22ʰ   21ʰ

Star brightness (magnitudes)

☆ 1st (−1.4 to 1.5)
● 2nd (1.6 to 2.5)
● 3rd (2.6 to 3.5)
· 4th (3.6 to 4.5)

Oct. 21   10 P.M.
Nov. 21   8 P.M.
Dec. 21   6 P.M.

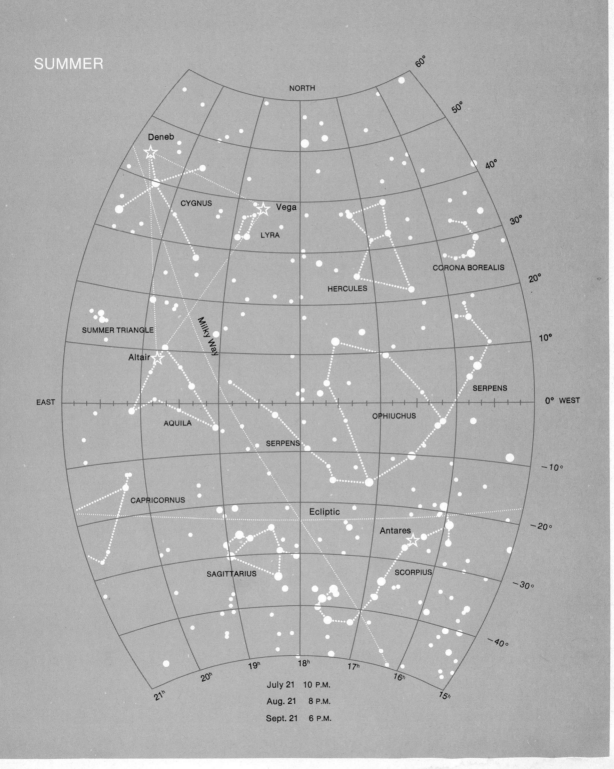

SUMMER

NORTH

60°
50°
40°
30°
20°
10°
0° WEST
−10°
−20°
−30°
−40°

EAST

Deneb

CYGNUS

Vega

LYRA

HERCULES

CORONA BOREALIS

SUMMER TRIANGLE

Milky Way

Altair

SERPENS

AQUILA

OPHIUCHUS

SERPENS

CAPRICORNUS

Ecliptic

Antares

SAGITTARIUS

SCORPIUS

21ʰ 20ʰ 19ʰ 18ʰ 17ʰ 16ʰ 15ʰ

July 21   10 P.M.

Aug. 21   8 P.M.

Sept. 21   6 P.M.

INDEX